ADVANCES IN PERINATAL THYROIDOLOGY

ADVANCES IN EXPERIMENTAL MEDICINE AND BIOLOGY

Recent Volumes in this Series

ADVANCES IN PERINATAL THYROIDOLOGY

Edited by

Barry B. Bercu and Dorothy I. Shulman

University of South Florida
College of Medicine
Tampa, Florida

PLENUM PRESS • NEW YORK AND LONDON

Library of Congress Cataloging-in-Publication Data

International Symposium on Advances in Perinatal Thyroidology (1990 :
Longboat Key, Fla.)
    Advances in perinatal thyroidology / edited by Barry B. Bercu and
Dorothy I. Shulman.
        p.   cm. -- (Advances in experimental medicine and biology ; v.
299)
    "Proceedings of an International Symposium on Advances in
Perinatal Thyroidology, held December 2-4, 1990, in Longboat Key,
Florida--T.p. verso.
    Includes bibliographical references and index.
    ISBN-13: 978-1-4684-5975-3   e-ISBN-13: 978-1-4684-5973-9
    DOI: 10.1007/978-1-4684-5973-9
    1. Thyroid gland diseases in pregnancy--Congresses. 2. Congenital
hypothyroidism--Congresses. 3. Thyroid hormones--Physiological
effect--Congresses.   I. Bercu, Barry B. II. Shulman, Dorothy I.
III. Title. IV. Series.
    [DNLM: 1. Perinatology--congresses. 2. Thyroid Diseases-
-etiology--congresses. 3. Thyroid Diseases--in infancy & childhood-
-congresses. 4. Thyroid Gland--physiology--congresses.   W1 AD559
v. 299 / WK 200 I601a 1990]
RG580.T47I58 1990
618.3'26--dc20
DNLM/DLC
for Library of Congress   91-24087
                                                               CIP

Proceedings of an International Symposium on
Advances in Perinatal Thyroidology,
held December 2-4, 1990, in Longboat Key, Florida

ISBN-13: 978-1-4684-5975-3

© 1991 Plenum Press, New York
Softcover reprint of the hardcover 1st edition 1991
A Division of Plenum Publishing Corporation
233 Spring Street, New York, N.Y. 10013

PREFACE

Perinatal problems in thyroid gland physiology are common
but complicated and present a diagnostic dilemma for the
primary clinician. In December 1990, an international
group of basic and clinical investigators gathered in
Longboat Key, Florida to address these issues. The
participants included internists, obstetricians, pedia-
tricians, neurologists, pathologists and basic scientists
in cellular metabolism, endocrine physiology, and molecular
biology. The presentations contained within this book
bring together their most current and vital research
related to the field of perinatal thyroidology.

This book is based on the dynamic and fruitful exchange
of the participants at the symposium. We are indebted
to these individuals whose valuable insights and efforts
are contained within this text.

Barry B. Bercu
Dorothy I. Shulman

CONTENTS

Session III

Session IV

MECHANISM OF THYROID HORMONE ACTION

Leslie J. DeGroot

Thyroid Study Unit, Department of Medicine
The University of Chicago
Chicago, IL

The pathway through which thyroid hormone controls metabolism begins with secretion of thyroxine and triiodothyronine by the thyroid, peripheral deiodination of about 1/3 of secreted T4 to T3, and uptake of the iodothyronines by passive and active mechanisms into the cell. Thyroid hormone in serum and in cell cytosol exists largely bound to proteins in a reversible equilibrium, with a tiny free fraction. It is this free fraction which equilibrates between serum and cells. Free hormone in the cell cytoplasm, and T3 generated from T4 within the cell diffuse into the nucleus and bind to thyroid hormone receptor proteins. These receptors may exist in the nucleoplasm, or, more likely, largely attached to DNA. Occupancy of receptors which are bound to specific sequences in DNA (thyroid response elements or TREs) leads to activation or repression of transcription, altered amounts of mRNA, and thus altered protein synthesis. It is the final alteration in amounts of specific proteins--enzymes, structural proteins, DNA binding proteins, which carry out the "physiologic" functions which we recognize as the increased metabolic activity seen as an individual moves from hypo-thyroidism to thyrotoxicosis. The key elements in this response are the thyroid hormone itself, the nuclear receptor proteins (TRs), and the TREs on hormone responsive genes.

Since 1972 presumed receptor proteins were known to exist in cell nuclei, especially in cells from tissues which respond to thyroid hormone,[1] and to have an affinity for thyroid hormone analogs which paralleled the in vitro effectiveness of these hormone analogs. The receptor proteins were known to be associated with DNA, to have required sulfhydryl groups, molecular weights of approximately 45,000 - 55,000, and to exist in two forms.[2] However, their exact structure escaped detection because of their low abundance and great difficulties in purification.[3]

Forms of Thyroid Hormone Receptor and Related Molecules

Thyroid hormone receptors were recognized to be members of a family of nuclear receptors in 1986.[4,5] Previous work established that estrogen and progesterone receptors were related to an oncogene v-erbA, which with v-erbB, can produce erythroblastosis in chickens.[6] Subsequent analysis has shown that estrogen, progesterone, gluco- corticoid, Vitamin D, retinoic acid, aldosterone, testosterone, and thyroid hormone receptors, are all closely related to v-erbA, and presumably evolved from a common precursor. The first human (h) TR, coded on chromosome 3, was cloned by Weinberger in 1986 and is now referred to as hTRβ.[4] Subsequently, other forms have been defined including hTRα1,[7] which is coded by a gene on chromosome 21, a related form hTRVα2,[8] which does not bind thyroid hormone, TRβ₂, which is a pituitary specific form of receptor in rats,[9] and the related molecule rev-erbA2.[10] The structural organization of the α and β forms of receptor is similar; an aminoterminal domain which may impart "specificity" to receptor function,[11] a DNA-binding "zinc- finger" domain, a hinge region which may provide nuclear localization signals, and a carboxy terminal ligand binding domain to which thyroid hormone or analogs bind.[12] Also in the carboxy terminal region is a domain in which leucine residues appear in every seventh position. These "heptad" repeats define a structure which has been referred to as a "leucine zipper" and may promote specific inter- actions between similarly constructed proteins.[37] hTRβ1 consists of 461 amino acids, has high similarity to α1 in the DNA binding domain, and identity in carboxy terminal 46 amino acids. hTRα1 and Vα2 are variants formed by differential splicing of exons coding for the carboxy terminal end of the molecule. α1 and α2 are identical through amino acid 370. After this there is total dis- similarity, and α2 is 80 amino acids longer than α1. hTRα1 is a functional thyroid hormone binding receptor. hTRα2 does not bind hormone and is, as will be discussed below, a "dominant negative" regulator of thyroid hormone receptor function. Since it does not bind thyroid hormone specifically, it is referred to as an hTR variant, or hTRVα2. The β2 form of the receptor is formed by differential splicing of the aminoterminal portion of beta mRNA and codes for a receptor which appears to be found specifically in rat pituitary.[9] It doubtless exists in human pituitaries as well.

Functional Domains of TRs

Crucial features of the domain structure described above include the DNA binding domain, the heptad repeats, and the ligand binding area. The DNA binding domain consists in all of the nuclear hormone receptor family, of two "zinc fingers" which are formed by loops of amino acids held in position by four cysteine molecules coordinated to zinc atoms.[14,15] It is thought that the two zinc-fingers, which make contact with the DNA, have different functions. The amino- terminal finger probably has more to do with specificity of binding

to the DNA, and the second finger may have more to do with spacing
between elements in the DNA recognition site[16] and may stabilize
DNA-receptor interaction. Specific mutations of the tips of the
zinc fingers alter the ability of the protein to bind to DNA, and
amino acids at the 3' base of the 1st finger seem to impart specificity for the cognate DNA sequence.[16]

Receptor molecules may exist free in nucleoplasm as monomers,
or possibly as dimers held together by their "leucine zippers".[17]
It has not been established that dimers exist apart from DNA, and
it has not been shown as yet that thyroid hormone binding to DNA
affects dimerization. The sites to which the receptors bind on
DNA, probably 6 - 8 nucleotides in length, the "response elements",
are described in more detail below. These TREs usually exist in
clusters as partial direct repeats, or palindromes. The palindromes
may be separated by zero to several nucleotides. It is believed
that binding of a single monomer of the receptor to a single TRE
does not constitute an effective signal for activation of a gene.
The idealized positive response element appears to be a palindrome
which binds two receptors, functioning as a dimer. It is believed
that TRs bind to the response element in the absence of ligand, and
there is no evidence that ligand binding actually increases the
affinity of receptor for the TRE. Receptors can bind to any DNA
but bind to the specific TRE with high affinity. Binding to the
TRE probably is augmented by a variety of other proteins.[18,19]

It is presumed, but not proven, that in the idealized "normal"
situation a palindromic TRE is occupied by a dimer made up of homo-
logous or heterologous TRs. Bound TR, without ligand, functions
on a _positive_ regulatory TRE as a negative response signal. Binding
ligand converts the negative signal into a positive transcriptional
signal. Receptors such as α2, or mutated receptors which cannot
bind ligand, function as inhibitors of transcriptional signals,
and are said to have a "dominant negative effect".[20,21] The interaction of α1, β, and α2 at "negative" TREs such as that on hTSH
alpha is less clear, but is probably very similar.

Thyroid Hormone Receptor Response Element(s) —TREs

Considerable effort has been made to define a thyroid hormone
response element, and proposals for a consensus sequence have been
made.[22-24] This work has involved careful analysis using progressive
deletion of promoter elements of responsive genes in transcription
trans-activation assay systems,[25] DNAse footprinting, gel shift
assays, and biotin labelled oligonucleotide affinity assays. Gel
shift assays appear to provide good specificity, although final
analysis requires proof of function in a transcriptional assay.
By analysis of the rGH promoter, a palindromic octamer or hexamer
(AGGTC/AA) was described as the idealized TRE.[22,23] A slightly
different TRE sequence has been described for malic enzyme.[24] rGH

probably includes three functional variants of the hexamer, including
a direct repeat and a palindrome. The malic enzyme probably includes
at least two tandem TREs.[24] Possibly palindromes are associated with
positive transcriptional effects and direct repeats with negative
regulation. It is clear that there is considerable variation in
the base sequences to which thyroid receptors can bind--the receptor
can be said to be "promiscuous". TREs which negatively regulate
transcription have been defined between the TATA box and the
transcriptional initiation site in hTSHα, and in the first intron
in hTSHβ.[26,27] Positive response elements are thought to function
by fostering binding of the receptor, along with other proteins,
which make up a transcriptional complex. The negative response
elements possibly work by binding TRs to DNA in the path of the
transcriptional complex and preventing its progression, at least
when the TR has bound ligand.

A variety of studies indicate that, for positive regulation,
at least two TRE hexamers, preferably present as a palindrome, are
best, but that spacing between the halves is not critical.[23] The
response element functions as an enhancer, since it can function at
a considerably different distance from the transcriptional initiation
site.[28] Whether the same TRE in a position 3' to the TATA box
functions as a negative regulator is not totally resolved, but
likely the "negative" TRE consists of one or more analogous hexamers.

Receptors in Human Tissues

The distribution of receptors in human tissues has been partially
clarified. Initial studies have shown β1, α1, and hTRVα2 mRNAs to
be present in large amounts in brain, prostate, and thyroid, and in
lower amounts in tissue such as heart, kidney, liver, spleen,
tonsil, and placenta.[29] It also appears, from initial studies with
specific antibodies, that the various forms of the receptor proteins
are all distributed in several of these tissues.[30] There appears
to be a significant discrepancy between the amount of mRNA present
and the apparent responsiveness of tissues to thyroid hormone,
suggesting that there are levels of control which have not yet been
recognized, or that mRNA levels are not directly equatable with
tissue concentration of receptor protein. In most of these tissues,
hTRVα2 mRNA exists in amounts equal or greater in concentration than
hTRα1 and β. Since hTRα2 functions as a dominant negative regulator
of the function of the other receptors, it may play a crucial role
in determining the amount of functional receptor available in
various tissues.

TR Expression and Processing

We are only beginning to learn about the control of receptor
expression and receptor processing. The level of functional receptor

is diminished in animals by starvation, by treatment with glucagon, and in seriously ill animals with growing tumors. Receptor content is low in the rat fetal liver and newborn rat and increases by age 15 - 25 days to normal adult levels.[31] β2 receptor in the rat pituitary is negatively regulated by thyroid hormone, and β1 receptor in the pituitary is positively regulated.[9] α1 and α2 message levels are negatively regulated by thyroid hormone in rat pituitary, heart, kidney, and liver.[22] Thyroid hormone regulation of receptor concentration probably involves binding of receptor to the TR gene promoter region in a direct feedback loop, since we have shown that a TRE is present in the flanking region of hTRβ. Receptor processing must play an important role in the regulation of receptor concentration and definition of subtypes in individual cells. Nothing is known about the control of splicing of β1 and β2, or of α1 and hTRVα2. It has been shown that a TR can be phosphorylated possibly by PKA or PKC,[33,34] and phosphorylation may be important in either TR binding to DNA or ligand binding.

Why Multiple Receptor Forms?

The reasons for the existence of multiple TREs remains speculative. Tissue or cell specificity of TR function is a logical assumption, but has only been suggested so far for the apparent role of β2 in the pituitary. In the human tissues examined other than pituitary, β, α1, and hTRVα2 are all present. We are just beginning to learn about differences in gene expression in individual cells, and this has not been examined in the human.

It is possible that specific receptors bind to specific genes, but in the systems examined both β1 and α1 appear to regulate the same genes, whether in a negative or positive manner.[36,37] We note that the functional receptors can regulate either positively or negatively; and hTRVα2 causes a "dominant negative" modulation of a negative TR signal as well as of a positive signal.

Differential regulation of receptor expression is another possible reason for different forms of the receptor. This appears to be a potential explanation for the different forms in the pituitary, since β2 is regulated in an opposite direction by thyroid hormone in comparison to β1.

It is also possible that the receptors have different affinities for ligands or different affinities for response elements, although these differences have not been fully examined. Some data suggest that hTRβ1 has a higher affinity for T3 and T4 than does hTRα1, and that hTRβ1 would be the predominant receptor in most tissues.[38] It also has been suggested that hTRα1 binds Triac less effectively than does hTRβ1, but the importance of this in normal physiology is obscure.[7]

Another, and we believe preferable, explanation for the
presence of multiple TREs is that the system effectively represents
a buffer system. Clearly the role of thyroid hormone in the human
is quite different from that of hormones such as estrogen or progesterone, which mediate very specific gene responses which are
precisely modulated in a rapid manner. Thyroid hormone has a more
ubiquitous role, maintaining a constant level of metabolic action,
probably, in the adult, by affecting many proteins. This system
is very carefully controlled around a physiologically established
set point, and many methods are provided to buffer the system
against change. These "defenses" include the provision of two forms
of hormone, metabolically controlled conversion of a less active to
a more active form of the hormone, a feedback system through the
pituitary, and a buffered system in responsive tissue with at least
two functional forms of receptor, and multiple response elements
in responding genes. Perhaps the multiple receptors and variety
of response elements are designed to prevent major shifts in the
level of response, rather than to foster rapid transcriptional
changes as designed into the estrogen/progesterone systems.

Potential Interaction of TR, RAR and ER

A recently recognized aspect of thyroid hormone physiology is
the interaction of thyroid hormone receptors and other receptors.
Retinoic acid receptors, which are important in developmental processes, can form dimers with thyroid hormone receptors.[39] Retinoic
acid receptor can bind to TREs and activate thyroid hormone
responsive genes, and can function additively with thyroid hormone.[40]
Thyroid hormone receptors can bind to retinoic acid response
elements[41] and to estrogen response elements,[22] although it has not
been shown that the thyroid hormone receptor activates these genes.
Thus the possibility exists for a positive stimulatory interaction
with retinoic acid, and a negative regulatory interaction at least
with estrogen receptors.

Abnormal Receptors in Human Pathophysiology

The role of abnormal receptor function in human disease is
just being explored. We have reported the presence of a mutant non-
T3 binding form of thyroid hormone receptor in a patient with
generalized resistance to thyroid hormone action,[42] and have shown
that this receptor can bind to DNA and inhibit the function of
normal receptors.[43] This aspect of our work will be covered in a
separate section of this survey. Obviously the possibility that
pituitary resistance to thyroid hormone could be related to a
specific abnormality in the β2 form of the receptor is tantalizing,
but so far this has not been established. Since thyroid hormone
provides a differentiating signal, loss of receptor might be related
to tumor induction or growth. In fact possible loss of expression
of one form of hTR has been reported in carcinomas of the colon[44]

and possible deletion of a TR gene has been observed in carcinomas of the lung.[45] Whether abnormalities of receptor function are present in individuals with other medical problems such as the euthyroid sick syndrome, prematurity, and extreme obesity, remains unexplored to date.

REFERENCES

1. J.H. Oppenheimer, D. Koerner, H.L. Schwartz, and M.I. Surks, Specific nuclear triiodothyronine binding sites in rat liver and kidney, J. Clin. Endocrinol. Metab. 35:330 (1972).
2. L.J. DeGroot, A. Nakai, A. Sakurai, and E. Macchia, The molecular basis of thyroid hormone action, J. Endocrinol. Invest. 12:843 (1989).
3. K. Ichikawa and L.J. DeGroot, Purification and characterization of rat liver nuclear thyroid hormone receptors, Proc. Natl. Acad. Sci. USA. 84:3420 (1987).
4. C. Weinberger, C.C. Thompson, E.S. Ong, R. Lebo, D.J. Gruol, and R.M. Evans, The c-erbA gene encodes a thyroid hormone receptor, Nature. 324:641 (1986).
5. J. Sap, A. Munoz, K. Damm, Y. Goldberg, J. Ghysdael, A. Leutz, H. Beug, and B. Vennstrom, The c-erbA protein is a high-affinity receptor for thyroid hormone, Nature. 324:635 (1986).
6. A. Krust, S. Green, P. Argos, V. Kumar, P. Walter, J.-M. Bornert, and P. Chambon, The chicken estrogen receptor sequence homology with v-erbA and the human estrogen and glucocorticoid receptors, EMBO J. 5:891 (1986).
7. A. Nakai, A. Sakurai, G.I. Bell, and L.J. DeGroot, Characterization of a third human thyroid hormone receptor coexpressed with other thyroid hormone receptors in several tissues, Mol. Endocrinol. 2:1087 (1988).
8. A. Nakai, S. Seino, A. Sakurai, I. Szilak, G.I. Bell, and L.J. DeGroot, Characterization of a thyroid hormone receptor expressed in human kidney and other tissues, Proc. Natl. Acad. Sci. USA. 85:2781 (1988).
9. R.A. Hodin, M.A. Lazar, B.I. Wintman, D.S. Darling, R.J. Koenig, P.R. Larsen, D.D. Moore, and W.W. Chin, Identification of a thyroid hormone receptor that is pituitary-specific, Science. 244:76 (1989).
10. N. Miyajima, R. Horinchi, Y. Shibuya, S.-I. Fukushige, K.-L. Mastubaya, K. Toyoshima, and T. Yamamoto, Two erbA homologs encoding proteins with different T3 binding capacities are transcribed from opposite DNA strands of the same genetic locus, Cell. 57:31 (1989).
11. L. Tora, H. Gronemeyer, B. Turcotte, M.-P. Gaub, and P. Chambon, The N-terminal region of the chicken progesterone receptor specifies target gene activation, Nature. 333:185 (1988).
12. R.M. Evans, The steroid and thyroid hormone receptor super-family, Science. 240:889 (1988).

13. Z.D. Horowitz, C.-R. Yang, B.M. Forman, J. Casanova, and H.H. Samuels, Characterization of the domain structure of chick c-erbA by deletion mutation: in vitro translation and cell transfection studies, Mol. Endocrinol. 3:148 (1989).

14. J.M. Berg, Proposed structure for the zinc-binding domains from transcription factor IIIA and related proteins, Proc. Natl. Acad. Sci. USA. 85:99 (1988).

15. T. Hard, E. Kellenbach, R. Boelens, B.A. Maler, K. Dahlman, L.P. Freedman, J. Carlstedt-Duke, K.R. Yamamoto, J.-A. Gustafsson, and R. Kaptein, Solution structure of the glucocorticoid receptor DNA-binding domain. Science. 249:157 (1990).

16. K. Umesono and R.M. Evans, Determinants of target gene specificity for steroid/thyroid hormone receptors, Cell. 57:1139 (1989).

17. B.M. Forman, C.-R. Yang, M. Au, J. Casanova, J. Ghysdael, and H.H. Samuels, A domain containing leucine-zipper-like motifs mediate novel in vivo interactions between the thyroid hormone and retinoic acid receptors, Molecul. Endocrinol. 3:1610 (1989).

18. A.L. O'Donnell, D.S. Darling and R.J. Koenig, T3 receptor mutations that impair transcriptional regulation also impair enhancement of DNA binding by JEG cell extracts, in: "Program and Abstracts," The Endocrine Society, 72nd Annual Meeting, Atlanta, GA, Abstract NO. 1329 (1990).

19. M.B. Murray and H.C. Towle, Identification of nuclear factors that enhance binding of the thyroid hormone receptor to a thyroid hormone response element, Molecul. Endocrinol. 3:1434 (1989).

20. K. Damm, C.C. Thompson, and R.M. Evans, Protein encoded by v-erbA functions as a thyroid-hormone receptor antagonist, Nature. 339:593 (1989).

21. R.J. Koenig, M.A. Lazar, R.A. Hodin, G.A. Brent, P.R. Larsen, W.W. Chin, and D.D. Moore, Inhibition of thyroid hormone action by a non-hormone binding c-erbA protein generated by alternative mRNA splicing, Nature. 337:659 (1989).

22. C.K. Glass, J.M. Holloway, O.V. Devary, and M.G. Rosenfeld, The thyroid hormone receptor binds with opposite transcriptional effects to a common sequence motif in thyroid hormone and estrogen response elements, Cell. 54:313 (1988).

23. G.A. Brent, J.W. Harney, Y. Chen, R.L. Warne, D.D. Moore, and P.R. Larsen, Mutations of the rat growth hormone promoter which increase and decrease response to thyroid hormone define a consensus thyroid hormone response element, Molecul. Endocrinol. 3:1996 (1989).

24. K.J. Petty, B. Desvergne, T. Mitsuhashi, and V.M. Nikodem, Identification of a thyroid hormone response element in the malic enzyme gene, J. Biol. Chem. 265:7395 (1990).

25. T.N. Lavin, J.D. Baxter, and S. Horita, The thyroid hormone receptor binds to multiple domains of the rat growth hormone 5'-flanking sequence, J. Biol. Chem. 263:9418 (1988).

26. V. Krishna, K. Chatterjee, J.-K. Lee, A. Rentoumis, and J.L. Jameson, Negative regulation of the thyroid-stimulating hormone

alpha gene by thyroid hormone: Receptor interaction adjacent to the TATA box, Proc. Natl. Acad. Sci. USA. 86:9114 (1989).

27. F.E. Wondisford, E.A. Farr, S. Radovick, H.J. Steinfelder, J.M. Moates, J.H. McClaskey, and B.D. Weintraub, Thyroid hormone inhibition of human thyrotropin beta-subunit gene expression is mediated by a cis-acting element located in the first exon, J. Biol. Chem. 264:14601 (1989).

28. G.A. Brent, J.W. Harney, D.D. Moore, and P.R. Larsen, Effects of varying the thyroid hormone DNA response element position within the rat growth hormone promoter: implications for positive and negative regulation by T3, in: "Program and Abstracts," The Endocrine Society, 72nd Annual Meeting, Atlanta, GA, Abstract No. 1331 (1990).

29. A. Sakurai, A. Nakai, and L.J. DeGroot, Expression of three forms of thyroid hormone receptor in human tissues, Molecul. Endocrinol. 3:392 (1989).

30. E. Macchia, A. Nakai, A. Janiga, A. Sakurai, M.-E. Fisfalen, P. Gardner, K. Soltani, and L.J. DeGroot, Characterization of 'site specific' polyclonal antibodies to c-erbA peptides recognizing human thyroid hormone receptors alpha-1, alpha-2, and beta and native T3 receptor and study of tissue distribution of the antigen, Endocrinology. 126:3232 (1990).

31. L.J. DeGroot, M. Robertson, and P.A. Rue, Triiodothyronine receptors during maturation, Endocrinology. 100:1511 (1977).

32. R.A. Hodin, M.A. Lazar, and W.W. Chin, Differential and tissue-specific regulation of the multiple rat c-erbA messenger RNA species by thyroid hormone, J. Clin. Invest. 85:101 (1990).

33. Y. Goldberg, C. Glineur, J.-C. Gesquiere, A. Ricouart, J. Sap, B. Vennstrom, and J. Ghysdael, Activation of protein kinase C or cAMP-dependent protein kinase increases phosphorylation of the c-erbA-encoded thyroid hormone receptor and of the v-erbA encoded protein, EMBO J. 7:2425 (1988).

34. C. Glineur, M. Bailly, and J. Ghysdael, The c-erbA alpha-encoded thyroid hormone receptor is phosphorylated in its amino terminal domain by casein kinase II, Oncogene. 4:1247 (1989).

35. D.J. Bradley, W.S. Young III, and C. Weinberger, Differential expression of alpha and beta thyroid hormone receptor genes in rat brain and pituitary, Proc. Natl. Acad. Sci. USA. 86:7250 (1989).

36. A. Rentoumis, V. Krishna, K. Chatterjee, L.D. Madison, S. Datta, G.D. Gallagher, L.J. DeGroot, and J.L. Jameson, Negative and positive transcriptional regulation by thyroid hormone receptor isoforms, Molecul. Endocrinol. (1990 -- in press).

37. A. Nakai, A. Sakurai, E. Macchia, V. Fang, and L.J. DeGroot, The roles of three forms of human thyroid hormone receptor in gene regulation, Molecul. Cellul. Endocrinol. 72:143 (1990).

38. P.A. Schueler, H.L. Schwarts, K.A. Strait, C.N. Mariash, and J.H. Oppenheimer, Binding of 3,5,3'-triiodothyronine (T3) and its analogs to the in vitro translational products of c-erbA protooncogenes: differences in the affinity of the alpha and

beta forms for the acetic acid analog and failure of the human
testis and kidney alpha-2 products to bind T3, Molecul. Endocrinol. 4:227 (1990).

39. C.K. Glass, S.M. Lipkin, O.V. Devary, and M.G. Rosenfeld,
Positive and negative regulation of gene transcription by a
retinoic acid-thyroid hormone receptor heterodimer, Cell.
59:697 (1989).

40. K. Umesono, V. Giguere, C.K. Glass, M.G. Rosenfeld, and R.M.
Evans, Retinoic acid and thyroid hormone induce gene expression
through a common responsive element, Nature. 336:262 (1988).

41. K. Umesono, V. Giguere, C.K. Glass, M.G. Rosenfeld, and R.M.
Evans, Retinoic acid and thyroid hormone induce gene expression
through a common responsive element, Nature. 336:262 (1988).

42. A. Sakurai, K. Takeda, K. Ain, P. Ceccarelli, A. Nakai, S.
Seino, G.I. Bell, S. Refetoff, and L.J. DeGroot, Generalized
resistance to thyroid hormone associated with a mutation in
the ligand-binding domain of the human thyroid hormone receptor
beta, Proc. Natl. Acad. Sci. USA.   86:8977 (1989).

43. A. Sakurai, T. Miyamoto, S. Refetoff, and L.J. DeGroot, Dominant
negative transcriptional regulation by a mutant thyroid hormone
receptor beta in a family with generalized resistance to thyroid hormone, Molecul. Endocrinol. (1990 -- in press).

44. S. Markowitz, M. Haut, T. Stellato, C. Gerbic, and K.
Molkentin, Expression of the erbA-beta class of thyroid
hormone receptors is selectively lost in human colon carcinoma,
J. Clin. Invest. 84:1683 (1989).

45. I.U. Ali, R. Lidereau, and R. Callahan, Presence of two members
of c-erbA receptor gene family (c-erbA beta and c-erbA2) in
smallest region of somatic homozygosity on chromosome 3p21-p25
in human breast carcinoma, J. Natl. Cancer Inst. 81:1815 (1989).

THYROID SYSTEM ONTOGENY IN THE SHEEP: A MODEL

FOR PRECOCIAL MAMMALIAN SPECIES

Delbert A. Fisher

Department of Ped᠎ ᠎ics
Harbor-UCLA Medical Center
Torrance, CA   90509

I -   Models For Study of Thyroid System Ontogenesis

Human thyroid system ontogenesis can be divided into
three phases which roughly parallel the three classic trimesters
of pregnancy. These include the period of embryogenesis during
the first trimester, a period of relative hypothalamic-pituitary
quiescence during the mid-trimester of pregnancy, and a final
phase of synchronized maturation of hypothalamic-pituitary
control of TSH secretion, thyroid hormone secretion, and peripheral
iodothyronine metabolism (1). These phases of thyroid ontogenesis
occur, as well, in other mammals and have been investigated
in considerable detail in two models, the developing rat
and the fetal sheep (1). In both models thyroid system development generally resembles human thyroid ontogenesis. However, in
the rat, an altricial species, the total period of thyroid
development encompasses about 50 days (21 fetal days + 28
postnatal days) and delivery occurs at approximately the midpoint
of development during the phase of relative system quiescence.
In the sheep, a precocial species, the period of thyroid system
maturation extends from conception to two weeks of postnatal
age, about 165 days. Birth occurs at approximately 150 days
gestation in this species.

The three species differ with regard to timing of thyroid
hormone dependent brain maturation relative to intrauterine
development (1,3). However thyroid dependent brain maturation
generally occurs during phase three of thyroid system development, and in humans it extends well beyond this period. In the
infant rat the major period of thyroid dependency of brain
maturation extends from a few days to about 3 weeks of postnatal
age.

Advances in Perinatal Thyroidology, Edited by B.B. Bercu and
D.I. Shulman, Plenum Press, New York, 1991

In the sheep, thyroid dependent brain maturation extends from
70 to 90 days to two postnatal weeks. In the human infant
the period of thyroid dependent brain maturation extends from
the perinatal period to about 2 years of postnatal age.

An additional difference in the species with regard to
thyroid development is the extent of placental permeability
to thyroid hormones. There are structural differences in
the placentas of the three species (4,5). In the sheep placenta,
an epitheliochorial placenta, the maternal and fetal circulations are separated by 6 tissue layers (maternal endothelium,
maternal connective tissue, maternal endometrial syncytium,
cytotrophoblast, fetal basement membrane, and fetal endothelium)
(4). The rat placenta is a haemotrichorial placenta with
4 tissue layers separating the maternal and fetal circulations
(chorionic epithelium, syncytiotrophoblast, cytotrophoblast,
chorionic connective tissue, and fetal endothelium). The
human placenta is haemomonochorial with 3 tissue layers between
the maternal and fetal circulations (syncytiotrophoblast,
chorionic connective tissue and fetal endothelium) (4,5).

Part of the placental barrier is metabolic. The placentas
of the three species contain an iodothyronine inner ring monodeiodinase which converts thyroxine (T4) to inactive reverse
triiodothyronine (rT3) and T3 to inactive diiodothyronine
(T2) (6). The net result of these structural and metabolic
factors in all three species is a marked maternal to fetal
thyroid hormone gradient. However, in the rat and human species, significant net maternal to fetal thyroid hormone transfer
occurs (7,8). In the sheep, in contrast, there appears to
be only limited transfer (1,9). There also is little or no
placental transfer of 3,5, dimethyl-3'-isopropyl-L-thyronine
(DIMIT) in the sheep; DIMIT is a non halogenated, more lipid
soluble thyronine which more readily traverses the rat placenta
than T4 (9,10).

Thus, in most mammalian species, and particularly in
the sheep, there is limited maternal-fetal transfer of thyroid
hormones. In addition the placenta is impermeable to TSH
(1,6). As a result the mammalian fetal thyroid system matures
relatively free of maternal thyroid influence.

II Thyroid System Maturation in the Sheep

Fetal hypothalamic-pituitary and thyroid gland embryogenesis occur during the first third of gestation in humans and
sheep (1,2). The pattern of selected critical events of thyroid
system maturation in the fetal sheep are shown in figure 1.
TRH and TSH are identifiable in hypothalamus and pituitary
gland, respectively, by 50 days gestation. This represents

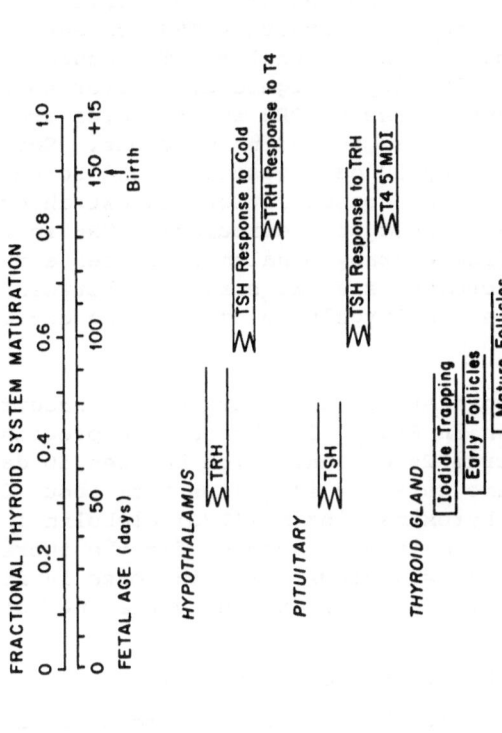

Figure 1. Approximate timing of selected events in thyroid system maturation in the fetal sheep.

about 0.3 of the total period of thyroid system maturation.
At this time the thyroid gland is well developed, thyroid
follicular cells trap radioiodine, and developing thyroid
colloid-filled follicles are identifiable (1,2).

The maturation patterns of serum and pituitary TSH, and
serum and tissue TRH concentrations are summarized in figure
2. Fetal serum TSH levels increase from low values at 50
days to peak concentrations at 90-100 days gestation and decrease
again thereafter to term. Pituitary TSH content increases
slowly and progressively until 100 days after which a more
rapid rate of increase is observed to term. This abrupt acceleration in rate of increase in pituitary TSH content is likely
due to the rapid increase in hypothalamic TRH concentration
which begins at 90 to 100 days (figure 2). Prior to that
time the major concentrations of TRH in the fetus are observed
in placenta and gut tissues, including pancreas. Serum TRH
levels are significant by 50 days gestation and serum levels
appear to parallel the TRH concentrations in extrahypothalamic
tissues. These data suggest that circulating TRH in the fetus
is derived largely from extrahypothalamic sources and support
the view that the increase in fetal serum TSH between 50 and
100 days gestation is due largely to circulating extrahypothalamic TRH production.

The maturation pattern of circulating T4 concentration
in the fetus is shown in Figure 3. There is a progressive
increase in fetal serum T4 concentration between 50 days and
term during which time total T4 increases from about 20 to
180 nmol/L. Serum thyroxine binding (TBG) globulin levels
increase in parallel and largely account for the increase
in total T4, but there is a progressive increase in free T4
concentration presumably as well driven by the increase in
serum TSH (figure 2).

III   Maturation of Thyroid Hormone Metabolism

The first step in metabolism of T4 is monodeiodination
to T3 or reverse T3 (rT3) (1,2,11). The monodeiodination
is mediated by at least 3 types of particulate cellular iodothyronine monodeiodinase (MDI) enzyme activities (11). Type I
MDI is a high Km enzyme, inhibited by propylthiouracil (PTU)
and stimulated by thyroid hormone; it catalyzes monodeiodination
of T4 to T3 and rT3 to T2. The preferred substrate for type
I MDI is rT3. Type II MDI is a low Km enzyme, insensitive
to PTU, and inhibited by thyroid hormone; it catalyzes T4
conversion to T3; the preferred substrate is T4. The Type
III inner ring MDI catalyzes conversion of T4 to rT3. The
maturation patterns of type I, type II and type III MDI activities in fetal ovine tissues are summarized in figure 4 (11-15).
As indicated in figure 4, the predominant fetal MDI activity

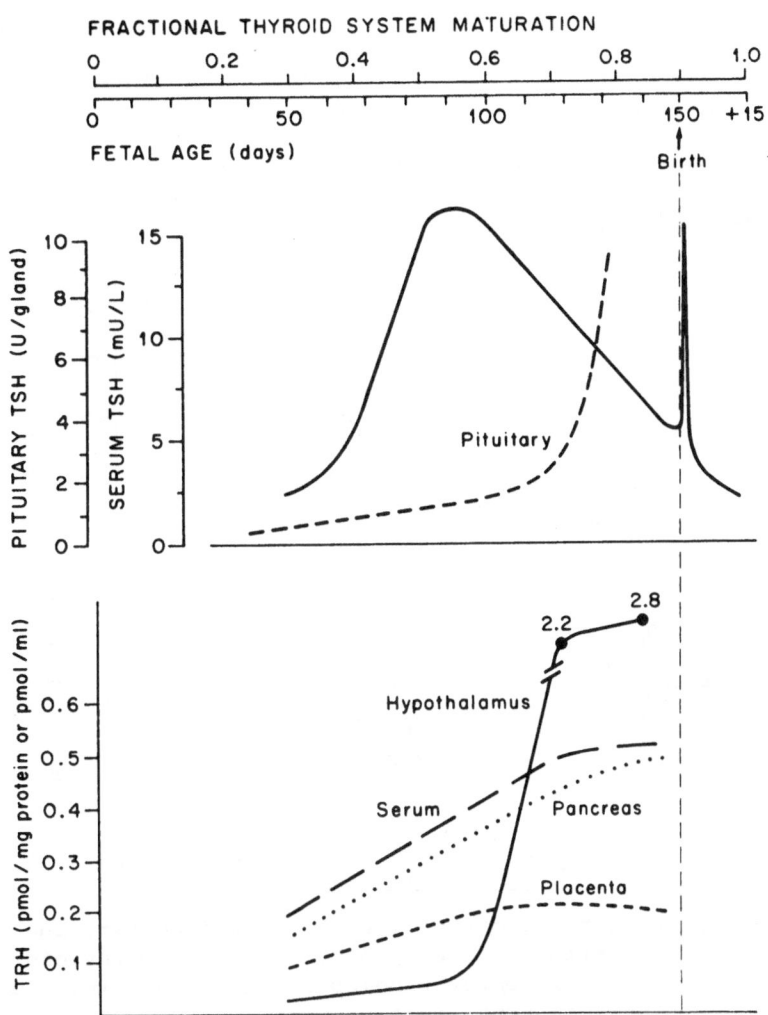

Figure 2. Patterns of maturation of pituitary and serum TSH
(upper panel) and of selected tissue and serum TRH
concentrations (lower panel) in the fetal sheep.

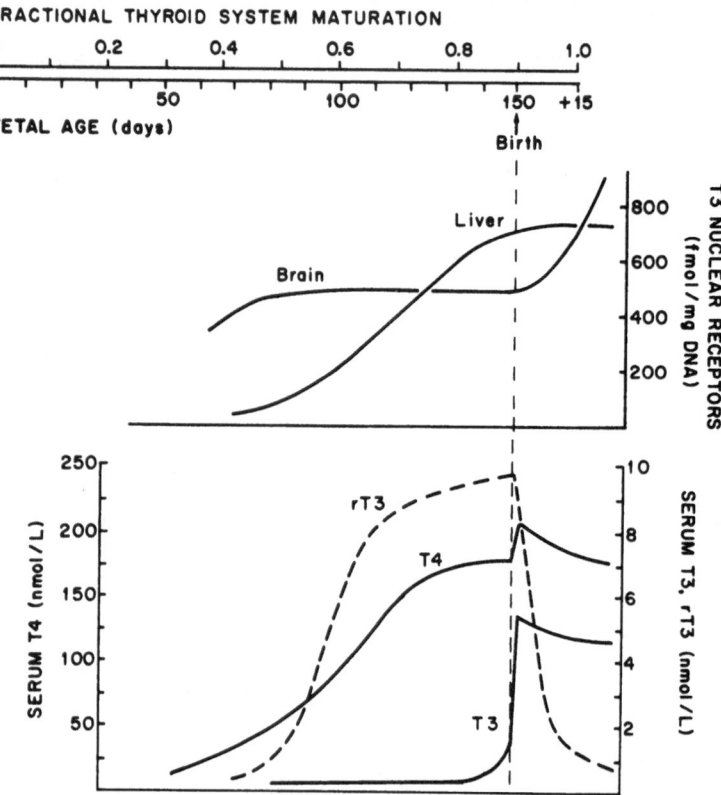

Figure 3.   Patterns of maturation of serum iodothyronine concentrations (lower panel) and nuclear thyroid
hormone receptor binding in brain and liver (upper
panel) of fetal sheep.

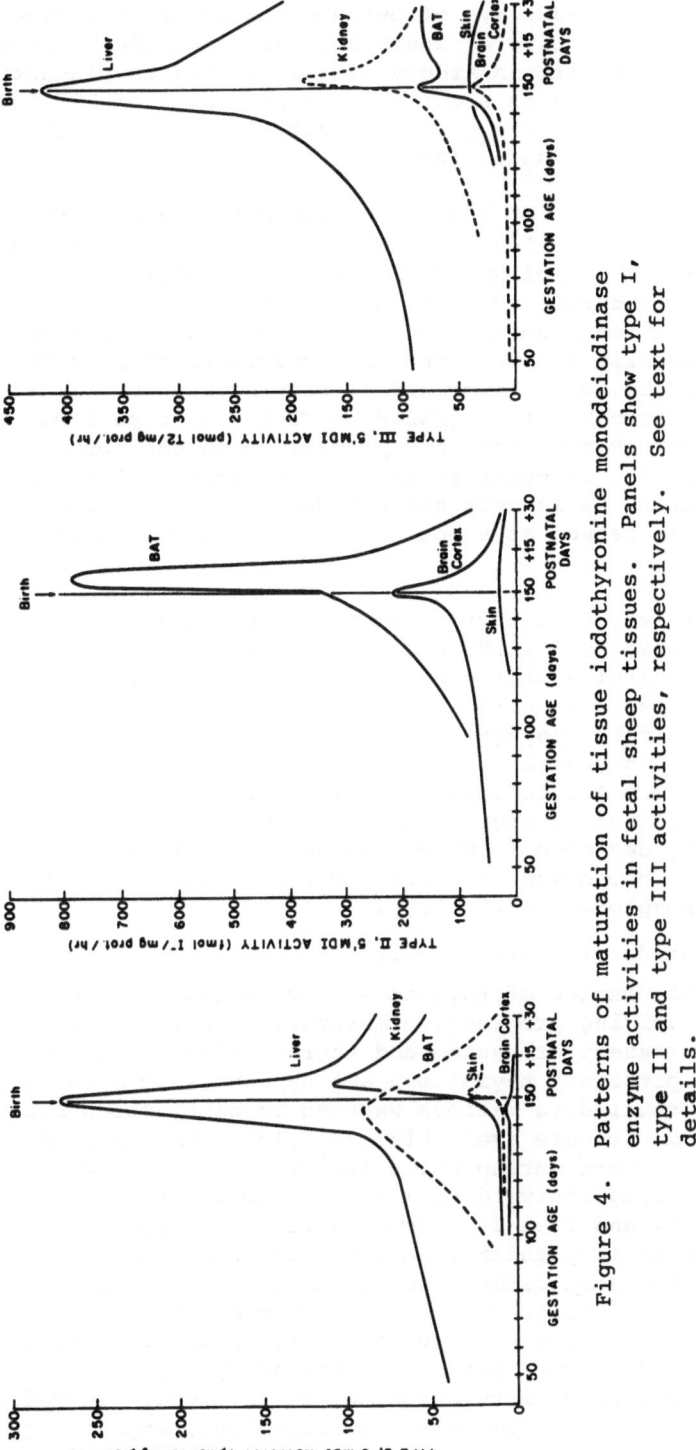

Figure 4. Patterns of maturation of tissue iodothyronine monodeiodinase
enzyme activities in fetal sheep tissues. Panels show type I,
type II and type III activities, respectively. See text for
details.

during the first 130 days of ovine gestation is the type III
activity in fetal liver. There also are significant levels
of type III activity in kidney (figure 4) and in placenta
(not shown). There is significant type I activity in fetal
liver and significant type II activities in fetal brain and
in brown adipose tissues (BAT).

The maturation of rT3 and T3 concentrations in fetal
serum are shown in figure 3. Reverse T3 is the predominant
product of T4 metabolism in the fetus and levels increase
progressively during the last half of gestation. (1,2). Serum
T3 concentrations increase only at term. Thus the major pathway
of T4 metabolism in the fetus is monodeiodination to rT3.
This occurs in fetal liver and perhaps kidney and placenta.
Reverse T3 probably is degraded to T2 by the type I MDI in
fetal liver. Brain, BAT, and probably skin can convert T4
to T3; this T3, in contrast to that produced in liver and
other tissues, is largely sequestered locally and inaccessible
to serum; it serves as a source of T3 for local thyroid actions
(1,6).

There are other thyroid hormone metabolic pathways including glucuronide and sulfate conjugations and progressive cleav-
age of the alanine side chain via decarboxylation and transamina-
tion (2). Thyroid hormones in the adult are excreted as glucuronide and sulfate conjugates in urine and stool. There is
no information regarding the maturation of these pathways
in the fetus. Acetic acid derivatives of the iodothyronines
have significant biological activity in vitro, but in adults
are rapidly degraded in vivo. Pyruvate and lactate derivatives
have little biological activity. These pathways too, remain
largely unexplored in the fetus.

IV Maturation of Thyroid Control

The maturation of thyroid system control is a complex
process involving progressive maturation of the hypothalamic-
pituitary transducer system and thyroid gland function. The
events of pituitary thyrotroph and hypothalamic maturation
have been studied to various degrees in rats, sheep and humans.
Recent summaries are available (2,17,18). In the sheep this
maturation occurs during phase III of thyroid system maturation.
There is a progressive decrease in serum TSH (figure 2) as
the serum T4 and free T4 concentrations increase (figure 3).
The hypothalamic-pituitary response to cold develops relatively
early but the somatostatin and dopamine pathways which modulate
pituitary TSH release develop only in the neonatal period
(2,17,18). The molecular events limiting the thyroid response
to TSH during thyroid maturation are not yet clear. Receptor
and/or postreceptor maturation may be involved. Thyroidal
autoregulation by iodide also develops late, explaining the

susceptibility of premature infants and lambs to iodide induced
hypothyroidism and goiter (2).

V Adaptation to the Extrauterine Environment

    The transition from the fetal environment to the neonatal,
extrauterine milieu involves a complex series of cardiovascular and metabolic alterations. With regard to the thyroid
system, there is a marked early neonatal TSH surge (figure
2) which evokes acute thyroidal release of T4 and T3 and increases
in serum T4 and T3 concentrations (2,12,19). The neonatal
increase in serum T4 is transient, but the increase in serum
T3 in the neonate is sustained beyond the period of TSH stimulation. There also is a marked increase in thyroid dependent
neonatal BAT thermogenesis acutely stimulated by catecholamines
(12,20).

    Neonatal BAT thermogenesis is critical for extrauterine
survival of the newborn lamb (1). It is not dependent on
the neonatal T3 surge, but is dependent on the prevailing
T4 concentration. The maturation of BAT function in the fetal
lamb is summarized in figure 5. This maturation is reflected
in progressive increases in BAT cell volume and norepinephrine-
stimulated thermogenesis during the last third of gestation
(21). In addition there are progressive increases in both
type I and type II MDI activities in BAT during this period
(15). Both MDI activities are thyroid hormone responsive
but in opposite directions; type I activity increases in response
to T4 while type II activity decreases (15). Both MDI activities probably are important to BAT function, providing a reliable source of T3 to maintain optimal uncoupling protein (thermo-
genin) concentrations for effective BAT thermogenesis (22).
In hypothyroidism, the type II activity predominates (15).
Both may contribute to circulating T3 levels in the neonatal
period (12).

    At term there are marked increases in the type I MDI
activities in liver and kidney and type II activity in BAT
(2,11,23). These increases are largely responsible for the
marked increase in T4 to T3 conversion and the increase in
serum T3 concentrations in the perinatal period (1,2,11).
In the sheep, the prenatal cortisol surge during the 7-10
days prior to parturition is at least partly responsible for
the increases in hepatic and renal type I MDI activities (24).
The type I MDI activity in BAT is thyroid hormone responsive
and the increase in activity in the newborn period may be
due to the increased thyroid hormone levels induced by the
TSH surge (15). The reason for the marked increase in BAT
type II activity is not clear; type II activity decreases
in response to thyroid hormone (15). Catecholamines may be

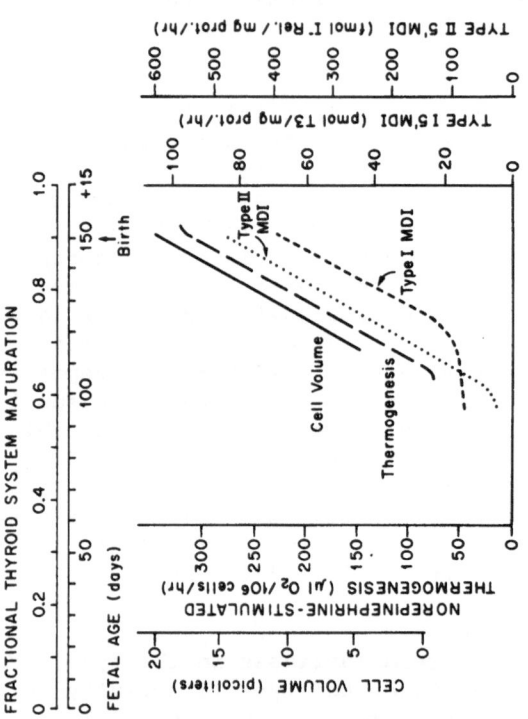

Figure 5.   Maturation of cell volume, norepinephrine-stimulated thermogenesis (measured as oxygen-consumption), and monodeiodinase enzyme activities in developing brown adipose tissue (BAT) in fetal sheep.

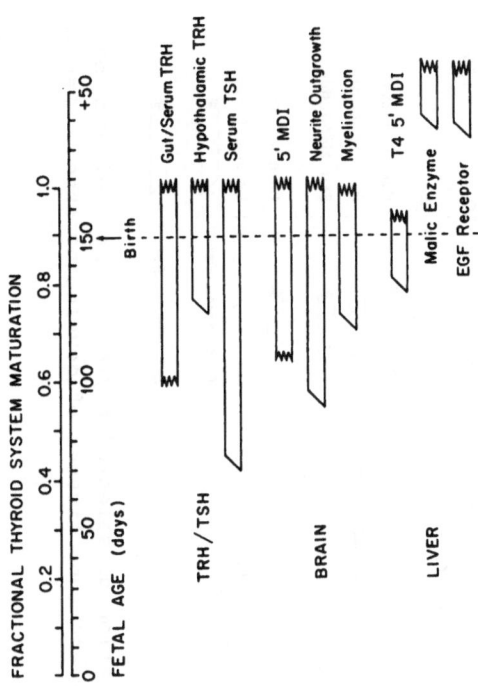

Figure 6. Maturation of selected thyroid hormone effects on brain and gut tissues of the developing sheep. See text for details.

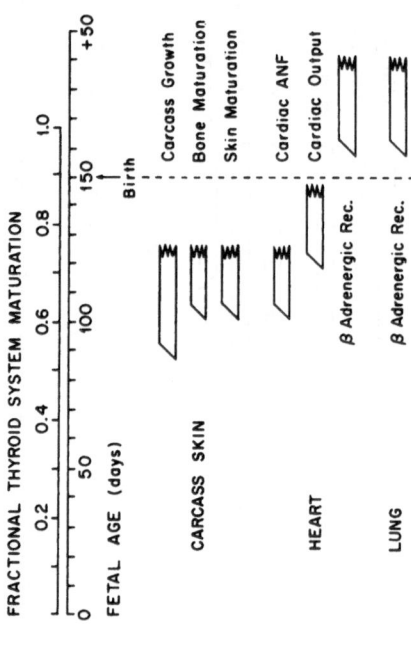

Figure 7. Maturation of selected thyroid hormone effects on carcass, skin, heart and lung tissues of developing sheep. See text for details.

involved in the early neonatal stimulation of type II activity
in BAT.

Postnatal thyroid dependent metabolism is largely mediated
by the increase in T3 production in the newborn. This increase
is shown in table I (24). The transient two-fold increase
in T4 production in the neonatal period is accompanied by
a five-fold increase in T3 production and a 3 to 4 fold decrease
in rT3 production (24). The increases in T3 production and
in serum T3 levels and the decrease in rT3 production and
serum rT3 levels persist indefinitely as a result of alterations
in the postnatal patterns of T4 monodeiodination (1,2).

VI   Maturation of Thyroid Hormone Actions

The maturation of thyroid hormone actions is summarized
in figures 6 and 7 and reviewed in detail in references 2
and 25. Figure 3 summarizes the patterns of maturation of
thyroid nuclear T3 receptors in brain and liver of the fetal
sheep (26,27). Levels in lung tissue approximate the levels
in fetal brain at midgestation (not shown) (26). Pituitary
TSH production and TRH concentrations in extrahypothalamic
and hypothalamic tissues are thyroid hormone responsive.
These responses to thyroid hormones are among the first to
appear and are manifest during the middle third of gestation
(25,28,29). Effects of thyroid hormone to inhibit brain 5'
MDI activity, stimulate neurite outgrowth, and stimulate myelination are observed at 90-100 days (10,25,29,30). Effects on
carcass growth and on skin and bone maturation are observed
at 90-100 days (19,23,24,30). The stimulatory effect of thyroid
hormone on cardiac ANF concentrations is observed early in
the last third of gestation (31). An effect of thyroid hormone
on cardiac output and the action to stimulate hepatic 5' MDI
activity are observed only near term (14,32). The stimulatory
action of thyroid hormone on hepatic malic enzyme activity,
hepatic EGF receptor binding, and on beta adrenergic receptor
binding in heart and lung tissues are delayed until the neonatal
period (2,25,33,34).

These effects of thyroid hormone are largely related
to effects on gene transcription or posttranscription events
in thyroid responsive cells. Whether the wide variation in
time of appearance of thyroid hormone actions is related to
variation in thyroid nuclear receptor maturation or to some
post-nuclear receptor events is not clear. The early appearance
of thyroid hormone effects on brain development in the fetal
sheep clearly relates to the early appearance of thyroid nuclear
receptors in brain tissue. The late (postnatal) appearance
of thyroid effects on selected hepatic proteins must be related
to post-receptor events.

VII Summary

There is now extensive data in two animal models to characterize the events of thyroid system maturation in mammalian species. Data from the rat have contributed importantly to our understanding of thyroid development in the altricial species while studies in the sheep have characterized thyroid ontogenesis in precocial mammals. More limited data are available from human studies and this information, with the animal data, indicate that the major features of thyroid system development are qualitatively similar in all three species. The combined data has provided an important overview of mammalian thyroid ontogenesis. However the picture remains unfinished. We have limited insights into the maturation of the molecular events of hypothalamic pituitary control, the maturation of TSH and thyroid nuclear receptors and postreceptor events, and the mechanisms of the thyroid effects on brain maturation. We can look forward to continuing contributions in these areas.

References

1.   Fisher DA, Dussault JH, Sack J, Chopra IJ. Ontogenesis of hypothalamic-pituitary-thyroid function and metabolism in man, sheep and rat. Rec Prog Horm Res 33:59-116 (1977).

2.   Fisher DA, Polk DH. Development of the thyroid, Bailliere's Clinical Endocrinology and Metabolism 3:627-657 (1989).

3.   Morreale de Escobar G and Escobar del Rey F. Brain damage and thyroid hormone, In: Burrow GN and Dussault JH, Eds, Neonatal Thyroid Screening, Raven Press, New York, pp 25-50 (1980).

4.   Kaufman P. Functional anatomy of the non-primate placenta. Placenta, Suppl. 1, 13-28 (1981).

5.   Ramsey EM. Functional anatomy of the primate placenta. Placenta, Suppl. 1, 29-42 (1981).

6.   Roti E, Gnudi A, Braverman LE. The placental transport, synthesis and metabolism of hormones and drugs which affect thyroid function. Endocrine Rev 4:131-149 (1983).

7.   Morreale de Escobar G, Calvo R, Obregon MJ, Escobar del Rey F. Contribution of maternal thyroxine to fetal thyroxine pools in normal rats near term. Endocrinology 126:2765-2767, (1990).

8.   Vulsma T, Gons MH, DeVijlder JJJM. Maternal fetal transfer of thyroxine in congenital hypothyroidism due to a total organification defect or thyroid agenesis. N Engl J Med 321:13-16 (1989).

9.   Bachrach LK, Kudlow JE, Silverberg H, Kent G, Burrow GN. Treatment of ovine cretinism in utero with 3,5,dimethyl-3'-isopropyl-1-thyronine. Endocrinology 111:132-135 (1982).

10. Mano MT, Potter BJ, Belling GB, Martin DN, Gragg BG,

Chavadej J, Hetzel BS. The effect of thyroxine, 3,5, dimethyl-3'-isopropyl-l-thyroxine and iodized oil on fetal brain development in the iodine-deficient sheep. Acta Endocrinologica 121:7-15 (1989).

11. Wu SY, Fisher DA, Polk D, Chopra IJ. Maturation of thyroid hormone metabolism, In: Wu SY, Hershman JM, Eds, Thyroid Hormone Metabolism: Regulation and clinical implications, Blackwell Scientific, New York, in press.

12. Polk DH, Wu SY, Fisher DA. Serum thyroid hormone and tissue 5' monodeiodinase activity in acutely thyroidectomized newborn lambs. Am J Physiol 21:E151-E155 (1986).

13. Wu SY, Polk DH, Fisher DA. Biochemical and ontogenic characterization of T4 5'monodeiodinase in brown adipose tissue from fetal and newborn lambs. Endocrinology 118:1334-1339,(1986).

14. Polk DH, Wu SY, Wright C, Reviczky AL, Fisher DA. Ontogeny of thyroid hormone effect on tissue 5'monodeiodinase activity in fetal sheep. Am J Physiol 254:E337-E341 (1988).

15. Wu SY, Merryfield ML, Polk DH, Fisher DA. Two pathways for thyroxine 5'monodeiodination in brown adipose tissue in fetal sheep: ontogenesis and divergent responses to hypothyroidism and T3 replacement. Endocrinology 126:1950-1958 (1990).

16. Ferreiro B, Bernal J, Morreale de Escobar G, Potter BJ. Preferential saturation of brain 3,5,3' triiodothyronine receptor during development in fetal lambs. Endocrinology 122:438-443 (1988).

17. Fisher DA. Development of fetal thyroid system control, In: Delong GR, Robbins J, Condliffe PG, Eds, Iodine and the Brain, Plenum Publishing Inc., New York, pp 167-176 (1989).

18. Roti E. Regulation of thyroid-stimulating hormone (TSH) secretion in the fetus and neonate. J Endocrinol Invest 11:145-158 (1988).

19. Klein AH, Oddie TH, Fisher DA. Effect of parturition on serum iodothyronine concentrations in fetal sheep. Endocrinology 103: 1453-1457 (1978).

20. Polk DH, Padbury JF, Callegari C, Newnham JP, Reviczky AL, Klein AH, Fisher DA. Effect of fetal thyroidectomy on newborn thermogenesis in lambs. Pediatr Res 21:453-457 (1987).

21. Klein AH, Reviczky A, Chou P, Padbury J, Fisher DA. Development of brown adipose tissue thermogenesis in the ovine fetus and newborn. Endocrinology 112:1662-1666 (1983).

22. Bianco AC, Silva JE. Intracellular conversion of thyroxine to triiodothyronine is required for the optimal thermogenic function of brown adipose tissue. J Clin Invest 79:295-300 (1987).

23.   Wu SY, Klein AH, Chopra IJ, Fisher DA. Alterations
      in tissue thyroxine 5'-monodeiodinating activity in
      the perinatal period. Endocrinology 103:235-239 (1978).
24.   Klein AH, Oddie TH, Fisher DA. Iodothyronine kinetic
      studies in the newborn lamb. J Devel Physiol 2:29-35
      (1980).
25.   Fisher DA and Polk DH. Maturation of thyroid hormone
      actions, In: Delange F, Fisher DA, Glinoer D, Eds, Research
      in Congenital Hypothyroidism, Plenum Publishing Inc.,
      New York, pp 61-77 (1989).
26.   Ferreiro B, Bernal J, Potter BJ. Ontogenesis of thyroid
      hormone receptor in foetal lambs. Acta Endocrinologica
      116:205-210 (1987).
27.   Polk DH, Cheromcha D, Reviczky A, Fisher DA. Nuclear
      thyroid hormone receptors: ontogeny and thyroid hormone
      effects in sheep. Am J Physiol 256:E543-E549 (1989).
28.   Polk DH, Reviczky A, Lam RW and Fisher DA. Thyrotropin
      releasing hormone in the ovine fetus: ontogeny and effect
      of thyroid hormone. Am J Physiol: in press.
29.   Hetzel BS, Mano MT. A review of experimental studies
      of iodine deficiency during fetal development. J Nutr
      119:145-151 (1989).
30.   Erenberg A, Omori K, Menkes JH, Oh W, Fisher DA. Growth
      and development of the thyroidectomized ovine fetus.
      Pediatric Res 8:783-789 (1974).
31.   Castro R, Polk DH, Lam RW, Leake RD, Fisher DA. Atrial
      natriuretic factor (ANF): effect of thyroidectomy on
      concentration in fetal sheep atria and ventricles.
      Pediatric Res 23:274A (1988).
32.   Breall JA, Rudolph AM, Heymann MA. Role of thyroid
      hormone in postnatal circulating and metabolic adjust-
      ments. J Clin Invest 743:1418-1424 (1984).
33.   Padbury JF, Klein AH, Polk DH, Lam RW, Hobel CJ, Fisher
      DA. The effect of thyroid status on lung and heart
      beta adrenergic receptors in fetal and newborn sheep.
      Devel Pharmacol Therapeut 9:44-53 (1986).
34.   Cheromcha DP, Callegari CC, Reviczky A, Polk DH, Fisher
      DA. Ontogenesis of thyroid hormone receptor and post-recep-
      tor effects in fetal ovine liver. Clin Res 35:232A
      (1987).

PATHOLOGIC STUDIES OF FETAL THYROID DEVELOPMENT

Douglas R. Shanklin

Departments of Pathology and of
Obstetrics and Gynecology
University of Tennessee, Memphis
Memphis, Tennessee 38163

Ordinarily the assignment of tissue change to the category of
lesion or to variation is straightforward, depending on the charac-
ter and extent of the change and its functional impact. The developmental process alters this relationship in large part because the
effect is often in the future, sometimes the distant future. As
such, what is "pathological" from the developmental point of view
requires more understanding of the coordination of developing systems, the sequences for obtaining functional competency, and the
effects of growth, maturation, and diseases in other organ systems.
Specified injuries may have different effects on the time scale of
development. Moreover, lesions in fetuses and newborns often have
unusual features. It should be kept in mind that much of the final
form, content, and location of many organs often depends on the removal of primordial structures which were earlier stages of those
tissues. For the thyroid gland the process in question is that of
folliculogenesis [Norris, 1914], through the formation of a primordium later converted to follicles, a unique vertebrate structure.

Embryology of the thyroid

Thyroid cells derive from foregut endoderm [Arey, 1956; Copp,
1969; Copp, et al., 1967], in a median diverticulum of the ventral
wall of the pharynx [Gray, 1948]. Contribution of cells from the
ultimobranchial bodies was suggested by Loewenstein and Wollman in
1970 with this distribution:

1. Centrilobular follicles and those adjacent to the
   parathyroid glands from the ultimobranchial bodies.

2. Peripheral follicles and those of the isthmus from the
   ventral pharyngeal pouch.

That this distinction may be more apparent than real is indicated by the observation that ultimobranchial bodies are merely the
final derivatives of the left and right 5th pharyngeal pouches with
possible contributions from the 4th pouches. They are enveloped by
the expansion of the lateral lobes of the thyroid primordium, derived from the 4th pouches [Gray, 1948]. Briefly, a tract remnant of
the migration of the thyroid primordium to mid-neck is termed the
thyroglossal duct which, if it persists, may yield lesions in the
pathway, especially the pyramidal lobe of the thyroid (a retrogres-
sive extension back to the hyoid bone) and various cysts, usually
small [Gray, 1948].

Nevertheless, we find lobular distinctions in follicles during
fetal life, possibly a zonal effect or different time thresholds of
functional maturation, part of the complex origin of thyroid [Fig-
ures 1, 2 and 13]. This will be described and discussed below.

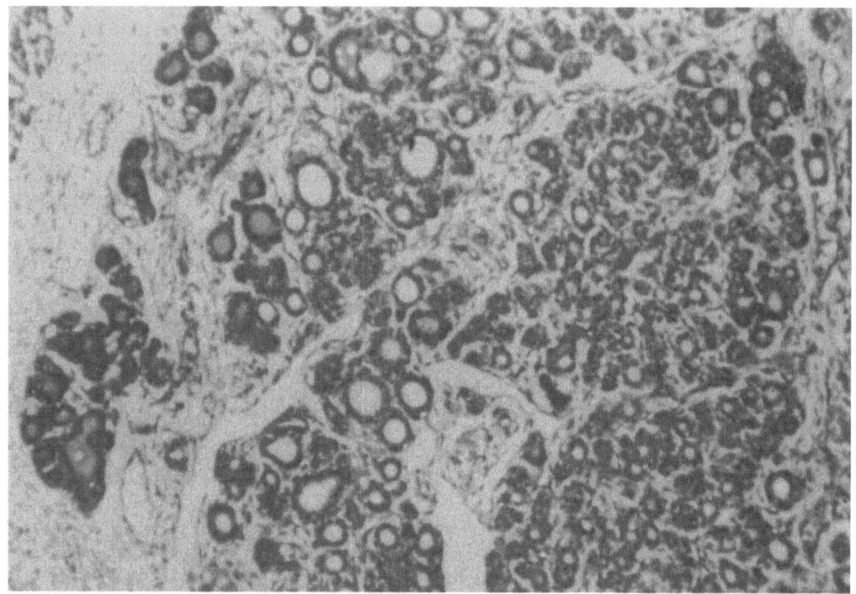

Figure 1. Thyroid folliculogenesis in the first half of human ges-
           tation, a period of organogenesis, relatively slow fetal
growth. Thyroid of fetus weighing 150 g, 13.0 cm crown-rump length
(CRL).   Minimal (1+) dilation of follicles, some without colloid
(possible artefact from very low protein content, i.e., no matrix
congealed by formalin [Shanklin, 1964; Morada, 1968]).   Loose con-
connective tissue; the thyroid lobules are not compact. Gestational
age (GA) equivalent to 18 weeks, menstrual age.   Hematoxylin and
eosin stain.   Original magnification, 100X.

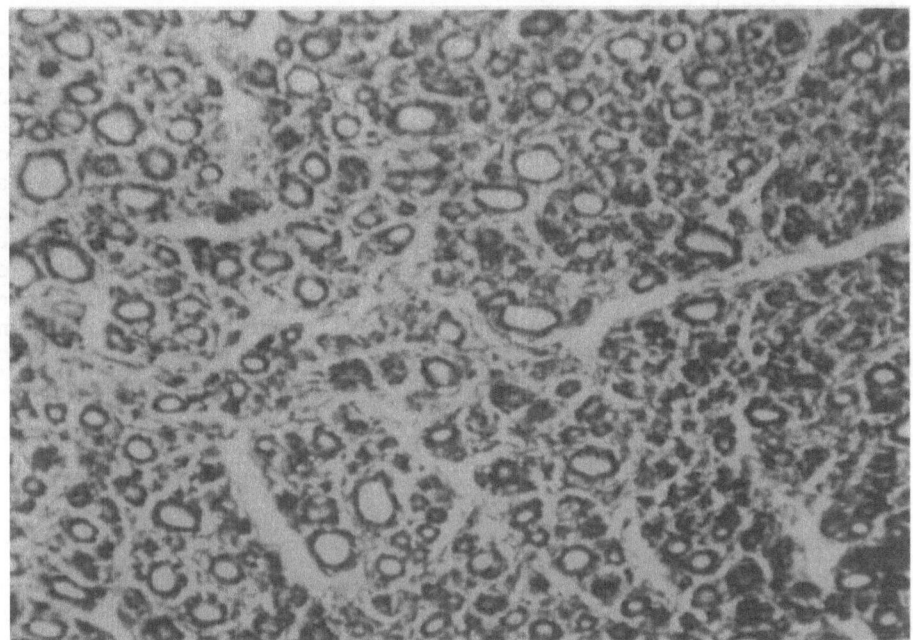

Figure 2. Thyroid folliculogenesis in the first half of human gestation, a period of organogenesis, relatively slow fetal
growth. Thyroid of fetus weighing 320 g, CRL = 17.0 cm. There are
more dilated yet "empty" follicles. GA = 20.5 weeks, menstrual age.
Hematoxylin and eosin stain. Original magnification, 100X.

Cytology of the thyroid

Thyroid cells share a functional attribute common to other en-
dodermally derived cells; they secrete bidirectionally: (1) toward
the lumen of their tissue structure (away from their blood vessels)
and (2) into those vessels [Pantic, 1974].   For the thyroid, the
follicular structure requires also that the stored material, thyroglobulin, be transported back across the epithelium for much of the
endocrine (vascular secretory) function [Pantic, 1974].   The possibility of distinguishing follicles which arise from ultimobranchial bodies [Gorbman, 1949; Sehe, 1966] comes from the observation
of Wetzel and Wollman [1969] of centrilobar follicles with unusual
cells and foamy colloid.

There is evidence for a second type of follicle from studies
on the C3H/HeN mouse   [Dunn, 1944; Gorbman, 1947a, 1947b,; Wetzel
and Wollman, 1966; Wetzel and Wollman, 1969; and Neve and Wollman,
1972].

Functional correlation

Correlation of changes in thyroid tissue with functional parameters is difficult.   None of the aforecited studies included con-
concurrent chemical information   [Arey, 1956; Copp, 1969; Copp, et
al., 1967; Loewenstein and Wollman, 1970].   Early work was based on
determinations on recently delivered infants from a range of gestational ages, without reference to the possible contribution of maternal or fetal disease influencing thyroid function   [Chapman, et
al., 1948; Hodges, et al., 1955; Fisher, et al., 1964; Fisher, et
al., 1970].   The work by Fisher and associates showed an increased
concentration of maternal thyroxine binding protein and an inherent

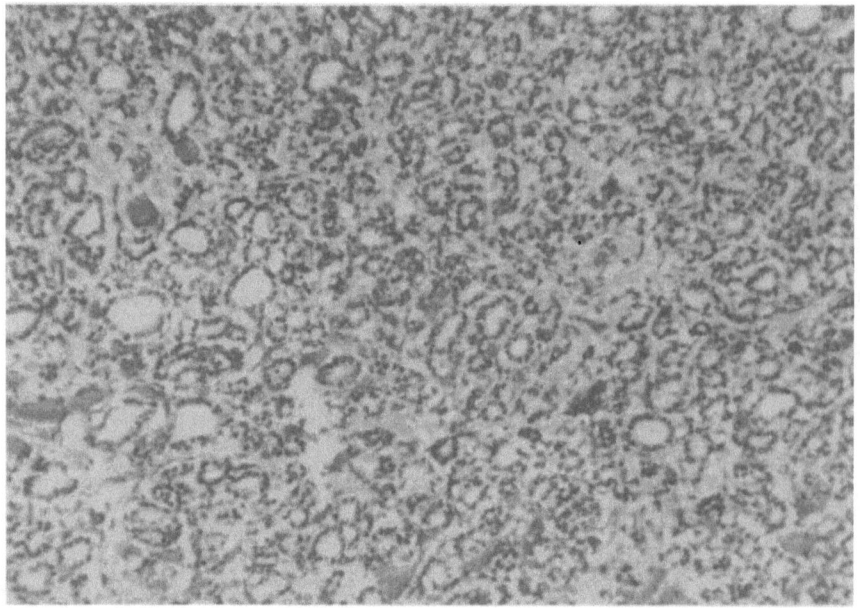

Figure 3.   Thyroid folliculogenesis in second half of human gesta-
            tion, a period of organ maturation and relatively rapid
fetal growth.   Gestation age range (GAR) 20-24 weeks. Birth weight,
638 g, lived 31 hours. Mixed lung disease, hyaline membrane formation, hemorrhage, and interstitial emphysema; matrix hemorrhage and
intraventricular rupture.   There is little difference here in follicular development from fetuses in the previous four week period.
Thyroid cells are larger.   An unanswered question is whether this
mix of follicles and cell size represents a pathological, hypofunc-
tional state, significant to the outcome.   Newborns from this maturity class, living only minutes to a few hours, have thyroids usually indistinguishable from fetuses of similar gestational age.
Hematoxylin and eosin stain. Original magnification, 100X.

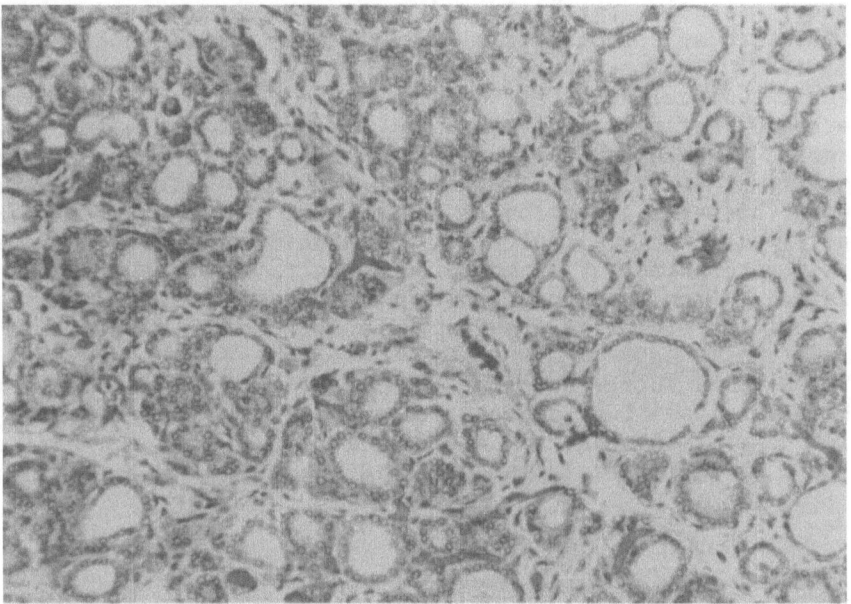

Figure 4.   Thyroid folliculogenesis in second half of human gesta-
            tion, a period of organ maturation and relatively rapid
fetal growth.   GAR 25-28 weeks.   Birth weight 640 g, lived 43 days.
Twinned with a grade 1 chorioma. Varied follicles with some retain-
ed colloid.   Lesser survivals in this maturity class often have
more colloid, suggesting a depletion or drawing down effect due to
the principal intercurrent and ultimately fatal disease. Ventilator
lung disease with multiple small and large lung abscesses.   Hematoxylin and eosin stain.   Original magnification, 100X.

impermeability to thyroxine as limiting factors in placental transport of thyroxine [1964] as well a possible early role for chorionic thyrotropin [1970].   Thyroid stimulating hormone turns over very
rapidly, with a half-life of 20 minutes or less [Sterling, 1979].
Although specific dynamic studies have not been done in fetuses or
newborns there seems little reason for major differences in this
from adult tissues, as the substance is the same and has comparable
effects on RNA and protein synthesis [Sterling].

    More recently, levels of human thyroid activity (rT3, T3, T4,
TSH) on postnatal preterm cord blood correlated with certain patho-
physiological states [Uhrmann, et al., 1978].   Studies during this
period on animals and human newborns led to the belief that thyroid
function remained at basal levels until midgestation [Fisher and
Klein, 1981], with implication for the extreme thyroid dysfunction
of agenesis or total organification defect [Vulsma, et al., 1989].

Most recently, chemical studies have been made on fetal human blood obtained by cordocentesis [Porreco and Bloch, 1990; Wenstrom, et al., 1990; Thorpe-Beeston, et al., 1991], without any companion studies of folliculogenesis. Cordocentesis is an invasive diagnostic method available to outpatients, carrying the very low risk of fetal death of 0.14 per cent [Weiner, 1988]. Moreover, the great sensitivity of new analytic methods for thyroid factors is shown by the recent neonatal detection of generalized resistance to thyroid hormone [Weiss, et al., 1990]. The significance of the value of precise determination of thyroid function is shown by a recent demonstration of the need for supplemental thyroxine in patients with primary hypothyroidism in pregnancy [Mandel, et al., 1990].

The data of Thorpe-Beeston, et al., comes from a large sample and the normative values adduced provide a useful developmental matrix for comparison, despite absence of histological comparisons. Their data show important increases in fetal TSH and rises in both

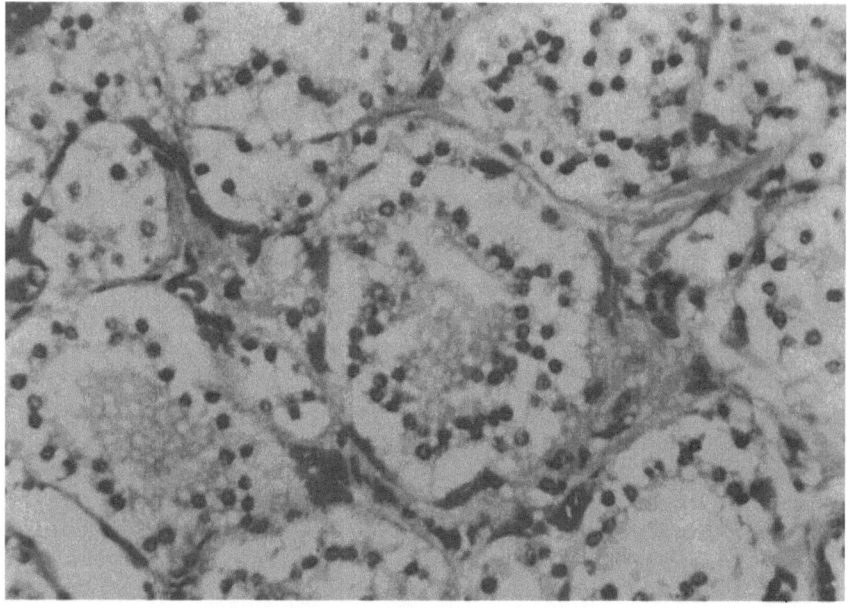

Figure 5. Thyroid folliculogenesis in second half of human gestation, a period of organ maturation and relatively rapid fetal growth. GAR 29-31 weeks. Birth weight 2240 g, lived 9 hours. Undiagnosed congenital toxic goiter with an unexpected early death. Thyroid gland was 15 times expected size. There was multifocal myocardial necrosis, congestive heart failure, mild hemolytic anemia. Hyperfunctional subnuclear vacuoles are prominent. Hematoxylin and eosin stain. Original magnification, 200X.

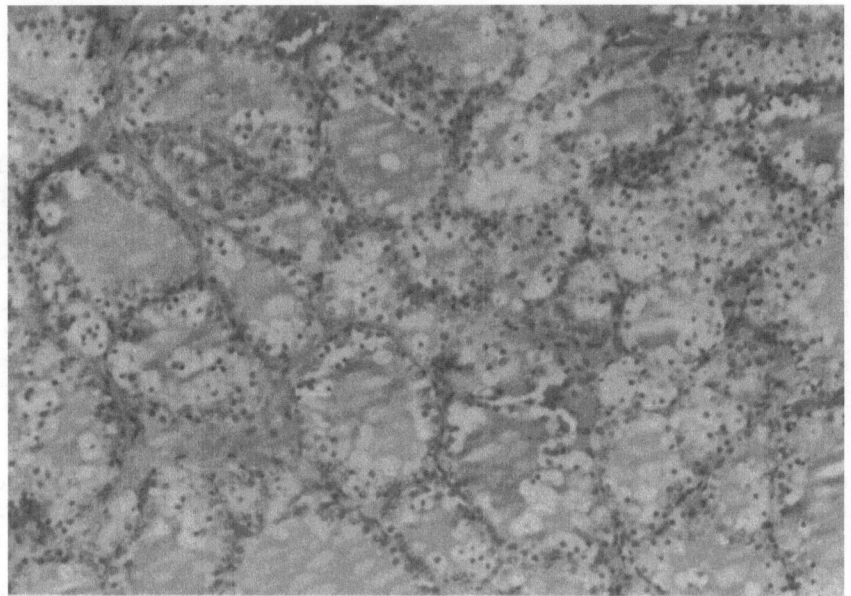

Figure 6.   Thyroid   folliculogenesis in second half of human gesta-
             tion, a period of organ maturation and relatively rapid
fetal growth. GAR 33-36 weeks. Birth weight 2170 g, lived only 55
minutes.   Hyperactive thyroid, colloid filled with many subnuclear
vacuoles. The cause of death was segmental tracheal aplasia. Hematoxylin and eosin stain. Original magnification, 100X.

free and total T4 between gestational weeks 12 and 20, the latter
despite much higher maternal levels during the same period.   Total
T3 also increases over this interval; free T3 does not until after
20 weeks.   These data support a progressive maturation of the system from earlier stages in human fetal development than previously
understood to be the case.   It is now useful to put this perspec-
tive into a direct examination of human fetal thyroid histology.

     Our studies of the fetal and neonatal human thyroid show that
three general conclusions are in order.   These must be put against
the variation in body length and mass by menstrual age when corrected for individual cycles with respect to the probable conceptual
gestational age. These three general conclusions are:

     1. Folliculogenesis is highly varied per gestational age,

     2. Folliculogenesis is highly varied per post natal age,
        seemingly independent of the gestational age at the
        time of birth, and

3. Follicular development is affected by other factors, including diseases of the mother, the placenta, and other parts of the fetus itself.

The objective of this report is to show our observations and data that support these general conclusions. In addition, preliminary correlations of thyroid folliculogenesis with developmental change in the brain, especially in the cerebral cortical mantle, is made by drawing compositely of the works of Marin-Padilla, Hulse, and Dorovini-Zis and Dolman. This presentation will begin with the growth of the thyroid gland, reflected in the organ weight.

The weight of the human thyroid gland

Thyroid weights from early gestation have been presented infrequently. The most thorough reports are those of Shepard, Andersen and Andersen [1964]. Their study of 79 very early fetuses is particularly useful because five were embryos at Streeter's horizon XXIII (27-31 mm crown-rump length) [Streeter, 1920; Streeter, et al., 1951]. The trend over the range of crown-rump length from 30 to 150 mm was biphasic to triphasic, indicating a hyperbolic growth relationship with the second departure point (on a triphasic plot) at about 110 mm crown-rump length and a thyroid weight of 25 mg.

Table One combines the data from Shepard, et al., with standard weights from Potter [1952], modified by the author's personal collection of perinatal autopsies over the period 1931-1991.

Contemporaneous brain development

de Vries, et al., [1982] recorded real time ultrasound assessments of fetal position, posture, and movements. By 15 weeks some sixteen different movements were identified. Before 16 weeks there was a preference for the supine position; afterwards a lateral position was dominant. The wide variety of neurological deficits in infantile hypothyroid [Hulse] has implications for every aspect of brain differentiation and maturation: neuroblastic multiplication, glial proliferation, dendritic growth, field connectivity, and myelination. Hulse has emphasized that some functional deficits might be due to faulty sequencing of these maturational steps. Marin-Padilla explored this with specific neuroarchitectural studies at the important gestational ages of 5 and 7 months (not defined as to whether calendar or lunar months).

His study was confined to the precentral gyrus (area 4); Golgi stains were performed; thick cut sections were examined. For this discussion, calendar months converted to weeks are used by which to examine the data; this comports well with brain maturation so well

TABLE ONE

Thyroid weights in first half of gestation (*)

Thyroid weight (mg)	Crown-heel length (mg)	Equivalent fetal age (weeks)/weight (gm)		Thyroid:body weight ratio
"0.0"	25	9.0	24	"0.00000"
	30	9.5	27	
1.0	35	10.0	30	0.00033
	40	10.4	33	
1.5	45	10.7	36	0.00042
	50	11.1	39	
2.5	55	11.5	43	0.00058
	60	12.0	47	
6.0	65	12.3	52	0.00115
	70	12.7	57	
12.0	75	13.2	62	0.00194
	80	13.6	70	
16.0	85	14.0	75	0.00213
	90	14.5	83	
21.0	95	15.0	92	0.00228
	100	15.4	100	
28.0	105	15.8	110	0.00255
	110	16.3	120	
44.0	115	16.8	136	0.00324
	120	17.2	150	
60.0	125	17.5	160	0.00375
	130	18.0	180	
80.0	135	18.2	190	0.00421
	140	18.5	207	
98.0	145	18.8	223	0.00439
	150	19.1	240	
120.0	155	19.4	280	0.00429
	160	19.7	305	
130.0	167	20.0	340	0.00382

* Data from Shephard, et al., in columns 1 and 2; remainder of data adapted from Potter and from the author's database of over 10,000 perinatal autopsies. The final column is a composite ratio; thyroid weight as per cent of body weight.

shown by Dorovini-Zis and Dolman. At 22 weeks only primary gyration
is present, outer granular cortex persists, mantle is incompletely
formatted, and there are no specific layers of neurones.   In future
layer 5 (internal pyramidal layer) some neurones are larger.   These
are designated as protopyramidal cells.   The size differential is
more impressive later.

At 31 weeks secondary gyration is advanced. Outer granulocytes
are largely gone and the cortical mantle is almost completely formatted. Few migratory neural elements remain.   Layering is recog-
nizable, best defined in layer 3-4; layer 4 is not easily distinguished from 3.   Pyramidal connectivity has developed in layer 5.
Myelination is found in the internal capsule.

Variation in development and maturation of cortex were shown
by Marin-Padilla.   A third fetus, rated at 7.5 calendar months, or
33.6 weeks, had a more mature layer 4, among other features.

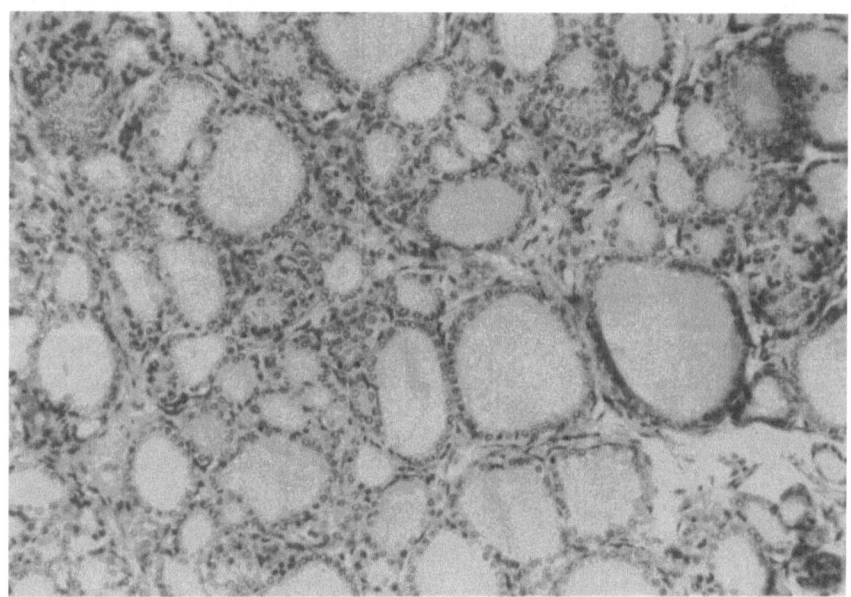

Figure 7. Thyroid folliculogenesis in second half of human gesta-
           tion, a period of organ maturation and relatively rapid
fetal growth. GAR 33-36 weeks. Birth weight 1450 g, lived 32 hours.
Abundant colloid storage with uneven activity of thyroid epithelium.   An infant with osteogenesis imperfecta, borderline lung hypoplasia and diffuse, severe hyaline membrane disease.   Hematoxylin
and eosin stain. Original magnification, 100X.

Observations on fetal and neonatal thyroids

The first twelve figures proceed in order of gestational age. Descriptive comments will be limited here, as the principal observations are shown in the photographs, which were selected from a larger number presented during the symposium. Human thyroid shows a few follicles late in the first half of gestation [Figures 1 and 2]. The essential difference over this 2.5 week span seems to be reduced interstitial connective tissue, as described by Shepard, et al., [1964b], but comparable change is found in various zones of a single gland. By contrast, follicles four weeks later may be less well-differentiated [Figure 3], given a birth weight twice as much. The lesion of hyaline membrane disease seems appropriate to a relative immaturity of follicles, as lower T3 has been correlated with hyaline membrane disease [Uhrmann, et al.]. Similar concerns on the effects of nonthyroidal disease on thyroid function were expressed by Chopra, et al., but not in newborns. The proportion of follicles with colloid was also examined by Shepard, et al., [1964b]; it was reported that 85-90% of follicles show the presence but not the accumulation of colloid when the crown-rump length is 12.0-14.0 cm. This distinction is important because periodic-acid-Schiff staining is required to find lumenal "colloid" at these early stages, whereas distension of the follicle toward the neonatal pattern, in our experience, begins at this time in development. Accordingly, it would seem Shepard, et al., [1964b] are indicating that an earlier cellular function leads to secretion which contains a PAS-positive moiety. Whether this represents storage of thyroglobulin or iodine remains uncertain. In fairness, the same report did note variations roughly comparable to those reported here. Their largest fetus had a crown-rump length of 20.1 cm; only 85% of the follicles contained colloid. Clearly, more attention needs to be paid to this aspect of the matter. Similarly, Figure 4 shows another 640 gram newborn living 43 days. Some, if not most, of the follicular enlargement is due to postnatal stimulation by TSH. Comparable follicular development is seen often at or near term [Figure 10], except for that infant the integrity of the hypothalamic-hypophyseal trophic system was in doubt. An increase in thyroid activity occurs in premature infants up to and including toxic hyperactivity [Figure 5]. Lesser follicular activity occurs, but in this infant relevance of thyroid function is doubtful, given the total ventilatory obstruction due to malformation [Figure 6]. There is no known causal relationship between earlier fetal thyroid function and this anomaly or vice versa. Further variation in the amount of colloid is shown in Figure 7 for an infant of the same gestational age as in Figure 6 but suffering from ventilatory insufficiency due to thoracic effects of osteogenesis imperfecta. However, a further correlation with the findings of Uhrmann, et al., may be pertinent to this example. The thyroid epithelium was low cuboidal, not flat like that in colloid goiter, but unstimulated. Very likely this was a factor in severe

hyaline membrane disease as a feature of premature birth. Certainly, not every newborn with osteogenesis imperfecta dies neonatally
or has hyaline membrane disease. The infant with massive cerebral
ependymoma [Figure 10] shows the possible effects of failure of the
hypothalamic-hypophyseal trophic system; Figure 8 is a more advanced example of this phenomenon. By 33-36 weeks the absence of TSH-
release hormone has not interfered with folliculogenesis or with
colloid accumulation. Pertinent is the absence of concurrent cellular secretion. Cases such as this can be interpreted as evidence
for fetal effects of maternal TSH greater than usually considered.
Alternatively, there may be persistence of chorionic TSH-like activity when the fetal pituitary fails. Since similar human malfor-
mations are rare, better understanding on this point might occur if

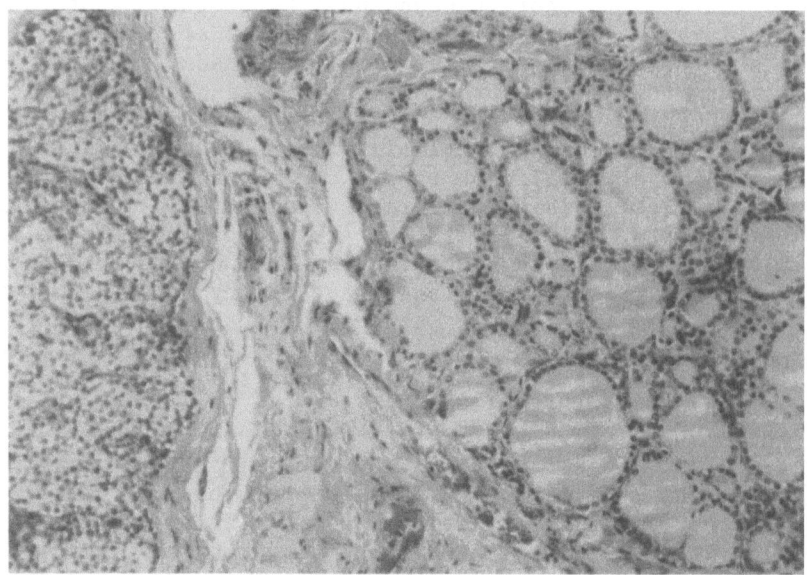

Figure 8. Thyroid folliculogenesis in second half of human gesta-
          tion, a period of organ maturation and relatively rapid
fetal growth. GAR 33-36 weeks. Birth weight 2380 g, lived 10 days.
Fairly uniform follicles, colloid filled but no cellular signs of
concurrent activity. The adjacent parathyroid gland (left) appears
normal for the age. From an infant with a global cerebral malformation, nonsyndromic hydranencephaly. There was no hypothalamus and
only dysplastic pons and medulla; the cerebellum was a gliofibril-
lary cyst. Absence of the hypothalamus challenges the functional
competency or timing of hypothalamic-hypophyseal regulation or modulation of secretion of thyroid stimulating hormone in fetal and
neonatal thyroid maturation (See also Figure 10 in this respect).
Hematoxylin and eosin stain. Original magnification, 100X.

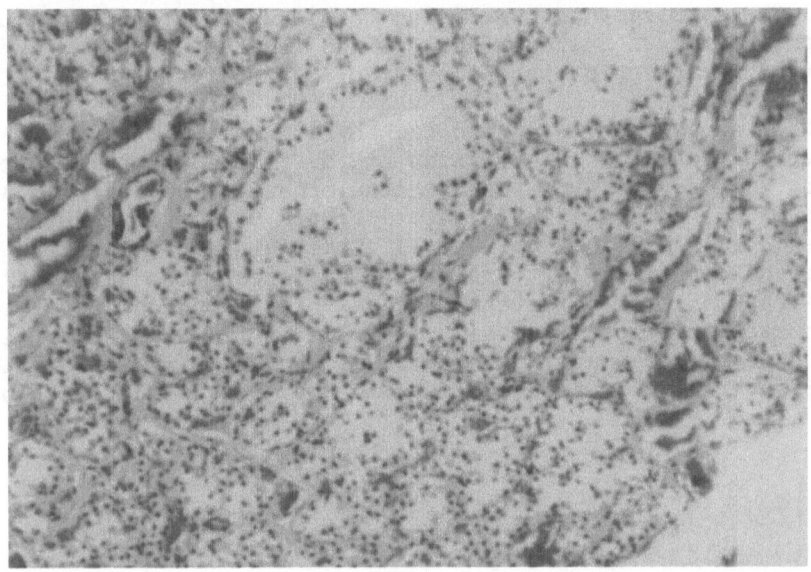

Figure 9. Thyroid folliculogenesis in second half of human gestation, a period of organ maturation and relatively rapid
fetal growth. GAR 37-40 weeks. Birth weight 3020 g, fetal death was
four hours before stillbirth. There was varied follicular colloid
content with overdistension like colloid goiters. The veins of the
umbilical cord insertion site were much dilated. There were numerous pathological signs of obstructive congestive heart failure and
chronic hypoxia. There was cephalic cyanosis, a ligature mark on
the neck, deep lung aspiration of amniotic content, and 1+ meconium
passage. Hematoxylin and eosin stain. Original magnification, 100X.

the animal models studied so intensively by Escobar and associates
[Ruiz de Ona, et al., 1988] were adapted by concurrent removal of
the hypothalamus. Reports on hydranencephaly have generally described the extent of the cerebral anomalies, with scant attention to
details in other tissues; chemical information under these circumstances does not presently exist. However, these massive cerebral
lesions call attention to the possibility that the relationship of
these factors has complexity beyond that presently understood. For
example, if chorionic TSH-activity is critical to early stages but
is superseded by fetal pituitary TSH, what is the signal for transfer? If the signal fails to arrive, will chorionic TSH continue?
Absent a pituitary source of TSH how is thyroid stimulated? While
the inferences derived from these individual examples cannot solve
the question of the earliest stimuli to follicle formation, autonomously directed genomic expression outside of hormonal effect seems

to fill the requirements. Remaining candidates for study are possible coordination with growth hormone and the turn down of chorionic gonadotropin just prior to the onset of thyroid growth.

The response of the developing thyroid to various pathophysiologic states in fetal life outside of the thyroid trophic system is another important part of thyroid development. In Figures 9 and 11 through 14 are some possibilities. Figure 9 is from a fetal death of brief duration before delivery. This fetus was near term after a period of considerable cardiovascular stress from umbilical venous obstruction. In this fetus the thyroid responded by noteworthy colloid accumulation. Massive destruction of the brain by infection and post inflammatory sequelae is another line of consideration [Figure 11]. In this child prolonged, massive cytomegalovirus

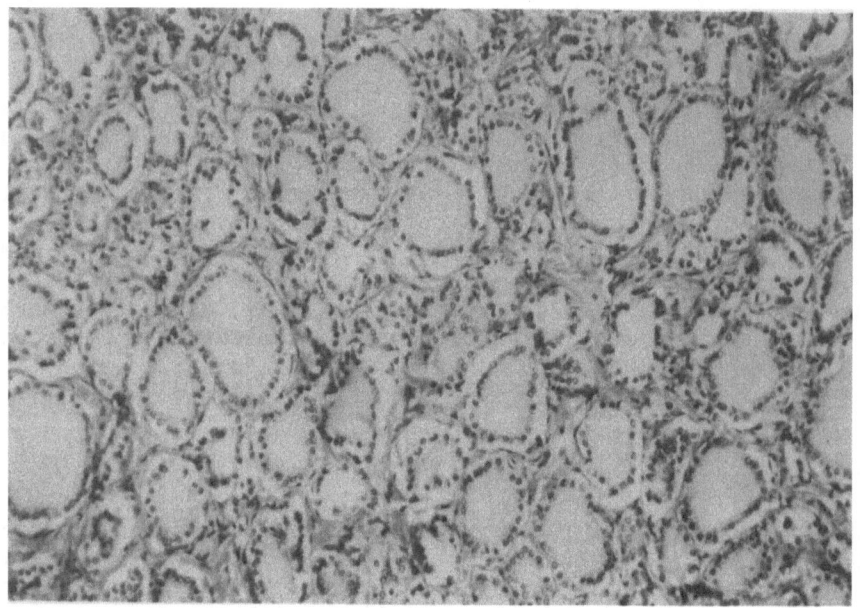

Figure 10. Thyroid folliculogenesis in second half of human gestation, a period of organ maturation and relatively rapid fetal growth. GAR 37-40 weeks. Birth weight 3850 g, lived 63 hours. Mediastinal and interstitial pulmonary emphysema. Distended follicles with very little current activity. This infant had massive cerebral ependymoma presenting as macrocrania without hydrocephaly. The tumor was bilateral and compressed the hypothalamus. Anterior hypophysis contained only small eosinophilic cells and small laminar calcifications, suggestive of diminished trophic stimulus (See also Figure 8 in this respect). The lungs showed ventilatory emphysema. Hematoxylin and eosin stain. Original magnification, 100X.

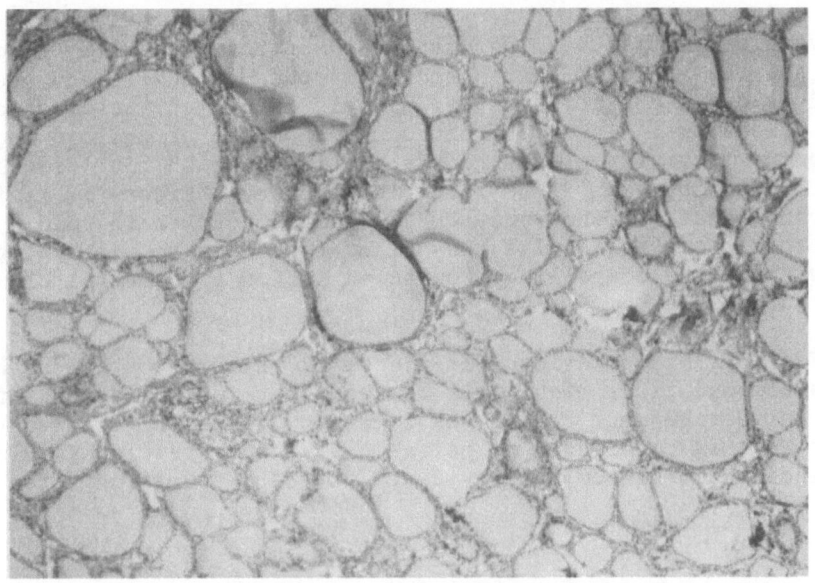

Figure 11. Thyroid folliculogenesis in second half of human gesta-
        tion, a period of organ maturation and relatively rapid
fetal growth. GAR 37-40 weeks. Birth weight 1560 g, lived 39 days.
The thyroid has a pattern of colloid goiter, 4+ distension of follicles and thin epithelium. The infant was born with severe systemic and cerebral infection by cytomegalovirus (CMV) and micro-
cephaly. There was extensive focal and massive segmental calcifi-
cation in the brain. Hematoxylin and eosin stain. Original magnification, 40X.

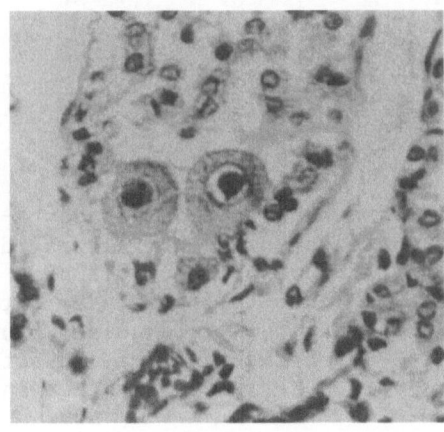

Figure 12. Case similar to that
        in Figure 11 living
just 92 hours. CMV inclusions in
thyroid epithelium. Hematoxylin
and eosin stain. Original magnification, 200X.

(CMV) infection of the brain largely reduced it to a walnut sized mostly calcific mass. The distortion of basal ganglia, hypothalamus, and the vascular arrangements of the brain were advanced and marked. Very large segments were simply calcified masses of centimeter dimensions. The near-goitrous collection of colloid suggests autonomous activity or aberrant stimuli. No CMV inclusions were found in the thyroid of the illustrated case. Figure 12 is from a different case, to show that CMV lesions occur in thyroid. Other infections come to involve the gland from time to time but generally without the compelling companion lesion of cerebral CMV noted in the above case.

Finally, midgestational responses of thyroid to various highly stressful conditions might well be such that considerations of what is normal development can be challenged. It must be allowed that a lot of what is considered to be the case about fetal physiology is from indirect or incomplete evidence. The information noted above

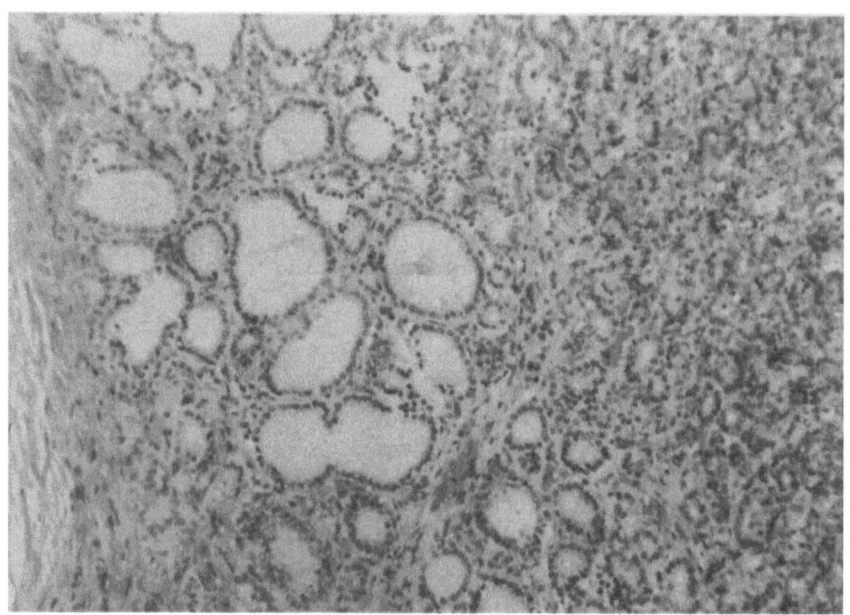

Figure 13. Thyroid folliculogenesis in second half of human gestation, a period of organ maturation and relatively rapid fetal growth. GAR 23 weeks. Donor twin in monochorionic, diamniotic, monozygotic parabiotic pair (see also Figure 14). Birth weight 318 g, lived 18 hours. Poorly differentiated thyroid gland with some subcapsular colloid accumulation. This twin had a small heart and developmentally normal sized glomeruli. Hematoxylin and eosin stain. Original magnification, 100X.

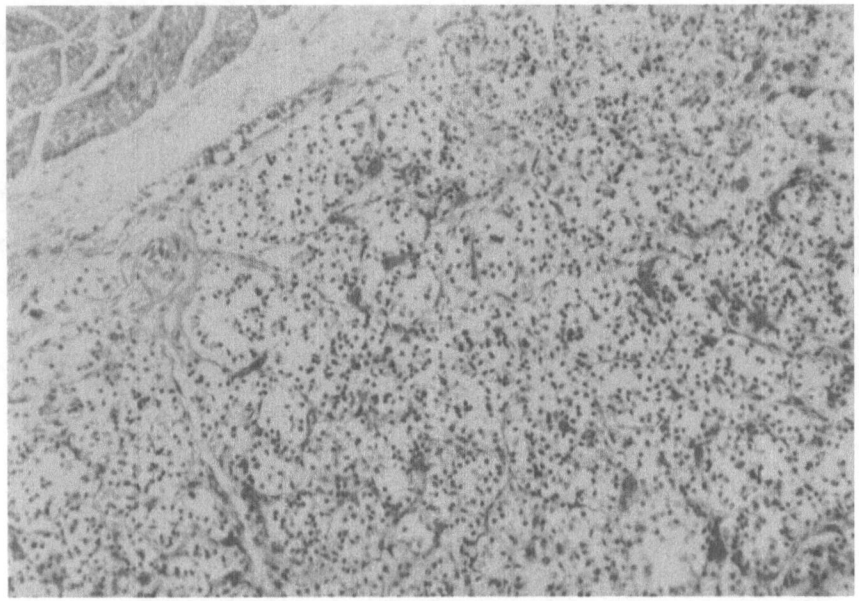

Figure 14.   Thyroid folliculogenesis in second half of human gesta-
              tion, a period of organ maturation and relatively rapid
fetal growth. GAR 23 weeks.   Recipient twin in monochorionic, diamniotic, monozygotic parabiotic pair   (see also Figure 13).   Birth
weight was 505 g, lived 11 hours. Hyperactive, moderately enlarged
(2+) thyroid gland.   This twin had an enlarged, dilated heart, systemic hypervolemia, and glomeruli which were twice expected normal
diameter and were more mature, with thin Bowman's epithelium. Hematoxylin and eosin stain.   Original magnification, 100X.

on the specifics of fetal thyroid function had to await a wider use
of cordocentesis; such values are clearly more representative than
those obtained by extrapolation from cord or early neonatal data.
The cases presented here strongly support this general concept by
showing high variability of folliculogenesis and colloid accumulation under a variety of late pregnancy conditions.   Figures 13 and
14, by contrast, are from 23 week twins with the unusual condition
of significant parabiotic transfusion.   Overtly this condition is
reflected in fluid volume differences, heart size, and glomerular
dimensions.   In general, the donor is in a state of reduced fluid
compartments and the recipient suffers from an excess.   The lesions
noted above are reflections of this general principal.   Figures 13
and 14 show the thyroid may be involved in parabiosis.   The donor
thyroid [Figure 13] is a poorly developed gland with zonal colloid
accumulation; by contrast, the recipient thyroid is diffusely over-
active with colloid depletion [Figure 14].   Thus, a pervasive fluid

redistribution can affect the endocrine system. The mechanisms for
this may be complex and no data is presently available on the subject.   Differential concentrations of TSH or of feedback T3/T4 are
possibilities; different general metabolic rates are another.

    In summary, various fetal disorders alter the development and
maturation of the thyroid. It may be reasonably inferred also that
various placental disorders do the same. Future examination of fetal values pertinent to thyroid function, whether obtained by cordocentesis or by other means, must account for these disorders and
the differing values to be found by sensitive methods.

    Hopefully, such data will elucidate the interrelationship be-
tween fetal thyroid competency and the growth, maturation, and the
functional level of individual organs or systems and what effects,
if any, the state of the thyroid has on the progress of fetal and
placental disorders and vice versa.

BIBLIOGRAPHY

Arey, L.B.: Developmental anatomy, 1956, Saunders, Philadelphia,
    p. 231
Chapman, E.M., G.W. Corner, D. Robinson and R.D. Evans: Collection
    of radioiodine by the human fetal thyroid. J.Clin.Endo. 8:717-
    720, 1948
Chopra, I.J., D.H. Solomon, G.W. Hepner and A.A. Morgenstein:
    Misleadingly low free thyroxine index and usefulness of reverse
    triiodothyonine measurement in nonthyroidal illnesses.
    Ann.Int.Med. 90:905-912, 1979
Copp, D.H.: Calcitonin and parathyroid hormone. Ann.Rev.Pharmacol.
    9:327-344, 1969
Copp, D.H., D.W. Cockcroft and J. Kuel: Calcitonin from ultimobran-
    chial glands of dogfish and chickens. Science 158:924-925, 1967
de Vries, J.I.P., G.H.A. Visser and H.F.R. Prechtl: The emergence
    of fetal behaviour. I. Qualitative aspects. Early Hum.Develop.
    7:301-322, 1982
Dorovini-Zis, K. and C.L. Dolman: Gestational development of brain.
    Arch.Pathol. 101:192-195, 1977
Dunn, T.B.: Ciliated cells of the thyroid of the mouse.
    J.Nat.Cancer Inst. 4:555-557, 1944
Fisher, D.A., C.J. Hobel, R. Garza and C.A. Pierce: Thyroid
    function in the preterm fetus. Pediatrics 46:208-216, 1970
Fisher, D.A. and A.H. Klein: Thyroid development and disorders of
    thyroid function in the newborn. New Eng.J.Med. 304:702-712, 1981
Fisher, D.A., H. Lehman and C. Lackey: Placental transport of
    thyroxine. J.Clin.Endocr. 24:393-400, 1964
Gorbman, A.: Functional and morphological properties in the thyroid
    gland, ultimobranchial body, and persisting ductus pharyngio-
    branchialis IV of an adult mouse. Anat.Rec. 98:93-101, 1947 (a)

Gorbman, A.: Thyroidal and vascular changes in mice following chronic treatment with goitrogens and carcinogens. Cancer Res. 7:746-758, 1947 (b)

Gorbman, A.: Tumorous growths in the pituitary and trachea following radiotoxic dosages of I131. Proc.Soc.Exp.Biol.Med. 71:237-240, 1949

Gray, H.: Anatomy of the human body, 25th ed., C.M. Goss, ed., 1948, Lea & Febiger, Philadelphia, pp. 1327-1329

Hodges, R.E., T.C. Evans, J.T. Bradbury and W.C. Keettel: The accumulation of radioactive iodine by human fetal thyroids. J.Clin.Endo.Metab. 15:661-667, 1955

Hulse, A.: Congenital hypothyroidism and neurological development. J.Child. Psychol.Psychiat. 24:629-635, 1983

Loewenstein, J.E., and S.H. Wollman: Mechanisms for abnormally slow release of some thyroid radioiodine; an autoradiographic study. Endocrinology 87:143-150, 1970

Mandel, S.J., P.R. Larsen, E.W. Seely and G.A. Brent: Increased need for thyroxine during pregnancy in women with primary hypothyroidism. New Eng.J.Med. 323:91-96, 1990

Marin-Padilla, M.: Prenatal and early postnatal ontogenesis of the human motor cortex: a golgi study. I. The sequential development of the cortical layers. Brain Res. 23:167-183, 1970

Morada, A.O.: Postmortem pulmonary edema; differential susceptibility in newborn and adult rabbits. Arch.Pathol. 85:468-474, 1968

Neve, P., and S.H. Wollman: Fine structure of a fifth type of epithelial cell in the thyroid gland of the C3H mouse. Anat.Rec. 172:37-43, 1972

Norris, E.H.: The morphogenesis of the follicles in the human thyroid gland. Amer.J.Anat. 20:411-448, 1916

Pantic, V.R.: The cytophysiology of thyroid cells. Int.Rev. Cytol. 38:153-243, 1974

Porreco, R.P. and C.A.Bloch: Fetal blood sampling in the management of intrauterine thyrotoxicosis. Obstet.Gynec. 76:509-512, 1990

Potter, E.L.: Pathology of the fetus and the newborn, 1952, Year Book Publishers, Chicago, p. 13

Ruiz de Ona, C., M.J. Obregon, F. Escobar del Rey, and G. Morreale de Escobar: Developmental changes in rat brain 5'-deiodinase and thyroid hormones during the fetal period: the effects of fetal hypothyrodism and maternal thyroid hormones. Pediat.Res. 24:588-594, 1988

Sehe, C.T.: Observations on the ultimobranchial gland in small wild and laboratory mammals, with special reference to the histochemical localization of polysaccharides. J.Morph. 120:425-441, 1966

Shanklin, D.R.: The influence of fixation on the histologic features of hyaline membrane disease. Amer.J.Path. 44:823-837, 1964

Shepard, T.H., H.J. Andersen and H. Andersen: The human fetal thyroid. I. Its weight in relation to body weight, crown-rump length, foot length and estimated gestational age. Anat.Rec. 148:123-128, 1964 (a)

Shepard, T.H., H. Andersen and H.J. Andersen: Histochemical studies
    of the human fetal thyroid during the first half of fetal life.
    Anat.Rec. 149:363-379, 1964 (b)
Sterling, K.: Thyroid hormone action at the cell level. New Eng.
    J.Med. 300:117-123, 173-177, 1979
Streeter, G.L.: Weight, sitting height, head size, foot length and
    menstrual age of human embryos. Carnegie Inst.* Wash. Pub.
    274, Contributions to Embryology (No. 55) 11:143-170, 1920
Streeter, G.L., C.H. Heuser, and G.W. Corner: Developmental
    horizons in human embryos. Contributions to Embryology (No.
    230) Reprint vol. IIs:166-186, 1951
Thorpe-Beeston, J.G., K.H. Nicolaides, C.V. Felton, J. Butler, and
    A.M. McGregor: Maturation of the secretion of thyroid hormone
    and thyroid stimulating hormone in the fetus. New Eng.J.Med.
    324:532-536, 1991
Uhrmann, S., K.H. Marks, M.J. Maisels, Z. Friedman, F. Murray, H.E.
    Kulin, M. Kaplan and R. Utiger: Thyroid function in the preterm
    infant: a longitudinal assessment. J.Pediat. 92:968-973, 1978
Vulsma, T., M.H. Gons and J.J.M. de Vijlder: Maternal-fetal trans-
    fer of thyroxine in congenital hypothyroidism due to total
    organification defect or thyroid agenesis. New Eng.J.Med.
    321:13-16. 1989
Weiner, C.P.: Cordocentesis. Obstet.Gynecol.Clin.N.Amer. 15:283-
    301, 1988
Weiss, R.E., S. Balzano, N.H. Scherberg and S. Refetoff: Neonatal
    detection of generalized resistance to thyroid hormone.
    J.A.M.A. 264:2245-2250, 1990
Wenstrom, K.D., C.P. Weiner, R.A. Williamson and S.S. Grant: Pre-
    natal diagnosis of fetal hyperthyroidism using funipuncture.
    Obstet.Gynec. 76:513-517, 1990
Wetzel, B.K. and S.H. Wollman: The fine structure of "foamy"
    follicles in thyroid glands of C3H mice. J.Appl.Physics
    37:3932. 1966 (abstract)
Wetzel, B.K. and S.H. Wollman: Fine structure of a second kind of
    thyroid follicle in the C3H mouse. Endocrinol. 84:563-578, 1969

THYROID HORMONE CONTROL OF BRAIN AND MOTOR DEVELOPMENT: MOLECULAR, NEUROANATOMICAL, AND BEHAVIORAL STUDIES

S.A. Stein,[1] P.M. Adams,[2] D.R. Shanklin,[3] G.A. Mihailoff,[4] and M.B. Palnitkar[1]

[1]Department of Neurology and
[2]Department of Psychiatry
University of Texas Southwestern Medical Center
Dallas, Texas
[3]Departments of Pathology and Obstetrics and Gynecology
University of Tennessee-Memphis
Memphis. Tennessee
[4]Department of Cell Biology and Anatomy
University of Mississippi School of Medicine
Jackson, Mississippi

DISORDERS OF THYROID HORMONE AND CEREBRAL PALSY, MENTAL RETARDATION, AND LEARNING DISABILITY

Thyroid hormones, T3 and T4, have been shown to play significant but poorly understood roles in development and differentiation of rodent and human brain(Lauder, 1989; Legrand, 1982-83; Stein et al, 1989a; 1991a,d; Eayrs, 1968; Morreale de Escobar et al, 1984; Garza et al, 1988; Ruiz-Marcos, 1989; Nunez et al, 1989). Hypothyroidism leads to molecular(Stein et al, 1989a,c; 1991a; Nunez et al, 1989; Hendrich et al, 1987), neuroendocrinological(Noguchi et al, 1986, Bakke et al, 1975, Stein et al, 1989b, Porterfield et al, 1981), neuroanatomical(Lauder et al, 1986; Lauder, 1989; Ruiz-Marcos, 1989; Eayrs, 1955; Garza et al, 1988; Morreale de Escobar et al, 1989; Marc et al, 1985; Legrand, 1982-83; Rami et al, 1986b; Narayanan et al, 1985; Marinesco, 1924; Lotmar, 1928; Rosman, 1975), behavioral and neuropsychological(Adams et al, 1989,1991; Anthony et al, 1991; Eayrs, 1968; Davenport et al, 1976; Klein, 1985; Rovet et al, 1987; Rovet, 1989; Man, 1971; Boyages et al, 1988; Pharoah,1984), and neurological abnormalities(Chaouki et al, 1989; Boyages et al, 1988; Delong et al, 1985; Nelson et al, 1986; Macfaul et al; 1978; Stein et al, 1991d, Rochiccioli et al, 1989) in the developing brain. Specifically, disorders of neuronal process growth and connectivity are noted neuroanatomically and motor syndromes involving motor cortex and pyramidal tracts are commonly observed in hypothyroid humans and rodents. These

Advances in Perinatal Thyroidology, Edited by B.B. Bercu and
D.I. Shulman, Plenum Press, New York, 1991

neurological and neuropathological abnormalities may be predicated on abnormalities in the cytoskeletal structures and in their molecular components. The cytoskeleton is a primary target for thyroid hormone in euthyroid and hypothyroid brain.

In the human, disorders of maternal and fetal thyroid function include maternal and secondary fetal iodine deficiency(Endemic myxedematous and neurological cretinism)(Delange et al, 1989), maternal hypothyroidism or hyperthyroidism(Davis et al, 1988, Lowe et al, 1990) , as well as disorders related to deficient fetal autonomous thyroid hormone secretion(Letarte el al, 1983; Fisher et al, 1981; Dumont et al, 1989), i.e., goiter or sporadic congenital hypothyroidism. Sporadic congenital hypothyroidism occurs in 1 out of 4000 live births(Fisher et al, 1981), while endemic cretinism effects at least 400 million people worldwide(Delange et al, 1989).

The timing, duration, and severity of fetal thyroid hormone deficiency is different for each of these disorders and directly linked to the neuroanatomical and clinical manifestations(Stein et al, 1989a,1991a) These disorders are identifiable causes of mental retardation(Klein et al, 1985; Wolter et al, 1979; Stein et al, 1991e; Rovet, 1989; Delong et al, 1985; Boyages et al, 1988), cerebral palsy(Klein, 1985; Rochiccioli et al, 1989; Wolter et al, 1979; Macfaul et al, 1978; Pharoah et al, 1984; Boyages et al, 1988; Delong et al, 1985), learning disabilities(Rovet et al, 1984, 1987; Rovet, 1989), and other significant neurological abnormalities(Klein, 1985; Wolter et al, 1979; Macfaul et al, 1978; Delong et al, 1985; Delong, 1989; Boyages et al, 1988).

Endemic cretinism represents one of the primary, if not the primary cause of cerebral palsy , as well as mental retardation worldwide(See Endemic Cretinism section below). Maternal thyroid hormone aberration represents one of the statistically significant risk factors that are associated with later cerebral palsy in the progeny. Untreated, late neonatal treated, and early neonatal treated sporadic congenital hypothyroidism can be common and consistent causes of motor disturbances that would be classed as cerebral palsy.

Significant unresolved issues relate to the pathophysiology of the neurological manifestations of these disorders and their reversibility. Despite early treatment for sporadic congenital hypothyroidism(Rovet et al, 1984,1987; Rovet, 1989) and positive results on the eradication of mental retardation(Glorieux et al, 1989; Rovet et al, 1987; New England Hypothyroidism Collaborative, 1990) in the industrialized world, some well controlled studies show a persistence of motor, mood, and neuropsychological abnormalities(Rovet, 1989; Birrell et al, 1987; Rochiccioli et al, 1989; Hulse, 1987). Despite early replacement of iodine prior to pregnancy, the neurological and mental burdens, i.e., cerebral palsy and mental retardation, of endemic cretinism are still significant and vary in severity in specific geographical regions(Chaouki et al, 1989; Boyages et al, 1989).

A potential understanding of the clinical disorders, their treatment, and their complete reversibility will only come when we have defined how, when, and where thyroid hormone works on specific developing fetal brain regions, particularly the motor system. Only then can we determine the nature of the

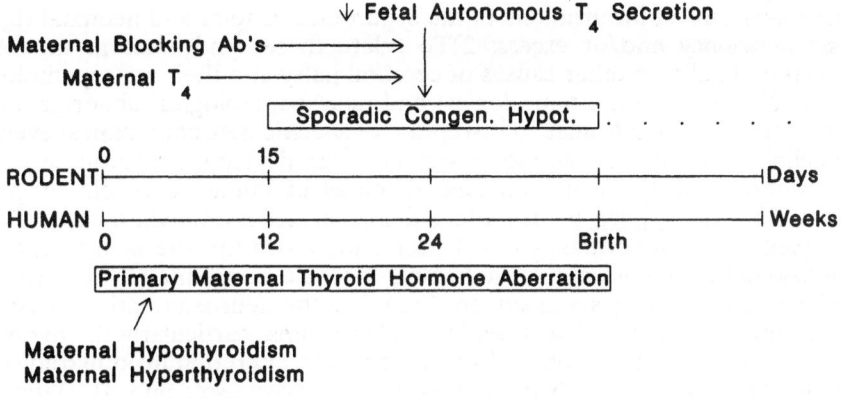

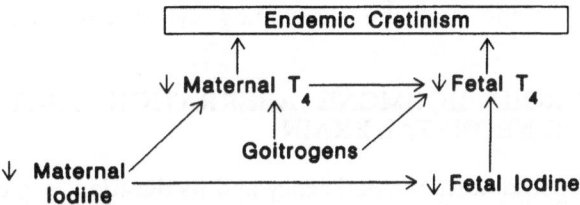

FIGURE 1. SYNDROMES OF THYROID HORMONE DEFICIENCY DURING PREGNANCY AND AFTER BIRTH: TIME AND DURATION OF OCCURRENCE AND POTENTIAL PATHOPHYSIOLOGY OF HYPOTHYROIDISM(Hypot. = Hypothyrodism; ···· = Temporal continuation of the hypothyroidism)

abnormalities and when these abnormalities can be reversed by thyroid hormone or iodine. Although no animal is completely the same as a human, rat, sheep and mouse models of these disorders(Beamer et al, 1981, 1982, 1987; Stein et al, 1989a,b,c, 1991a-e; Legrand et al, 1982-3, Morreale de Escobar et al, 1983, Hendrich et al, 1984; Narayanan et al, 1982; Van Middlesworth et al, 1980; Jianquan et al, 1985; Noguchi, 1988; Mills et al, 1988) have been particularly useful in trying to understand the sensitivity of the fetal brain to thyroid hormone aberration and by that, the pathophysiology of the clinical conditions and their manifestations.

The purposes of this paper are: 1)To describe the motor abnormalities and their neuroanatomical localizations in the syndromes of fetal and neonatal thyroid hormone deficiency and/or excess; 2)To relate these syndromes as models of human cerebral palsy to other causes of cerebral palsy and their neuropathological substrates; 3) To relate the neurological and neuropsychological abnormalities in the thyroid hormone syndromes to disruption of specific neuroanatomical events in certain cellular locations of the motor system in the developing nervous system at specific times; and 4) To use an animal model of human sporadic congenital hypothyroidism, the hyt/hyt mouse, which demonstrates prominant motor reflexive and complex motor abnormalities, to better understand the the neurological and neuropsychological abnormalities in human sporadic congenital hypothyroidism. This animal model provides a means to: 1) Define the neuroanatomical targets for thyroid hormone action; 2) The specific cellular groups, particularly the pyramidal neurons and other motor system cell groups, that are altered in hypothyroidism; 3) Define the process growth abnormalities in hypothyroidism; and 4) Define the potential molecular mechanisms underlying these process growth abnormalities, as they relate to thyroid hormone and the developing cytoskeleton and their components.

SYNDROMES OF THYROID HORMONE ABERRATION THAT MAY AFFECT THE FETAL AND NEONATAL BRAIN

The timing of the thyroid hormone deficiency and its duration can be used to classify specific thyroid hormone disorders in humans and in animal models of these disorders(FIGURES 1 and 2). Endemic cretinism is related to iodine deficiency and subsequent maternal hypothyroidism and secondary fetal iodine deficiency and hypothyroidism. This disorder usually spans the whole pregnancy and encompasses the period of maternal thyroid hormone dependence(first trimester), as well as autonomous fetal thyroid hormone secretion(second and third trimesters). The maternal and fetal iodine deficient rat and its progeny are models of this disorder(Jianquan et al, 1985, Van Middlesworth et al, 1980, Morreale de Escobar et al, 1989). Primary maternal hypothyroidism can span the pregnancy or occur for varying periods of time at particular times during pregnancy; this disorder involves a fetal hypothyroidism that is secondary to the maternal hypothyroidism. The maternally hypothyroid rat, produced by maternal thyroidectomy prior to conception, and its progeny simulate this human condition(Porterfield et al, 1981; Morreale de Escobar et al, 1983, 1985; Hadjzadeh et al, 1990). Both the iodine deficient rat and the maternal hypothyroid rat must be distinguished from the maternally hypothyroid rat that is produced by pharmacological intervention with PTU or methimazole or radioactive iodine(Legrand et al, 1982-3). This model does

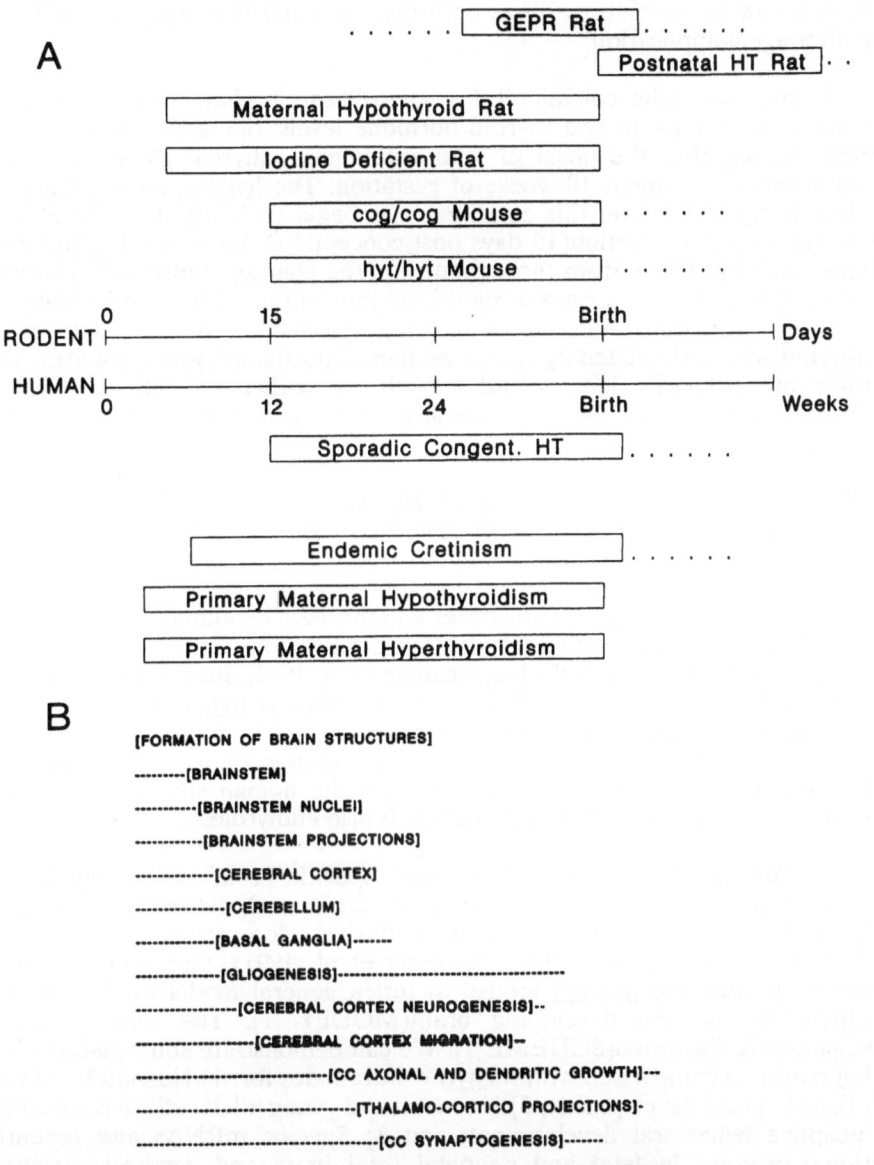

FIGURE 2. THE COMPARISON BETWEEN RODENT MODELS OF HUMAN FETAL AND NEONATAL HYPOTHYROIDISM AND HUMAN THYROID HORMONE SYNDROMES(PANEL A). TEMPORAL RELATIONSHIPS BETWEEN HUMAN AND RODENT SYNDROMES OF THYROID HORMONE DEFICIENCY AND NORMAL SEQUENTIAL DEVELOPMENT OF THE FETAL AND NEONATAL BRAIN(PANEL B). (HT = Hypothyroidism; ¨¨ = Temporal Continuation)

not have a human correlate and has both maternal and fetal hypothyroidism that begin after agent application.

Human sporadic congenital hypothyroidism is characterized by normal maternal thyroid function and thyroid hormone levels, but deficient fetal thyroid function starting after the onset of fetal autonomous thyroid gland secretion of thyroid hormones at about 10 weeks of gestation. The hyt/hyt and cog/cog mice and their progeny simulate this condition and begin their hypothyroidism at the time of autonomous secretion(15 days post conception). While the hyt/hyt mouse continues with hypothyroidism throughout life, the cog/cog mouse has a transient hypothyroidism that encompasses a significant part of the adult period, lasting until 10 months postnatally(Adkison al, 1990). Human neonatal and juvenile hypothyroidism are simulated by rodent models of postnatal hypothyroidism. These disorders are primarily induced after birth by artificial agents, i.e. PTU or methimazole, or may occur naturally, i.e. genetically epilepsy prone rat(GEPR)(Mills et al, 1988).

The HYT/HYT MOUSE IS A MODEL OF SPORADIC CONGENITAL HYPOTHYROIDISM: MODELS OF THYROID HORMONE ACTION USING THE HYT/HYT MOUSE

The hyt/hyt mouse demonstrates an inherited hypothyroidism that begins after the onset of fetal autonomous thyroid hormone secretion at 15 days postconception(Stein et al, 1989a,b,c; Beamer et al, 1982; Shanklin et al, 1988. By the use of a hyt/hyt hypothyroid father and a euthyroid hyt/+ mother, relatively equal numbers of hyt/hyt hypothyroid progeny and hyt/+ euthyroid progeny can be generated(Adams et al, 1989). The hyt/+ littermates act as an internal control group. The euthyroid hyt/+ mother simulates the human situation in sporadic congenital hypothyroidism where the mother is also euthyroid.

The hyt/hyt mouse demonstrates specific brain and cerebral cortex molecular and neuroanatomical abnormalities(Stein et al, 1989a, 1991a, Noguchi, 1988) and delayed and abnormal reflexive, locomotor, and adaptive behavior(Adams et al, 1989, 1991; Anthony et al, 1991). Our multidisciplinary approach has used the hyt/hyt mouse to test a general model of the effects of hypothyroidism on the developing brain(MODEL 1). The scheme for our investigations is illustrated(SCHEME 1). We can demonstrate abnormalities in the hyt/hyt mouse in comparison with its hyt/+ littermates for: 1) Neonatal and young adult thyroid gland development; 2) Neonatal and young adult reflexive, locomotor, and adaptive behavioral development; and 3) Specific mRNAs and potentially functional proteins in fetal and neonatal total brain and cerebral cortex and maturing optic nerve. This mouse has direct relevance for understanding the neurological and neuropathological abnormalities, particularly in the motor system, in human thyroid disorders and the potential molecular substrates of these abnormalities.

CHARACTERIZATION OF THE HYPOTHYROIDISM IN THE HYT/HYT MOUSE: THE INHERITED DEFECT IS IN THE THYROID GLAND

The hyt/hyt mouse(Beamer et al, 1981,1982) has a severe and pervasive inherited hypothyroidism(Stein et al, 1991d,1989b) that is characterized by 5-10

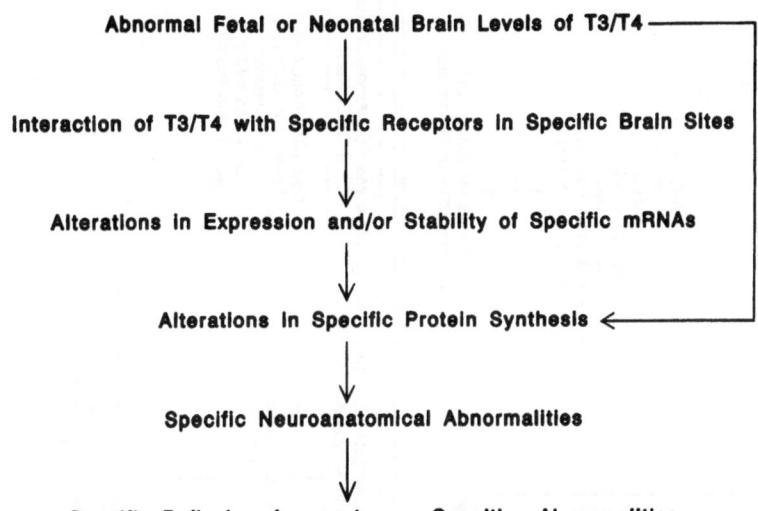

Abnormal Fetal or Neonatal Brain Levels of T3/T4

Interaction of T3/T4 with Specific Receptors in Specific Brain Sites

Alterations in Expression and/or Stability of Specific mRNAs

Alterations in Specific Protein Synthesis

Specific Neuroanatomical Abnormalities

Specific Reflexive, Locomotor, or Cognitive Abnormalities

MODEL 1. MODEL OF THYROID HORMONE ACTION USING THE HYT/HYT MOUSE. Our multidisciplinary approach has used the hyt/hyt mouse to test a general model of the effects of hypothyroidism on the developing brain(MODEL 1). Alterations in fetal serum and brain T3/T4 levels, following interaction with specific thyroid hormone receptors in specific brain sites or with specific proteins or the translational apparatus in specific sites, may lead to alterations in expression and/or stability of specific mRNAs or specific proteins. These mRNAs and proteins may have particular functional roles. These mRNA or protein abnormalities in specific sites and at specific times of gestation or the neonatal period may lead to specific neuroanatomical abnormalities. The neuroanatomical abnormalities affected depend on the normal events that are sequentially occurring at the time of the molecular abnormalities and the hypothyroidism.These neuroanatomical abnormalities in combination with other neuroanatomical abnormalities may contribute to some of the abnormalities in reflexive, locomotor, or adaptive behavior that are observed in hypothyroidism.

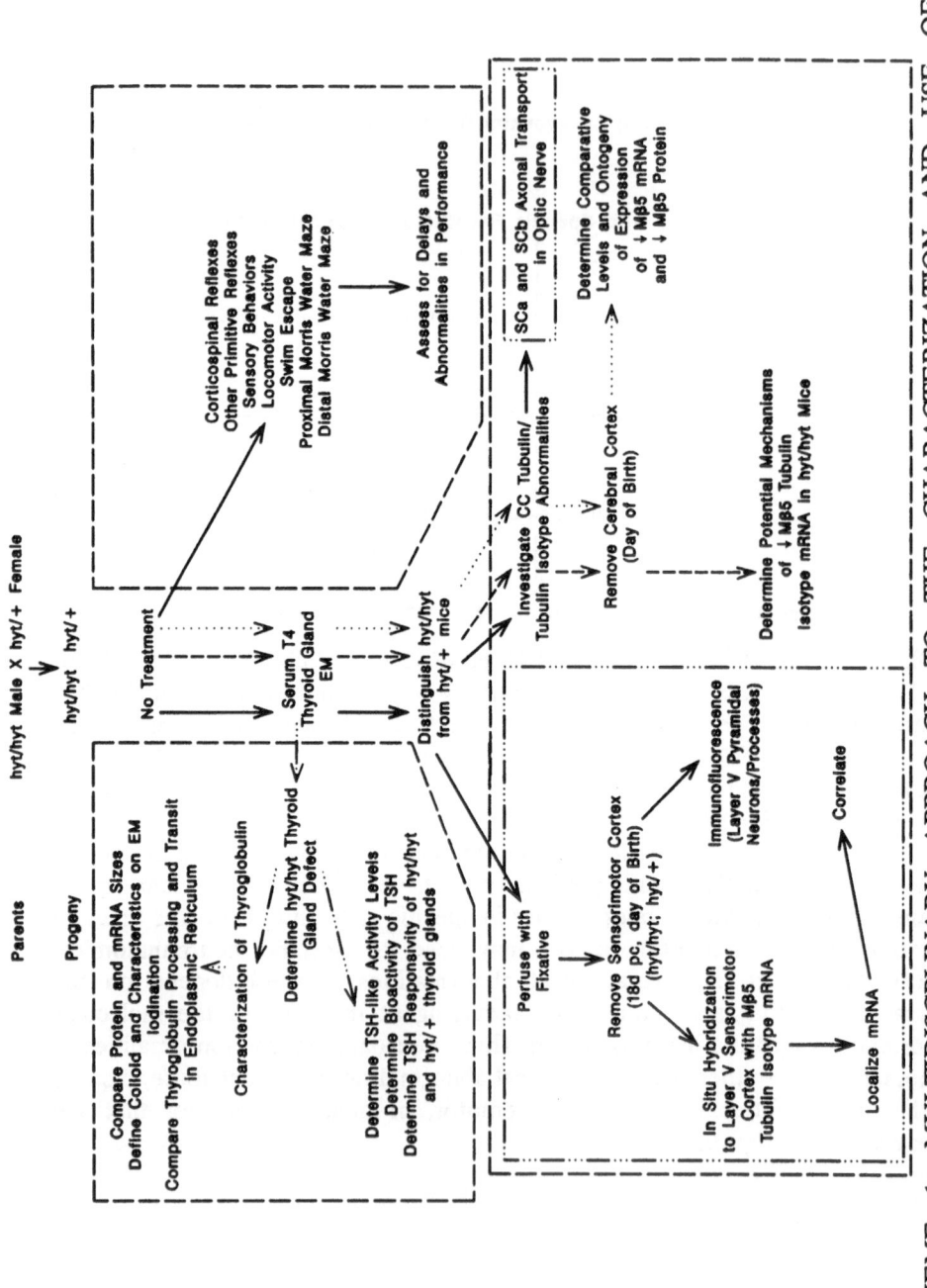

SCHEME 1. MULTIDISCIPLINARY APPROACH TO THE CHARACTERIZATION AND USE OF THE HYPOTHYROID hyt/hyt MOUSE AND ITS hyt/+ EUTHYROID LITTERMATES.

fold reductions in serum thyroxine(T4)(Beamer et al, 1981, Stein et al, 1989b)(FIGURE 3, Panel A), a 16 fold dimunition in triiodothyronine(T3)(Stein et al, 1989b) a 100 fold elevation in TSH-like activity(Stein et al, 1989b)(FIGURE 3, PANEL B), increased numbers of pituitary thyroid stimulating hormone(TSH) granules(Noguchi et al, 1986), and reduced hypothalamic TRH(Noguchi et al, 1986), when compared to euthyroid hyt/+ mice and progenitor strain BALB/cBY +/+ mice. Our studies indicate that the hyt/hyt mouse produces a biologically active TSH-like molecule that is capable of stimulating a heterologous thyroid gland in a mouse bioassay(Stein et al, 1991b). These results suggest that the TSH molecule is not the site of the inherited defect in the hyt/hyt mouse. Taken together with the elevation in serum TSH-like activity and in pituitary TSH granule number, our findings support the concept that the hyt/hyt mouse has a normal physiological feedback response of the hypothalamus and pituitary to the diminished T4 production by the hyt/hyt thyroid gland. The dimunition in TRH in the hyt/hyt hypothalamus(Noguchi et al, 1986) may reflect a hyperstimulation of TRH synthesis and/or increased release, related to reduced thyroid hormones feeding back on the hypothalamus. As evidence for the latter, exogenous T4 increases TRH levels in the hypothalamus(Noguchi, 1988). These results also suggest that the defect in the hyt/hyt mouse is a primary hypothyroidism at the level of the thyroid gland, rather than secondary or tertiary hypothyroidism(Stein et al, 1989b,1991b).

THE POTENTIAL SITE OF THE DEFECT IN THE HYT/HYT MOUSE: THE INHERITED DEFECT DOES NOT RESULT FROM A STRUCTURAL ABNORMALITY IN THE THYROGLOBULIN MOLECULE

The sequence of normal T4/T3 production by the thyroid gland, the effectors that stimulate T4/T3 synthesis, and their sites of action are illustrated(MODEL 2). A number of steps in the synthesis of T4/T3 are altered in the hyt/hyt mouse(MODEL 2).

On ultrastructural evaluation, the hyt/hyt gland demonstrates ballooning and irregular dilation of the endoplasmic reticulum(Stein et al, 1989b). This appearance of the endoplasmic reticulum resembles what is observed in human (Ketelbant-Balasse et al, 1975, Lissitzky et al, 1975b) and certain inherited mouse(cog/cog)(Beamer et al,1987; Mayerhofer et al, 1988; Adkison et al, 1990) and cattle(Ricketts et al, 1987) cases of congenital goiter with depressed T4. These cases, that demonstrate primary hypothyroidism, appear to be due to molecular defects in the thyroblobulin gene(Taylor et al, 1987; Ricketts et al, 1987) and thereby, in the mRNA and protein(Ricketts et al, 1987; Beamer et al, 1990). This is not the case in the hyt/hyt mouse. hyt/hyt thyroglobulin protein of a similar molecular size to euthyroid mice is made(Stein et al, 1989b). Unlike the cog/cog mouse(Adkison et al, 1990), the thyroglobulin in the hyt/hyt mice is normally transported and processed in follicular rough endoplasmic reticulum(Stein et al, 1989b). The hyt/hyt thyroglobulin is iodinated(Stein et al, 1989b) and the iodine is incorporated into hyt/hyt and hyt/+ iodotyrosines(Stein and Taurog, unpublished observations). These results suggest that neither a defect in cellular processing of thyroglobulin nor a potential thyroglobulin protein defect are present in the hyt/hyt mice. Therefore, the dilated endoplasmic reticulum can not be related to a major

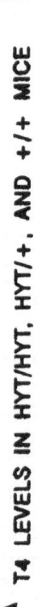

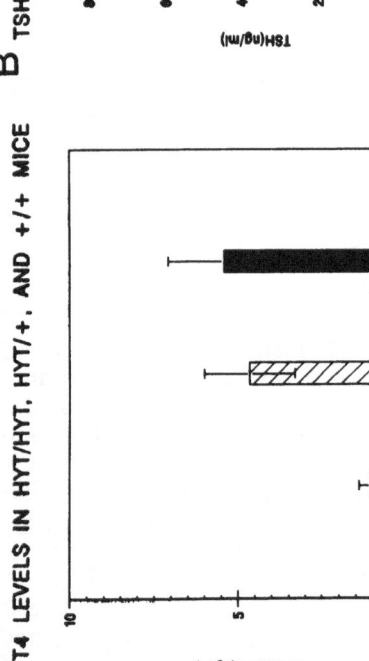

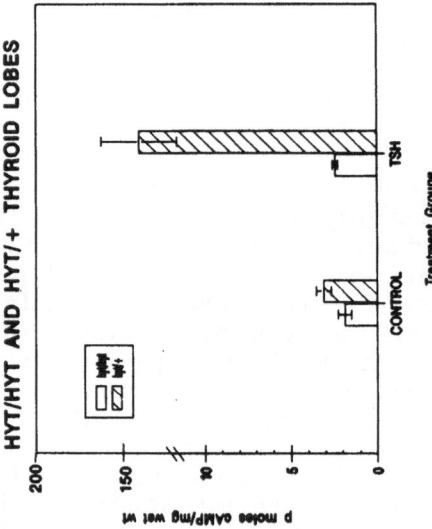

A T4 LEVELS IN HYT/HYT, HYT/+, AND +/+ MICE

B TSH LEVELS IN HYT/HYT, HYT/+, AND +/+ MICE

C EFFECTS OF TSH ON cAMP IN HYT/HYT AND HYT/+ THYROID LOBES

FIGURE 3. CHARACTERIZATION OF THE PRIMARY HYPOTHYROIDISM IN THE hyt/hyt MOUSE. PANEL A: T4 Levels in hyt/hyt, hyt/+, and +/+ Mice. Sera from 219 mice was used to determine serum T4 levels by radioimmunoassay(Regard et al). The hyt/hyt mice had a 5 fold reduction in T4 levels compared to hyt/+ and +/+ mice. PANEL B: TSH-LIKE ACTIVITY IN hyt/hyt, hyt/+, and +/+ MICE. hyt/hyt mice demonstrated a 100 fold elevation in TSH-like activity compared to hyt/+ and +/+ mice in a rat TSH immunoassay(Stein et al, 1989b). The TSH present in hyt/hyt mice is bioactive based on a modified McKenzie bioassay(Stein et al, 1991b). PANEL C: BASAL cAMP LEVELS AND THE cAMP RESPONSE OF INDIVIDUAL hyt/hyt, hyt/+, and +/+ THYROID LOBES TO EXOGENOUS TSH. hyt/hyt mice had significant reductions in basal cAMP and markedly diminished production of cAMP in response to exogenous TSH(Stein et al,1991b). The cAMP levels were determined using an immunoassay kit(Incstar Corporation, Stillwater, MN).

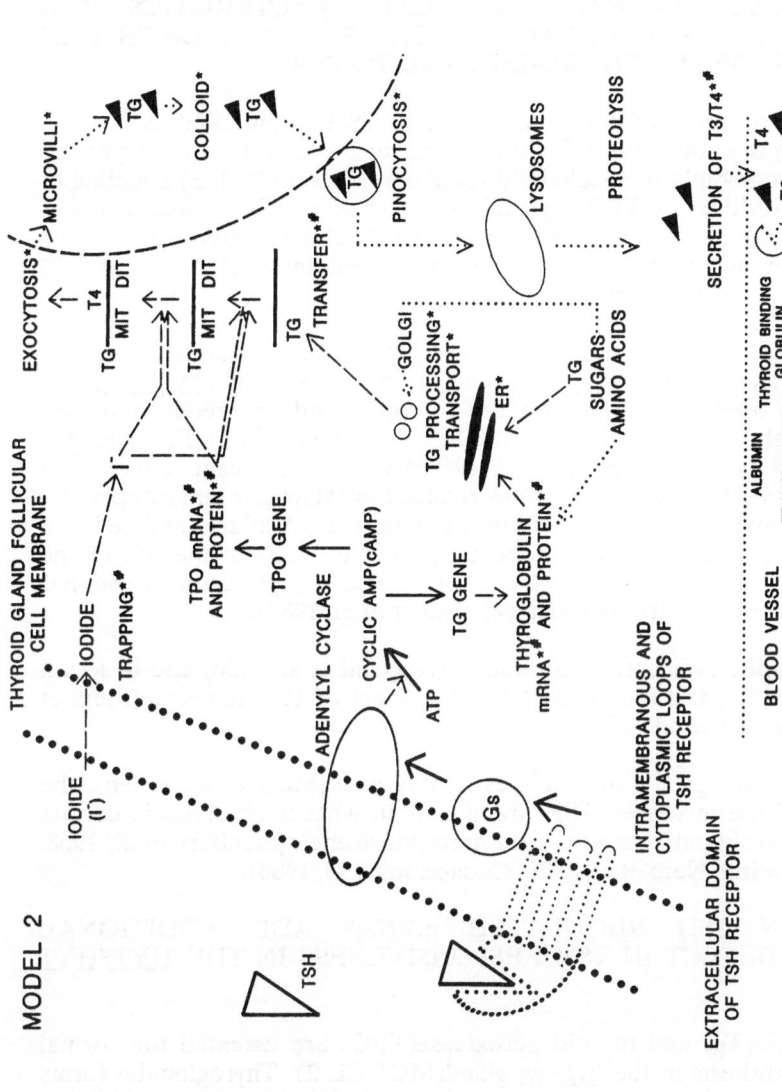

MODEL 2. MECHANISMS OF NORMAL PRODUCTION OF T4 AND T3 BY THE THYROID GLAND. The steps that are regulated by TSH(*) and TSH-cAMP(#) are indicated. The hyt/hyt mouse demonstrates abnormalities at the level of the TSH receptor and has decreased serum T3 and T4, intrathyroidal T4, cAMP, and iodide uptake,decreased levels of Thyroglobulin and TPO mRNAs, delayed and diminished production of endoplasmic reticulum(ER), reduced numbers of microvilli, and decreased amounts of colloid, G protein and adenylyl cyclase related responses and thyroglobulin protein size and processing are normal. as are the cAMP responses to agonist stimulations of hyt/hyt thyroid gland prostaglandin and noradrergic receptors[linked to the Gs(G stimulatory) proteins](Stein et al, 1991b).

structural abnormality in thyroglobulin, but rather a non-specific response of various tissues to hypothyroidism(Stein et al, 1989b). The additional point that is difficult to reconcile with a primary abnormality in thyroglobulin is that the hyt/hyt mouse has a hypoplastic thyroid gland, despite chronic stimulation by elevated TSH levels(Stein et al, 1989b).

THE hyt/hyt MOUSE DEMONSTRATES THE CHARACTERISTICS OF A HYPOTHYROIDISM THAT MIGHT BE RELATED TO A DEFECT IN RESPONSIVENESS OF THE THYROID GLAND TO TSH

The normal response of the thyroid gland to TSH stimulation(reviewed in Dumont et al, 1989) is an increase in: 1) iodide uptake; 2) iodination of the tyrosine residues in the thyroglobulin molecule by thyroid peroxidase(TPO); 3) coupling of these iodotyrosine residues by TPO; 4) synthesis of specific mRNAs and proteins, including thyroglobulin and thyroid peroxidase; 5) colloid production; 5) numbers and surface area of microvilli; 6) production of endoplasmic reticulum; and 7) glandular size during development(MODEL 2).

Several lines of evidence suggest that the hyt/hyt mouse has an inherited reduction in responsiveness of the thyroid gland to TSH. First, in the hyt/hyt gland, many of the TSH mediated effects on the thyroid gland are reduced or are abnormal(Stein et al, 1989b,1991b)(See FIGURE 3 and MODEL 2). For example, the hyt/hyt gland is hypoplastic(Stein et al, 1989b) with a significant reduction in weight(Stein et al, 1991b), and demonstrates reduced intraluminar and cytoplasmic colloid, reduced formation and numbers of endoplasmic reticulum, and reduced numbers of microvilli from the early neonatal period throughout the life of the animal(Stein et al, 1991d; Shanklin et al, 1988). The hyt/hyt mouse also has reduced iodide uptake and intrathyroidal T4(Beamer et al, 1982).

Second, despite an increase in pituitary(Noguchi et al, 1986) and in serum TSH(Stein et al, 1989b), and the production of a bioactive TSH molecule(Stein et al, 1991b), there is a reduction of T4.

Third, the histological abnormalities in the gland(Stein et al, 1989b), the diminished T4, and the increased TSH are similar to what is observed in certain congenitally hypothyroid patients with TSH unresponsiveness(Stanbury et al, 1968, Job et al, 1969, Medeiros-Neto et al, 1979, Codaccioni et al, 1980).

REDUCTIONS IN TSH REGULATED mRNAS ADD ADDITIONAL SUPPORT TO A DEFECT IN TSH RESPONSIVENESS IN THE HYT/HYT GLAND

Thyroglobulin(Tg) and thyroid peroxidase(TPO) are essential for normal thyroid hormone synthesis in the hyt/hyt gland(MODEL 2). Thyroglobulin forms the backbone for the synthesis of thyroid hormone, while TPO functions to iodinate hormonogenic tyrosine residues and to couple these iodotyrosine residues on the thyroglobulin. The synthesis of thyroglobulin and TPO mRNAs and their proteins is regulated and stimulated by TSH(MODEL 2). As further evidence of a defect in TSH responsiveness of the hyt/hyt gland, by solution hybridization, the levels of thyroglobulin and TPO mRNAs are reduced in a statistically significant fashion in the hyt/hyt versus littermatched hyt/hyt mice(p <.01, p<.05, respectively)(Stein et al, 1991b). Subsequent reductions in the protein products of these mRNAs may underly the reductions in serum T4 in the hyt/hyt mouse.

THE MOLECULAR DEFECT IN THE HYT/HYT MOUSE RESULTS IN A REDUCTION IN cAMP AND MAY INVOLVE AN INHERITED ABNORMALITY OF THE TSH RECEPTOR BUT NOT THE G PROTEIN-ADENYLYL CYCLASE SYSTEM

The normal molecular mechanisms of TSH effects on the thyroid gland provide potential insights into the pathophysiology of the inherited defect in the hyt/hyt mouse, while also emphasizing the defect in responsivity to TSH. Many of the effects of TSH are mediated and mimicked by cAMP(Dumont et al, 1989)(MODEL 2). The levels of TPO and thyroglobulin mRNAs and their proteins are regulated by TSH and by TSH through cAMP. As such, the levels of these mRNAs may be viewed as markers of TSH stimulation of the thyroid gland as mediated by cAMP. The reductions of these mRNAs, as observed in the hyt/hyt thyroid gland, might reflect on decreased levels of cAMP. Therefore, it is not suprising that the hyt/hyt gland shows a statistically significant reduction in basal levels of cAMP(p <.025) and that the hyt/hyt gland shows a reduced response to exogenous TSH compared to hyt/+ littermate and +/+ glands(p <.0001)(FIGURE 3, PANEL C)(Stein et al, 1991b). The increase in cAMP that is mediated by TSH stimulation of the thyroid gland normally occurs through a number of steps(MODEL 2). The TSH molecule binds to the TSH receptor. This binding triggers the activation of the G stimulatory protein(Gs)(Gilman et al, 1989; Ross et al, 1989); the Gs protein stimulates adenylyl cyclase which converts ATP into cAMP(MODEL 2). Then, cAMP exerts a variety of effects on the thyroid follicular cell(MODEL 2). Since the hyt/hyt TSH molecule demonstrates normal bioactivity, the inherited defect in the hyt/hyt mouse could be at the level of the extracellular region of the receptor, or the intramembranous or the intracytoplasmic regions of the receptor, the communication of the latter two regions with the Gs protein, the Gs protein itself, or the adenylyl cyclase molecule. In the hyt/hyt mouse, the defect is not present at the level of the Gs protein or adenylyl cyclase(Stein et al, 1991b). Our results suggest that the defect may be present within the TSH receptor molecule in terms of its ability to bind TSH or to transduce conformational changes to activate the Gs protein. Given the fact that the hyt/hyt produces a TSH receptor mRNA that is similar in size to hyt/+ and progenitor strain +/+ mice(Stein et al, 1991b), the inherited abnormality in the hyt/hyt mouse may reflect on a more subtle defect in the sequence and/or structure of the TSH receptor molecule. These findings suggest that the hyt/hyt mouse has a defect in TSH responsivity due to an inherited defect in the thyroid gland TSH receptor molecule(Stein et al, 1991b).

MARKERS RELATED TO THE PATHOPHYSIOLOGY OF THE HYT/HYT DEFECT PROVIDE A MEANS TO DISTINGUISH HYT/HYT FROM HYT/HYT LITTERMATES: SERUM T4 AND ANATOMICAL CRITERIA

After 6 days of age, hypothyroid hyt/hyt mice can be distinguished from hyt/hyt littermates and +/+ progenitor strain mice by thyroid hormone levels. The hyt/hyt mice have levels of less than 1.5 ug/dl, while the hyt/+ mice have levels greater than 4.5 ug/dl(Stein et al, 1989b). This distinction is corroborated by the time of eye opening and ear raising, which are delayed in the hyt/hyt mouse as compared to hyt/hyt littermates. Prior to 6 days of age, T4 levels can not be used alone. T4 levels and qualitative and quantitative indices of thyroid gland hyporesponsiveness to TSH can be used in combination for differentiation of hyt/hyt and hyt/+ mice(Stein et al, 1991d, Stein et al, 1991c). Prior to 6 days of age, maximally different littermates, as assessed by these criteria and by the ratio of T4 levels(hyt/+ to hyt/hyt > 3.0) provide the basis for animal choice as hyt/hyt and hyt/+ for molecular and neuroanatomical studies(see below). Despite some of the limitations of the distinction prior to 6 days, the hyt/hyt and hyt/+ mice differ primarily in T4 level from 15 days post conception throughout life. Therefore, each can be compared for behavioral, neuroanatomical, and molecular studies(SCHEME 1).

MOTOR SYNDROMES, NEUROLOGICAL SIGNS, AND NEUROPATHOLOGY OF HUMAN THYROID DISORDERS: IMPLICATIONS FOR PATHOPHYSIOLOGY AND REVERSIBILITY

ENDEMIC CRETINISM: MOTOR AND MENTAL SIGNS

In addition to varying degrees of deafness and deafmutism, mental retardation, learning disabilities, speech disturbances, and oculomotor palsy(Delange et al, 1989; Delong et al, 1985, Delong, 1989; Boyages et al, 1988; Pharoah et al, 1985; Fierro-Benitez et al, 1989), a specific motor syndrome(Delong et al, 1985, Delong, 1989; Boyages et al, 1988) is generally noted in endemic cretinism. This involves proximal spasticity with tone changes, altered reflexes, pathological reflexes, and/or hemiparesis(Delong et al, 1985; Boyages et al, 1988). In some cases, pes cavus, and hypotonia are noted(Delong et al, 1985). The syndrome also commonly includes proximal/axial rigidity, flexion dystonia, mask-like facies, and varying inability to walk. Slowness of initiation of action and apraxia are also observed. These abnormalities suggest dysfunction at the level of the cochlea, areas 4 and 6 and the supplemental motor area of the cerebral cortex, the subcortical aspects of the pyramidal tracts, and also, in some patients, dysfunction at the anterior horn cell level. In addition, these findings would suggest involvement of the globus pallidus and perhaps basal ganglia-cortical pathways. The cerebellum would appear to be spared clinically(Boyages et al, 1988), but may not be neuropathologically(Lotmar, 1929).

In these children with endemic cretinism, the cognitive abilities and motor competence of children ten to fifteen years of age are significantly correlated with maternal T4 levels rather than iodine or T3 during pregnancy(Pharoah et al, 1984). These results suggest that maternal T4, presumably transported through the placental to the fetus, is crucial for normal prenatal brain development. Further, these results along with data on sporadic congenital hypothyroidism(Rovet et al, 1987) and prenatal(Hadjzadeh et al, 1990; Hendrich et al, 1987; Davenport et al, 1976) thyroid hormone deficiency in rodents implies that aberration of thyroid hormone in utero can have lasting effects on later neurological and neurobiological function in the progeny.

In controlled trials when iodine is given prior to pregnancy or in the first trimester(Boyages et al, 1989; Chaouki, 1987; Fierro-Benitez et al, 1989), there has been an improvement in intellectual performance in school children and motor competence. Despite this therapy, the neurological and neuropsychological abnormalities have not been completely eliminated(Boyages et al, 1989; Chaouki, 1987). These observations are reflective of the complexity of the problem, where iodine, environmental toxins, i.e., cassava, and genetic substrate may contribute to the occurrence and reversibility of the clinical abnormalities.

HYPOTHYROIDISM AND HYPERTHYROIDISM IN HUMAN PREGNANCY: RISK FACTORS FOR CEREBRAL PALSY AND MENTAL RETARDATION

Maternal thyroid disorders during pregnancy are common but disregarded(Davis et al, 1988; Lowe et al, 1990) disorders that may have prominent neurological effects on the progeny. Children of hypothyroxinemic women may have poor fine motor coordination, varying degrees of mental retardation, and specific signs of cerebral palsy , including spasticity, tone abnormalities and hemiparesis (Man et al, 1971, Nelson et al, 1986). Clinically this is of concern because the hypothyroxinemia may occur only during pregnancy, and may go undiagnosed without screening or assiduous treatment of pregnant women(Davis et al, 1988; Lowe et al, 1990). Based on analysis of the progeny of the National Collaborative Perinatal Project, as well as other studies, both maternal hypothyroidism and hyperthyroidism prior to and during pregnancy are associated with increased risk(at a statistically significant level to as much as 20 times) for the development of cerebral palsy in progeny(Nelson et al, 1986).

SPORADIC CONGENITAL HYPOTHYROIDISM: MOTOR AND MENTAL RETARDATION SYNDROMES

If untreated or if treated after 1-3 months after birth(Klein et al, 1985), sporadic congenital hypothyroidism leads to severe mental retardation and profound neurological deficits , including cerebral palsy. The latter is defined by gross and fine motor clumsiness(Wolter et al, 1979), spasticity and other pyramidal tract signs, poor coordination, basal ganglia movement disorders, cerebellar ataxia , hearing loss, and squint(Wolter et al, 1979; Macfaul et al, 1978). Because of the success and use of early blood screening programs for T4, TSH, or both at birth, early diagnosis and treatment before 1 month of age have eliminated significant mental retardation (Klein, 1985; Rovet, 1989; Glorieux et al, 1989). Some have suggested that these children may be entirely normal(Fisher et al, 1989; New England Collaborative Group, 1990), in part due to maternal transplacental transport of thyroid hormones to the fetus(Vulsma, 1989; Larsen, 1989). However, the initial studies showing total prevention of mental retardation and of cerebral cortical neuropsychological dysfunction by early hormone therapy were marred by insufficient patients, inappropriate tests, inadequate follow-up times, inadequate control groups and limited neurological evaluations(reviewed in Stein et al, 1991e).

Despite early diagnosis and treatment, some children may be abnormal neurologically and neuropsychologically. Some studies have reported that IQs in early treated hypothyroids are still lower but in the normal range(reviewed in Stein et al, 1991e); in some very well constructed studies, some of these early treated children have learning disabilities, hearing deficiencies, aphasia, hyperkinesis and deficits in perceptual, fine motor, and language abilities(Rovet et al, 1984,1987) or delayed but eventually normal neuropsychological development(New England Collaborative Group, 1990; Glorieux et al, 1983). Whether these effects are due to inadequate postnatal T4 dosage(Fisher et al, 1989) , inadequate compliance(New England Collaborative Group, 1990), or irreversible fetal neuropathological changes is unresolved.

Specific abnormalities in motoric behavior, referrable to the motor cortex, corticospinal tracts, and other regions are common in sporadic congenital hypothyroidism. Patients treated within the first three months after birth have slowed motor performance(Wolter et al, 1979), possible cerebellar ataxia(Wolter et al, 1979, Macfaul et al, 1978), fine motor incoordination(Wolter et al, 1979), unilateral and bilateral pyramidal signs(Macfaul et al, 1978), and overt spasticity(Wolter et al, 1979). However, motor examinations of patients treated before two weeks of age may show no abnormalities(New England Collaborative Group, 1990) or significant numbers of patients with motor abnormalities(Birrell et al, 1987; Hulse, 1987). By quantitative neurological examination, coordination, fine hand and finger manipulative abilities and gross motor function are altered through 12 years of age in comparing hypothyroid to euthyroid children(Rochiccioli et al, 1989).

CEREBRAL PALSY AND MOTOR SYNDROMES IN ENDEMIC CRETINISM, SPORADIC CONGENITAL HYPOTHYROIDISM, AND MATERNAL HYPO/HYPERTHYROIDISM

Motor abnormalities are seen in all human thyroid disorders. Cerebral Palsy can be defined as "a disorder of movement and posture due to a defect or lesion of the immature brain."(Bax, cited in Weinstein et al, 1989). According to others(Freeman et al, 1988), "cerebral palsy is a fixed(static) deficit of motor function of early onset." By these criteria, the motor signs and neurological examination of patients with thyroid hormone disorders of prenatal onset would be classified as cerebral palsy. The deficit may affect one or more parts of the nervous system with resulting variation in symptoms. The major types are: "pyramidal, i.e., spastic quadriplegia, commonly associated with mental retardation and epilepsy; diplegia(more commonly in premature infants), or hemiplegia; extrapyramidal, including the dystonic and choreoathetoid types; and the mixed varieties involving both pyramidal and extrapyramidal systems."(Freeman et al, 1988). These specific forms of cerebral palsy are also observed in the thyroid hormone disorders of prenatal onset.

Motor function is dependent on a complex interaction between a number of systems of which the corticospinal tracts are primary effectors , as well as the layer V pyramidal neurons in motor and sensory cortex(Davidoff, 1989; Benecke,

1990; Kuypers, 1985). Other systems involved include the basal ganglia and extrapyramidal system, cerebellum, spinal cord, parietal lobes and premotor cortex. Tone and motor function may be regulated by the corticospinal tracts, the reticulospinal tracts, and a number of other central and peripheral systems.

The abnormalities observed in sporadic congenital hypothyroidism, endemic cretinism, and maternal hypothyroidism can be attributed to specific cortical regions and the corticospinal tracts(Roland et al, 1987; Freund, 1987; Kuypers, 1985; Porter 1985; Benecke, 1990). Clinically and neuropathologically, these patients have dysfunction of a number of anatomical regions. These include the corticospinal tracts, parietal regions involved in motor behavior, and the supplementary motor area, premotor cortex, and the primary motor area(MI). Distal fine motor abnormalities, observed in sporadic congenital hypothyroidism, may relate directly to premotor cortex abnormalities(Freund, 1987). Also, the medial and mediolateral precentral gyrus and the basal ganglia, particularly in the endemic cretinism, would appear to be involved.

The cerebral cortex, pyramidal neurons, pyramidal tracts, and cerebellum are particular targets for thyroid hormone deficiency. Although the numbers of human autopsies of endemic cretinism and sporadic congenital hypothyroidism are minimal cerebral atrophy(Lotmar, 1928, 1929), gyral abnormalities(Lotmar, 1928, 1929; Beierwaltes, 1959), reduced neuronal number of pyramidal and betz neurons in layer V(Jia-Liu et al, 1989; Jianquan et al, 1985, Marinesco, 1924; Lotmar, 1928, 1929), abnormalities in pyramidal neuron axonal and dendritic development(Jianquan et al,1985; Marinesco, 1924; Lotmar, 1928), and diminished size of the pyramidal tracts may be observed in cerebral cortex and brainstem(Rosman, 1975). These pathological lesions in hypothyroidism are similar to those observed in cerebral palsy(Christiansen et al, 1967; Gross et al, 1968; Malamud et al, 1964).

The types, sites, and patterns of the neurological, behavioral, and neuropathological abnormalities, generally, and involving the motor system, are also similar in rodent and human thyroid hormone disorders(reviewed in Stein et al, 1991f). Postnatal hypothyroid rodents, congenitally hypothyroid hyt/hyt mice and progeny of maternally hypothyroid rats and of fetal hypothyroid and iodine deficient rodents demonstrate delays and persistent abnormalities in reflexive(suckling, placing, and grasping), locomotor, and later developing adaptive behaviors with motor components(Adams et al, 1991a,b; Anthony et al, 1991; Davenport et al, 1972,1976; Davenport, 1976; Hendrich et al, 1984; Johanson, 1980; Narayanan et al, 1982; Rastogi et al, 1976; Schalock et al, 1979; Comer et al, 1985; Eayrs et al, 1968; Schapiro et al, 1970; Strupp et al, 1983). In the hypothyroid disorders, the rodent behavioral and human neurological signs reflect on anatomic abnormalities in specific regions of the cerebral cortex. These include the motor cortex and its corticospinal projections, parietal lobes, temporal lobe associative cortex, as well as hippocampus and brainstem. Because of these observations, animal models of human thyroid disorders have relevance for understanding the neurological and neuropathological abnormalities in these human thyroid disorders.

[ONSET OF AUTONOMOUS T4 SECRETION

[HYPOTHYROIDISM BEGINS IN THE hyt/hyt MOUSE

[ONSET OF NOVEL CEREBRAL CORTICAL mRNA EXPRESSION
INCLUDING CERTAIN TUBULIN ISOTYPE mRNAS

FETAL AND NEONATAL DEVELOPMENT OF RODENT CEREBRAL CORTEX

PRENATAL										POSTNATAL				
10	11	12	13	14	15	16	17	18	19	Birth	1	3	5	7	9

Neurogenesis(by Layers)

[------------------------------------(I,V,VI)---][----------II,III,IV--]

[----------------Migration----------------------]

Arrival of Thalamic Afferents to White

Matter   [--------------]

Arrival of Thalamic Afferents to IV   [---------]

Arrival of Callosal Afferents to Cerebral Cortex to III   [---]

Axonal Development of Corticospinal Tracts from V

[Emergence]   [DI][MB][PD]   [CEC][TC][LC][SC]

[Grey Mat. Pene....

Apical Dendrites(V) to I   [----

Appearance of Basal Dendrites and Branching of Apical and

Basal Dendrites(from V)   [-----------

(from III and other layers)   [-----------

Appearance of Dendritic Spines   [-]

Neuroglial Maturation   [-----

Legend (Sites Reached): DI = Diencephalon; MB = MidBrain; PD = Pyramidal Decussation;

CEC = Cervical Spinal Cord; TC = Thoracic Cord; LC = Lumbar

Cord; SC = Sacral Cord; Grey Mat. Pene. = Grey Matter

Penetration of Spinal Cord.

FIGURE 4. TIMING AND SEQUENTIAL DEVELOPMENT OF THE LAYER V PYRAMIDAL NEURON(SENSORIMOTOR CORTEX) AND CORTICOSPINAL TRACTS IN THE RODENT. The onset of autonomous thyroid hormone function begins at 15 days post conception(d pc) and the hyt/hyt mouse demonstrates deficient T4 production starting at this time. This particular period of time from 15 d pc onward corresponds with new cerebral cortex mRNA synthesis, dendritic development of layer V pyramidal neurons, and elongation and ingrowth of corticospinal axons in the brainstem and spinal cord.

NORMAL FETAL AND POSTNATAL BRAIN DEVELOPMENT IS AFFECTED IN THE SPECIFIC HUMAN AND ANIMAL MODELS OF THYROID HORMONE DISEASE: PATHOPHYSIOLOGIC AND NEUROANATOMIC UNDERPINNINGS OF THESE NEUROLOGICAL AND BEHAVIORAL DISORDERS

Several trends in the normal development of the nervous system and its alteration by thyroid hormone can be identified. These trends are relevant for understanding the neurological, behavioral, and neuropathological abnormalities in the rodent and human thyroid hormone disorders. First, in the human and the rodent, neurodevelopmental events(neurogenesis, migration) occur in a defined and logical sequence(FIGURE 2B, FIGURE 4). In a population of specific neurons, neurogenesis is followed by migration. Migration is followed by axonal outgrowth and elongation and by dendritic outgrowth, elongation and branching. These events are followed by synaptogenesis and subsequently, myelination(Sidman et al, 1982). This can be broken down into more specific events related to the cerebral cortex and corticospinal tracts of the mouse and rat(FIGURE 4). In motor cortex, all of the above events, except myelination, occur principally during the prenatal period(FIGURE 2B, FIGURE 4)(Crandall et al, 1989a,b; Demyer, 1967; Donatelle, 1977; Jones, 1981; Jones et al, 1982; Krist, 1978; Meller et al, 1968; Miller, 1987a,1988; Pinto Lord et al, 1979; Rice, 1975; Shoukimas et al, 1978; Schreyer et al, 1982; Wise et al, 1979). Myelination continues through the second year of life in the human or through the first several postnatal months in the rodent(Sidman et al, 1982).

Compared to the rodent, a similar temporal relationship for development is seen in the human cerebral cortex(Marin-Padilla, 1970,1988; Sidman et al, 1982; Lemire et al, 1975) for the motor cortex of the precentral gyrus between 9 weeks and 24 weeks of gestation. Layer V pyramidal neurons that will give rise to the human corticospinal tracts arise and migrate to the primitive cortical plate between 7-10 weeks of gestation(Marin-Padilla, 1970,1988), which corresponds to 14 days post conception(d pc) in the mouse. Initial axonal outgrowth and the main apical dendritic trunk to layer I with some initial basal dendrite formation are noted between 11-15 weeks(Sidman et al, 1982; Marin-Padilla, 1970,1988). The apical and basal dendrites become more profuse and branched from 15 weeks onward with acquisition of significant numbers of dendritic spines by 28 weeks of gestation(Marin-Padilla, 1970,1988; Purpura, 1975). The internal capsule reaches adult form by about 10-12 weeks, The corticospinal tract is intact prior to 33 weeks when it starts to be myelinated(Lemire et al, 1975). The onset of spontaneous motor behavior in the fetus suggests that these pathways may be intact early in the second trimester(Devries et al, 1982) when spinal cord synapses appear(Okado, 1982). During the first 24 months of postnatal age, particularly during the first three months of age, the axonal and dendritic branches of pyramidal neurons in the precentral gyrus increase, These processes are refined and pruned in number through at least six years of age(Conel, 1939,1941, 1947, 1951, 1955, 1959, 1963, 1967). The normal sequence of cerebral cortex and motor system development contributes to the normal sequence of development of motor control and integration and of cognitive function.

Second, the periods of brain development can be correlated with specific phases of thyroid influence. T3 and T4 are detected in the brain at 10 weeks of gestation in the human and at 21 d(days) pc(post conception) in the rat(Morreale de Escobar et al, 1985). Thyroid hormone receptors are noted in human brain at 7 weeks of gestation(Bernal, 1984) and at 12 d pc in the mouse brain(Stein et al, 1989a) and rat brain(Strait et al, 1990). Three phases of thyroid hormone influence can be defined: 1) early in utero development (prior to 15 d pc in rodents and 10-12 weeks in humans) when the only thyroid hormone exposure is via maternal hormones that are transported transplacentally(Morreale de Escobar et al, 1985; Vulsma et al, 1989). 2)in utero development subsequent to 15 d pc in rats (12 weeks humans) when the fetal thyroid gland is functional and fetal brain T3-T4 exposure could be from both fetal and maternal hormones(See Morreale de Escobar et al in this volume) and 3)postnatal development when brain exposure is dependent on endogenous hormone production. These periods encompass specific events of normal brain development of the human and the rodent(FIGURES 2 and 4).

Third, these neuroanatomical events may be irreversibly or reversibly altered by a variety of different effectors, including thyroid hormone. In the human and rodent, thyroid hormone can affect neurogenesis, cell migration, axonal and dendritic differentiation, synaptogenesis, and myelination. In the motor cortex and other areas of the motor system, abnormalities are noted in human and rodent thyroid disorders involving these neuroanatomical events. The different types of thyroid hormone disorders(FIGURES 1 and 2) temporally correspond with certain neuroanatomical events and can alter these events based on the timing of the thyroid hormone disorders.

Fourth, the nature and extent of the neurological and neuroanatomical effects of alteration in thyroid hormone level relate to the type, timing, severity, and duration of this alteration and the neuroanatomical events that are occurring during those times. Alteration in thyroid hormone level, both elevations and depressions, can occur for different durations, at different times, and of different severity in the human and rodent disorders of thyroid function.

Fifth, the timing of a thyroid hormone level insult can have multiple and different effects on different brain regions and cell populations within those regions. This is because different brain regions, i.e. brainstem or cerebral cortex are developing at different times and because different events are occurring in cellular populations in the same regions at the same time. As examples of this, the brainstem develops earlier in gestation than the cerebral cortex. The cerebellum develops later than the cerebral cortex and brainstem. Differentiation is occurring in the brainstem, while neurogenesis is occurring in the cerebral cortex. Layer V pyramidal neurons are differentiating, while Layer III pyramidal neurons are migrating. Although thyroid hormone can alter each of the events of development in different regions, it is the timing and severity of the insult that will dictate which neuroanatomical events are altered and their location. With a discretely timed early postnatal thyroid hormone insult, delayed migration of cerebellar granular neurons(Lauder et al, 1986; Legrand, 1982-83) and altered process growth in the cerebral cortex layer V neurons(Ruiz-Marcos, 1989) Similarly, the differentiation of pyramidal neurons in hippocampus and cerebral cortex motor regions will be affected by a similarly timed thyroid hormone insult. These patterns have relevance for understanding the neuroanatomical burden of the specific thyroid hormone insults in the human and rodents disorders of thyroid hormone.

In summary, abnormalities in the specific human and rodent thyroid hormone disorders can be related to the timing and severity of the period of hypothyroidism, and duration of the thyroid hormone abnormality. These disorders can also be related to the specific neuroanatomical events that are altered by the thyroid hormone insults and the subsequent disruption of those neuroanatomical events that normally follow the originally disrupted events (FIGURES 2 and 4).

Earlier, more severe, and/or longer insults lead to more severe neuroanatomical disruption and subsequent behavioral and neurological abnormalities. Severe cerebral palsy and mental retardation are the results of these insults. Mild cerebral palsy and/or learning disability may be the result of milder thyroid hormone insults. As an example of this, children suffering from congenital iodine deficiency, in utero and perinatally, and secondary hypothyroidism show more severe neurological impairment than children with sporadic congenital hypothyroidism. The greater severity of neurological impairment seen in iodine deficiency states than in sporadic congenital hypothyroidism is likely a result of the additive effects of maternal and fetal hypothyroidism. Because of the timing and severity of the hypothyroidism, in endemic cretinism, there may be interference with brainstem(which develops in the first trimester) and cerebral cortex development, while in sporadic congenital hypothyroidism, the cerebral cortex may be primarily affected. Based on the sequence and timing of normal development of the motor systems , including the layer V pyramidal neurons and the corticospinal tracts, the neurological, behavioral, and neuropathological abnormalites in human and rodent thyroid disorders may be related to disruption of these events by thyroid hormone deficiency in the fetal and postnatal brain. Based on the sequence of brain development, sporadic congenital hypothyroidism may lead to alterations in fetal cerebral cortex development and if not treated early in the postnatal period, may lead to abnormalities in postnatal refinement and myelination of the cerebral cortex and motor systems.

THE PERIODS OF THYROID HORMONE SYNTHESIS, OF CORTICOSPINAL TRACT AND LAYER V PYRAMIDAL NEURON DIFFERENTIATION AND OF NORMAL DEVELOPMENT OF MOTOR FUNCTION: THE TIME OF ONSET OF THE HYPOTHYROIDISM IN THE HYT/HYT MOUSE AND THE HYT/HYT MOUSE ARE RELEVANT FOR UNDERSTANDING THESE EVENTS

The late gestational and early neonatal hypothyroid hyt/hyt mouse and its euthyroid hyt/+ littermates provide a means to dissect out thyroid hormone effects during the narrow period of initial differentiation of the layer V pyramidal neurons and their corticospinal tracts(FIGURE 4). The hyt/hyt mouse is specifically relevant for motor behavior, because the onset of hypothyroidism(15 d pc) and its continuation into the early neonatal period encompasses cerebral cortex neuronal differentiation of pyramidal neurons, corticospinal tract development, and new specific mRNA and protein synthesis in pyramidal neurons and others cells of the cerebral cortex(Stein et al, 1989a, 1991a,c; Tucker et al, 1989; Nunez et al, 1988;

TABLE 1. SOMATIC AND BEHAVIORAL DIFFERENCES IN THE
HYT/HYT MOUSE All observations in the hyt/hyt mouse were compared to
littermatched hyt/+ mice and +/+ mice(Adams et al, 1989a, 1991; Anthony et
al, 1991).

MEASURE	AGE AT TESTING	DIFFERENCE NOTED
Eye Opening	14-18	Delayed
Ear Raising	15-21	Delayed
Body Weight	21,40	Reduced
Body Length	21	Reduced
Cliff Avoidance Response	6	Delayed
Negative Geotaxis Response	5	Delayed
Locomotor Activity	14	Elevated
Locomotor Activity	21,40	Decreased
Swimming Escape Learning	30-50	Impaired
Morris Maze		
Proximal	80-110	Impaired
Distal	80-110	Impaired

Miller et al, 1987b; Matus, 1988; Bond et al, 1983; Lewis et al, 1986). The timing of the hypothyroidism corresponds with the period of active growth of the corticospinal tracts from the cervical spinal cord to other caudal levels(Demyer, 1967; Schreyer et al, 1982; Jones et al, 1982), the periods of penetration of this tract into the grey matter at all levels of the spinal cord(Schreyer et al, 1982; Jones et al, 1982; Donatelle, 1977), the initial and continued apical and dendritic growth of these pyramidal neurons(Wise et al, 1979; Caviness et al, 1988; Crandall et al, 1984a,b; Miller, 1987a,1988; Jones, 1981; Rice, 1975), and the appearance of subsequent primitive motoric reflexes(after 5 days after birth)(Donatelle, 1977; Adams et al, 1989,1991) that may be dependent on the occurrence of these specific neuroanatomical events(FIGURE 4). We have investigated reflexive, locomotor, and adaptive behavior and their potential neuroanatomical and molecular foundations with the hyt/hyt hypothyroid mouse and its euthyroid littermates using a multidisciplinary approach(SCHEME 1, MODEL 1).

BEHAVIORAL STUDIES OF THE HYT/HYT MOUSE AND THEIR POTENTIAL RELATIONSHIP TO THYROID HORMONE REGULATED DEVELOPMENTAL NEUROANATOMICAL EVENTS

THE HYT/HYT MOUSE SHOWS DELAYS IN NORMAL MOTOR REFLEXIVE BEHAVIOR(TABLE 1, SCHEME 1)

Delays in cliff avoidance and negative geotaxis were found in the hyt/hyt mice compared to their euthyroid littermates(Adams et al, 1989). For cliff avoidance the hyt/hyt animals were significantly delayed at 6 days of age compared to the normal animals. For negative geotaxis, a similar significant delay was seen for the hyt/hyt animals versus the normal hyt/+ littermates at 5 days of age.   As the hyt/hyt animals grow older, the time required to make the criterion response became similar to the normal animals(Adams et al, 1989).

The results of the assessment of early motor reflexes , including placing, grasping and crossed extensor reflex indicate delays in forepaw and hindpaw grasping(TABLE 1) between the hyt/hyt and their euthyroid littermates , as well as a BALBcBY +/+ control group. Forepaw grip strength was significantly lower in the hyt/hyt mice at 8 and 9 days after birth and then reached a normal level at 10 days after birth(Adams et al, 1991). The corticospinal tracts , as well as other pathways may be involved in placing, grasping, and crossed extension. These potential corticospinal tract mediated reflexes appear normally soon after birth. The delays in forepaw grasping suggest a slower development of the corticospinal neurons and of rostral-caudal motor function maturation. In humans, altered or persistent or assymetric grasp reflexes may reflect on altered motor function or hemiparesis.

The timing of the hypothyroidism in the hyt/hyt mouse precedes and spans the period of emergence and refinement of these reflexes and their neuroanatomical substrates (FIGURE 4). Further, these reflexes and early behaviors may be sensitive to the function and anatomical development of the developing cerebellum, brainstem and vestibular system. The anatomical development of each of these systems is altered by prenatal and early postnatal hypothyroidism(Uziel, 1986; Narayanan et al, 1985; Legrand, 1982-83; Dememes,

1986). As further evidence, other studies have demonstrated delays in normal onset of of primitive cranial nerve and brainstem mediated reflexes in the mid-late gestation progeny of maternally hypothyroid rat(Narayanan et al, 1985).

Despite eventual normal performance in grasping and forepaw grip strength in hyt/hyt mice, the behavioral and anatomical delays, observed in hypothyroidism, may be particularly important because nervous system development involves multiple and complex events that are precisely timed. The normal performance and timing of performance of these motor reflexes may be required for the normal development of more complex motor behaviors in the rodent, i.e. swim escape and Morris maze activity. Therefore, in the hyt/hyt mice, the delays in these early reflexes may contribute to the abnormalities that are later observed for swim escape and the Morris maze.

MOTOR ACTIVITY AND LEARNED MOTOR BEHAVIORS ARE IMPAIRED IN THE HYT/HYT MOUSE

Locomotor activity was assessed by activity counts in an automated activity monitor(Adams et al, 1989). hyt/hyt mice demonstrated reductions in activity compared to euthyroid mice at days 21, 40, and 120; the hyt/hyt mice were hyperactive at day 14(Anthony et al, 1991)(Table 1).

Learned motor behavior was evaluated by the swim escape paradigm(Adams et al, 1989) and the proximal and distal Morris maze (Morris et al, 1982; Sutherland et al, 1988) tasks. Swim escape was assessed by determination of the time required by mice to find and climb a ladder on the opposite side of a water tank for five consecutive days between 45 and 50 days of age. The hyt/hyt mice showed no ability to learn this task(Anthony et al, 1991).

In the proximal Morris maze task(FIGURE 5), the hyt/hyt mice were significantly slower than the euthyroid mice(hyt/hyt and +/+) in solving the task, but showed some improvement in response time over the total period of testing(Anthony et al, 1991). For the distal Morris water maze testing, again, the hyt/hyt group showed little ability to learn the task, compared to both euthyroid groups of mice(Anthony et al, 1991).

ANATOMICAL LOCALIZATION OF MOTOR AND SPATIAL LEARNING AND THEIR RELATIONSHIP TO SENSORY BEHAVIORS

The performance on spatial learning tasks has been related to the number of pyramidal neurons in CA1 and CA3 layers of the hippocampus(Sutherland et al, 1988). The size of the hippocampus is reduced by hypothyroidism that starts in late gestation(Rabie et al, 1979). This size reduction may relate to the observed effects of hypothyroidism typified by reduced numbers of pyramidal and granule cells in the hippocampus(Rami et al, 1986a) , as well as by reductions in pyramidal and granule neuron process growth(Rami et al, 1986b, Lauder, 1989; Lauder et al, 1986). In fact, in the hyt/hyt mouse, hippocampal size is reduced and hippocampal process growth is impaired(Crawford and Stein, unpublished observations). It is conceivable that the abnormalities in spatial learning may be related to cellular abnormalities in the hyt/hyt hippocampus.

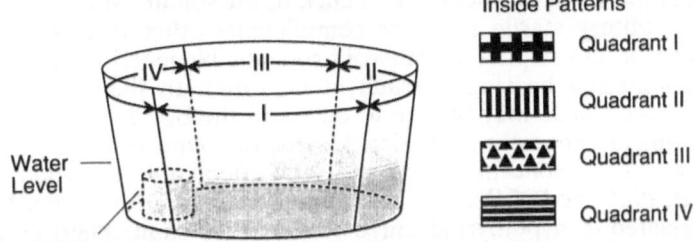

Inside Patterns

Quadrant I

Quadrant II

Quadrant III

Quadrant IV

FIGURE 5. THE MORRIS WATER MAZE PARADIGM (Anthony et al, 1991, Sutherland et al, 1988). In the proximal morris maze and the distal morris maze, a washbasin, filled with water, was used. A small platform was placed at one end of the washbasin and an individual animal was put in the water at the other end of the washbasin. In the proximal task, mice from 80-110 days of age were assessed for their ability to find the platform and climb out of the water, as measured by time from placement into the water until finding the platform. This testing was repeated 4 times/day for eight consecutive days. The hyt/hyt mice were significantly slower than the euthyroid mice(hyt/hyt and +/+) in solving the task, but showed some improvement in response time over the total period of testing. Again, the hyt/hyt group showed little ability to learn the task, compared to both euthyroid groups of mice. Distal Morris maze testing employed the same apparatus as used above for the proximal testing. However, the platform was submerged under the water and its location was obscured by coloring the water with opaque non-dairy creamer. Clear plastic sheets with specific patterns were affixed to the sides of the wash basin and these acted as spatial cues for the mouse to locate the platform. This testing began after the proximal testing and was conducted for 8 consecutive days with four trials per day.

Spatial learning involves not only motor and associative processes but also visual function. Visual stimuli must be received and processed at various levels of the visual system, including visual cortex. These inputs must then be integrated with the hippocampus and the motor systems. The pyramidal neurons of the visual cortex(Ruiz-Marcos et al, 1989) and other components of the optic pathway(Stein et al, 1991a) may be impaired in hypothyroid animals and humans(see Axonal Transport section below, FIGURE 8, TABLE 2).

The effects of hypothyroidism are not limited to hippocampus, the visual system or the motor systems. Similarly, other sensory behaviors, i.e., olfactory orientation and auditory startle, were found to develop more slowly in the hyt/hyt than for the euthyroid mice(TABLE 1)(The auditory startle was defined as positive if a reflexive response was observed to a click of set volume at a set distance). The delay in the auditory startle response compliments other results relating to the auditory system in developing hypothyroid rodents. Neonatal hypothyroid rodents demonstrate impaired hearing(Dussault et al, 1989). In the hyt/hyt mouse, the delays in auditory startle may be due to delays in the development of ear raising. The ear flap might represent a physical barrier to sound transmission before ear raising. However, abnormal prenatal and neonatal development of the cochlea(Uziel, 1986) and of the pyramidal neurons in auditory cortex(Ruiz-Marcos et al, 1983) related to hypothyroidism(reviewed in Dussault, 1989) may contribute to the abnormalities in the hyt/hyt startle response.

MOTOR AND BEHAVIORAL ABNORMALITIES IN THE HYT/HYT AND OTHER HYPOTHYROID RODENTS: POTENTIAL PATHOPHYSIOLOGY AND RELEVANCE FOR HUMAN SPORADIC CONGENITAL HYPOTHYROIDISM

Thus, the hyt/hyt mice show delays in the normal development of reflexive behavior, as well as significant impairment in more complex learned motor behavior that does not improve with time. These hyt/hyt delays resemble delays and non-persistent abnormalities that resolve with time and that have been observed in humans with sporadic congenital hypothyroidism that was treated prior to one month of age(Glorieux et al, 1983; New England Collaborative Group, 1990). These observations may reflect delayed but normal neuroanatomical development or differences in normal neuroanatomical development that can be overcome by later development. These functions require the normal development and integration of function of a variety of brain regions and cells, including those comprising the motor system, that are targets for thyroid hormone.

Permanent impairments in cognitive and complex motor functions are also seen in the hyt/hyt mouse. The hyt/hyt mouse simulates untreated human sporadic congenital hypothyroidism. Untreated or late treated human congenital hypothyroidism(after 2-3 months of age) also demonstrates permanent cognitive and neurological deficits(Klein, 1985; Macfaul et al, 1978; Wolter et al, 1979).

Future studies are required to define the degree, timing, and mechanisms of reversibility of the hyt/hyt abnormalities in swim escape and spatial learning by T4. In both the human and the rodent, reversibility of the behavioral, neuropsychological, and neurological abnormalities may or may not be complete and may be linked to the time of T4 replacement and the complexity of the behavioral or neurological task that is being assessed. Animals studies of this sort may have relevance for the treatment of human sporadic congenital hypothyroidism.

Congenitally hypothyroid hyt/hyt mice, their euthyroid hyt/+ littermates, and progenitor strain, +/+ mice, are useful models for looking at normal cerebral cortex neuroanatomical and molecular development. The purposes of our studies have been to try to better understand the neuroanatomical and molecular basis for the hypothyroid related neuropsychological and neurological abnormalities, particularly those related to abnormalities in normal movement and motor behavior. The scheme for our neuroanatomical and molecular studies is illustrated(SCHEME 1).

THYROID HORMONES, PROCESS GROWTH, MICROTUBULES, AND CYTOSKELETAL PROTEINS: COMMON GROUND IN HUMAN AND ANIMAL STUDIES

One of the most consistent abnormalities related to fetal and hypothyroidism in both rodents and humans, particularly sporadic congenital hypothyroidism, are persistent abnormalities in process growth and connectivity. In the developing hypothyroid brain, disorders of neuronal process growth (Morreale de Escobar et al, , 1983; Eayrs, 1968; Lauder et al, 1986; Legrand, 1982-1983) are observed in both central and peripheral neurons. These abnormalities are manifested by diminished axonal(Lauder, 1986, Lauder et al, 1989; Legrand, 1982-83; Eayrs, 1955; Eayrs, 1968; Marinesco, 1924) and dendritic outgrowth and elongation(Lauder, 1986, Lauder et al, 1989; Legrand, 1982-3; Ruiz-Marcos, 1989) and branching(Marinesco, 1924; Ruiz-Marcos, 1989; Garza et al, 1988; Rami et al, 1986b). These results are consistent with the ability of thyroid hormone to induce pathway growth(Hoskins et al, 1983) and neurite outgrowth and elongation(Puymirat et al, 1983; Hargreaves et al, 1988). The diminished complexity of the hypothyroid brain, as measured by reduced connections between neural cell populations, (Morreale de Escobar et al, 1983; Marinesco, 1924) may be linked to these effects of hypothyroidism on process growth, on direct alteration of synaptogenesis (Ruiz-Marcos, 1989), on alterations in the number and distribution of dendritic spines (Ruiz-Marcos, 1989), and on the subsequent myelination of these processes(Noguchi et al, 1984; Sarlieve et al, 1984; Almazan et al, 1975; Allpress et al, 1986). These alterations in process growth and connectivity are observed in premotor regions of cerebral cortex and visual and auditory cortex, as well as hippocampal pyramidal and granule neurons, cerebral cortical neurons, and cerebellar purkinje neurons (Lauder et al, 1986, Garza et al, 1988; Eayrs, 1955; Ruiz-Marcos, 1983,1989; Rami et al, 1986b; Marinesco, 1924). This is further manifested by reduction in cerebrocortical white matter(Lotmar, 1928, 1929; Rosman, 1975; Beierwaltes, 1959) and by reduced thalamo-cortical fibers and corticocortical fibers(Marinesco, 1924) and reduced size of the pyramidal tracts(Rosman, 1975). The sites of these abnormalities in process growth and connectivity may be related to the motor, memory, and visuomotor abnormalities observed in human and rodent thyroid syndromes.

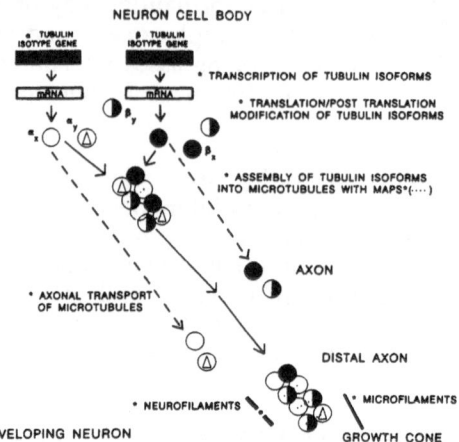

MODEL 3. POTENTIAL MODEL OF PROCESS GROWTH IN THE DEVELOPING AXON, ITS RELATIONSHIP TO CYTOSKELETAL STRUCTURES AND THEIR MOLECULAR COMPONENTS, AND POTENTIAL THYROID HORMONE REGULATED EVENTS(*). Normal process growth is dependent on the function of microtubules, neurofilaments, and microfilaments. Microtubules are 25 nm cytoskeletal fibers, built from dimers of α- and β-tubulin, and having side arms made up of microtubule associated proteins(MAPs) (Mandelkow et al, 1989; Matus, 1988). α and β tubulin isotypes and different MAPs are transcribed in the nucleus of the neuronal cell body. Distinct α and β tubulin isoforms as well as different MAPs, i.e. adult and fetal tau, MAP1A, and MAP1B are produced in the developing neuron(See text) versus the mature neuron(Lewis et al, 1988; Cowan et al, 1987; Sullivan et al, 1987; Bloom et al , 1983; Matus, 1988). Each isoform(Lewis et al, 1984; Burgoyne et al, 1988) and each MAP(Tucker et al, 1989; Matus, 1988; Bloom et al, 1983, 1984; Nunez, 1988) has a characteristic tissue and cellular distribution. The tubulin isoforms differ primarily in short segments of their carboxyterminus regions(Sullivan et al, 1987). Following translation and post-translational modifications, equal quantities of α and β isoforms are assembled(Nunez et al, 1989) together with MAPs into brain microtubules(MT). The MAPs may bind with different affinity to the distinct carboxyterminus regions of the tubulin isoforms of the MT. Certain MAPs may lead to greater(adult tau) or lesser(fetal tau) stability of the microtubule(Nunez et al, 1989). After assembly, these MTs are transported through the axoplasm of the axonal process to the distal axon and the growth cone(Hammerschlag et al, 1988). The transported microtubules may be added to the distal end of growing MTs and may be in dynamic equilibrium with free tubulin isoform and MAP subunits. Neurofilaments, which are 10 nm in diameter and which may complex with MTs, are the major determinants of axonal caliber (Hoffman et al, 1985) and are comprised of three subunits: NFH, NFM, and NFL. Neurofilament side arms, formed by the carboxy- terminal regions of NFH and NFM, contain repeated sequences that can be heavily phosphorylated (Lee et al. 1988). The degree of phosphorylation of the neurofilament and the numbers of neurofilaments relative to microtubules may favor process stabilization. Microfilaments are 4-6 nm fibrils consisting of actin subunits, which couple with microtubules in the growth cone. Microfilaments are particularly important in axonal growth cones and process growth (Yamada et al, 1971, Smith, 1989). Actin and neurofilament components are also synthesized in the nucleus and transported by axonal transport. Alteration in thyroid hormone level may affect process growth and maintenance by altering the cytoskeleton at multiple levels(*, See text, See MODEL 4).

A clue to some of the potential foundations of the abnormalities in process growth and maintenance and connectivity comes from the facts that these events are dependent on normal microtubule number and function, as well as the function and number of other neuronal cytoskeletal structures, including neurofilaments and microfilaments. Normal process growth during development depends on the synthesis of cytoskeletal proteins, their assembly into cytoskeletal structures, i.e. microtubules, neurofilaments, and microfilaments, and on the delivery of these proteins via axonal transport to the distal axon and growth cone(FIGURE 8, TABLE 2, MODEL 3)(Smith, 1988; Mitchison et al, 1988; Lasek, 1988; Brady, 1988; Hammerschlag et al, 1989; Yamada et al, 1971). Alterations in thyroid hormone level may alter each of these steps that contribute to normal process growth(See FIGURE 4, MODELS 3 and 4). In a specific neuron at a specific time, these effects of thyroid hormone may be operating simultaneously or individually.

ALTERATIONS IN MICROTUBULE NUMBER AND FUNCTION AND THEIR MOLECULAR COMPONENTS MAY CONTRIBUTE TO PROCESS GROWTH ABNORMALITIES

Dimunition in process growth could be related to the observed reductions in microtubule number and density in axons and dendrites of developing neurons related to hypothyroidism. (Faivre et al, 1983; Sarafian et al, 1983; Marc et al, 1985). Microtubules consist of dimers of α- and β-tubulin complexed with microtubule associated proteins, or MAPs (Mandelkow & Mandelkow, 1989; Matus, 1988). Thyroid hormone and specifically thyroid hormone deficiency, may influence microtubules by a number of mechanisms. First, hypothyroidism leads to reductions in the assembly of MAPs and tubulin isoforms into microtubules(reviewed in Nunez et al, 1989). These MAPs are thought to function in microtubule assembly and in microtubule interactions with other cytoskeletal structures (Nunez, 1988; Nunez et al, 1989). A hallmark property of most MAPs is that they also act in vitro as stimulators of tubulin polymerization and microtubule stability (Nunez et al, 1989).

Reduced microtubule assembly may be predicated on observed alterations in the pattern, site, level, and timing of expression of certain MAPS, including adult and fetal tau, MAP2, and MAP1B in fetal, neonatal, and juvenile brain(Nunez et al, 1989, Benjamin et al, 1988). In the developing brain, there is a delay in the transition of tau from juvenile to adult forms in hypothyroid animals (Nunez et al, 1989). This is potentially significant, because adult tau is a more potent stimulator of tubulin assembly than fetal tau (Nunez et al, 1989). Hence, a preponderance of fetal tau at the expense of adult forms of the protein could result in decreased neurite outgrowth. MAP2 is also altered in hypothyroid animals (Nunez, personal communication). Specific MAPs which may favor microtubule assembly are also regulated by thyroid hormone (Hargreaves et al, 1981) and reduced in hypothyroidism (Nunez et al, 1989; Nunez, personal communication).

Second, reductions in the level of total tubulin in different brain regions in fetal and neonatal brains , as well as maturing optic nerve(Stein et al, 1991a) are observed. This reduction in tubulin might contribute to decreased neurite formation, because the degree of microtubule assembly in cells may be proportional to the size of the tubulin pool (Drubin et al, 1985,1988). Total tubulin protein is also regulated by thyroid hormone in fetal cerebral cortex(Takahashi, 1983). However, in some cases, fetal and neonatal total tubulin mRNA and protein may not be altered by hypothyroidism(Stein et al, 1991c). This observation affirms the selective aspects of thyroid hormone action on the tubulins and microtubules in certain regions and cells.

Third, the velocity and the amount of axonal transport of microtubular components, i.e. tubulin, and thereby, microtubules are reduced in the maturing hypothyroid optic nerve axon(Stein et al, 1990c)(FIGURE 8, TABLE 2 and See below).

Fourth, hypothyroidism may lead to reduced expresssion of certain tubulin isoform mRNAs and their translation products at certain developmental times in certain neurons. This may alter the composition, stability, function, and assembly of microtubules in those neurons(MODELS 3 and 4).

THE UTILITY OF THE HYT/HYT MOUSE FOR STUDIES OF TUBULIN ISOFORM EXPRESSION, THYROID HORMONE REGULATION, AND PROCESS GROWTH ABNORMALITIES IN PYRAMIDAL NEURONS IN THE CEREBRAL CORTEX

The potential importance of these tubulin isoform mRNAs lies in the facts that: 1) Their protein products, tubulin isotype proteins, are the major components of tubulin protofilaments, which make up microtubules; and 2) The synthesis of these proteins and microtubule associated proteins are required for axonal outgrowth, elongation, and dendritic outgrowth, elongation, and branching. These findings can be correlated with: 1) The potential contribution of microtubule abnormalities to observed process growth abnormalities noted in hypothyroidism in motor cortex; and 2) The potential regulation of microtubule number and tubulin levels by thyroid hormone. Because of these observations,, we investigated the expression of the tubulin isoforms in the hyt/hyt, hyt/+ and +/+ fetal and neonatal cerebral cortex , as well as the transport of tubulins by axonal transport in the hyt/hyt and euthyroid optic nerves(SCHEME 1).

These mice can be used to evaluate the effects of alteration in thyroid hormone level on molecular events related to the cytoskeleton in the late fetal and neonatal cerebral cortex. These analyses are fostered by the use of hyt/hyt and hyt/+ littermate pairs, which provide a means to address molecular differences related only to differences in thyroid hormone level. We have used these mice to look at specific tubulin isotypes in the developing cerebral cortex to try to determine how this relates to process growth abnormalities in hypothyroidism.

For the tubulin isotypes, particularly Mβ5, the aims of our studies were: 1) To define the normal timing, pattern and level of expression of tubulin isotype mRNAs and their translation products and total α tubulin mRNAS and protein; 2) To establish when and whether specific tubulin isotype mRNAs and proteins and total β tubulin mRNA and protein are altered by hypothyroidism during a specific period of cerebral cortex differentiation; and 3) To determine if reductions in specific tubulin isotype mRNAs might underly reductions in specific tubulin proteins and total tubulin in hypothyroidism.

CERTAIN TUBULIN AND OTHER CYTOSKELETAL mRNAS AND THEIR PROTEINS MAY BE EXPRESSED DURING A NARROW WINDOW OF CEREBRAL CORTEX DEVELOPMENT IN THE MOUSE

Our studies have revealed several points related to the Mβ5, Mα1, and Mβ2 isoforms of tubulin(Stein et al, 1989a, 1991c).

First, in developing euthyroid mice, the mRNA for Mβ5 was shown to follow a developmental pattern of expression in cerebral cortex which was quite distinct from that of whole brain. The peak of expression was at 17 days pc in the total brain and on the day of birth in cerebral cortex. Hence, different brain regions must employ distinct programs for regulating the levels of Mβ5 mRNA during early development.

Second, the levels of Mβ5, Mβ2, and Mα1 mRNAs were expressed in fetal cerebral cortex, peaked on the day of birth, and declined to low levels after 7-16 days after birth. This low level of expression persisted through the adult period. Similarly, the translation products for Mβ5 peaked at one day and then declined significantly over the next six days.

Thus, Mβ5 and the other tubulin isotype changes demonstrate that the composition of microtubules in the fetal and neonatal developing cerebral cortex normally change related to constitutive patterns of expression, i.e. turning on, or turning off of certain tubulin isotypes. These patterns of expression are generally important for the developing brain, because similar constitutive patterns of expression in mice and rats are noted for other important cytoskeletal molecules; these molecules include other tubulin isotype mRNAs(Lewis et al, 1984; Bond et al, 1983; Farmer et al, 1986), total tubulin(Bond and Farmer, 1983), MAPs (Nunez et al, 1989; Nunez, 1988; Matus, 1988; Tucker et al, 1989; Lewis et al, 1986), and actin mRNAs (Bond et al, 1983) and other molecules important for axonal development, i.e., epidermal growth factor(EGF)(Stein et al, 1989a), and GAP43(Benowitz et al, 1987). Taken together, the above findings suggest that common endogenous signals or constitutive developmental programs, operating at the transcriptional level(Sullivan, 1988; Havercroft et al, 1984), may exercise tight control of mRNA and protein levels for a number of mRNAs and their translation products that are essential for process growth in the developing brain. These observations suggest that any analysis of thyroid hormone regulation of specific genes and their products must be interpreted with consideration of the endogenous programs simultaneously operating on those molecules.

Mβ5 TUBULIN mRNA AND PROTEIN ARE SIGNIFICANTLY REDUCED IN FETAL AND EARLY NEONATAL HYT/HYT CEREBRAL CORTEX

Compared to euthyroid hyt/+ littermates, Mβ5 mRNA and its translation product were significantly reduced(1.5-6 fold) in cerebral cortex on and shortly after the day of birth in hyt/hyt mice(Stein et al, 1989c,1991c). The specificity of these tubulin isotype mRNA abundance changes was suggested by the lack of change in total brain and cerebral cortex Mβ4 and Mα2 tubulin, total β tubulin, 9A6 mRNA(Stein, 1985,1988; Stein et al, 1989a,1991c), Poly A+ RNA, and ribosomal RNA in hyt/hyt versus hyt/+ mice. This compliments data related to earlier reductions in Mα1 in day of birth and 5 day old total hyt/hyt brain(Stein et al, 1989a). Dimunitions in thyroid hormone level lead to reductions in specific mRNAs, i.e. Mβ5, EGF(Stein et al, 1989a), and Mβ2(Stein, unpublished observations) in cerebral cortex, Mα1 in cerebral cortex and total brain(Stein et al, 1989a) and certain translation products, i.e., Mβ5, at specific brain times in specific regions, rather than causing global changes on all cerebral cortex mRNAs or RNAs.

The mechanisms of regulation for Mβ5 by thyroid hormone are not clear. Given the above facts, the depression in Mβ5 isotype mRNA and its translation product at birth and the rate of fall in abundance of these molecules in fetal and neonatal cerebral cortex(Stein et al, 1991c) could represent a complex regulation by thyroid hormone of the Mβ5 mRNA and its protein at the level of transcription(Samuels et al, 1989), mRNA stability(Fernyhough et al, 1989; Narayan et al, 1985; Cleveland, 1989), or both interacting(Diamond et al, 1985) with developmentally timed, genetic programs of expression(MODEL 4).

POTENTIAL RELEVANCE OF Mβ5 AND THYROID HORMONE FOR MICROTUBULES, AND PROCESS GROWTH.

Mβ5 or Class I tubulin isoform represents a ubiquitous isotype that is made in a variety of organ systems(Lewis et al, 1985; Cowan and Dudley, 1983) across the phylogenetic scale(Sullivan, 1988). In the fetal and neonatal mouse(Lewis et al, 1985) and rat(Bond et al, 1984) brain, this isoform is abundant and common. During this period, Mβ5 would be expected to contribute as a major isotype to microtubules. Further, the reduction of this isoform and other isoforms may have effects on the microtubule and subsequently, on process growth.

A clear illustration of the potential role of a certain isotype for the production of microtubules of specific function relates to the independent functions of β2 and β3 tubulin isoforms in Drosophila testis(Hoyle and Raff, 1990). The precedent for a linkage between a certain isotype and process growth comes from studies of Mβ2. Mβ2 mRNA(Hoffman et al, 1988) and Mβ2 protein(Jeantet et al, 1981), as opposed to other isotypes, are selectively increased in regenerating neurons(Hoffman et al,1988) and cells with extending neurites(Jeantet and Gros, 1981). Mβ2 is selectively expressed in differentiating but not dividing neurons in developing brain(Lee et al, 1990). In further support of this linkage, rat Tα1 mRNA is induced by NGF in cells with elongating neurites(Miller et al, 1987b).

The evidence for the potential linkage for Mβ5, microtubules, and process growth comes from a number of points. First, in the cerebral cortex, Mβ5 and other tubulin isotypes, Mβ2 and Mα1, are expressed(starting at 15 d pc) and peak(day of birth) during a narrow period of cerebral cortex development that correlates temporally and may contribute molecularly to the initial axonal outgrowth from layer III and layer V pyramidal neurons. During this period, the outgrowth, elongation, and branching of basal and apical dendrites for these neurons and the ingrowth of afferent thalamocorticortical fibers are also

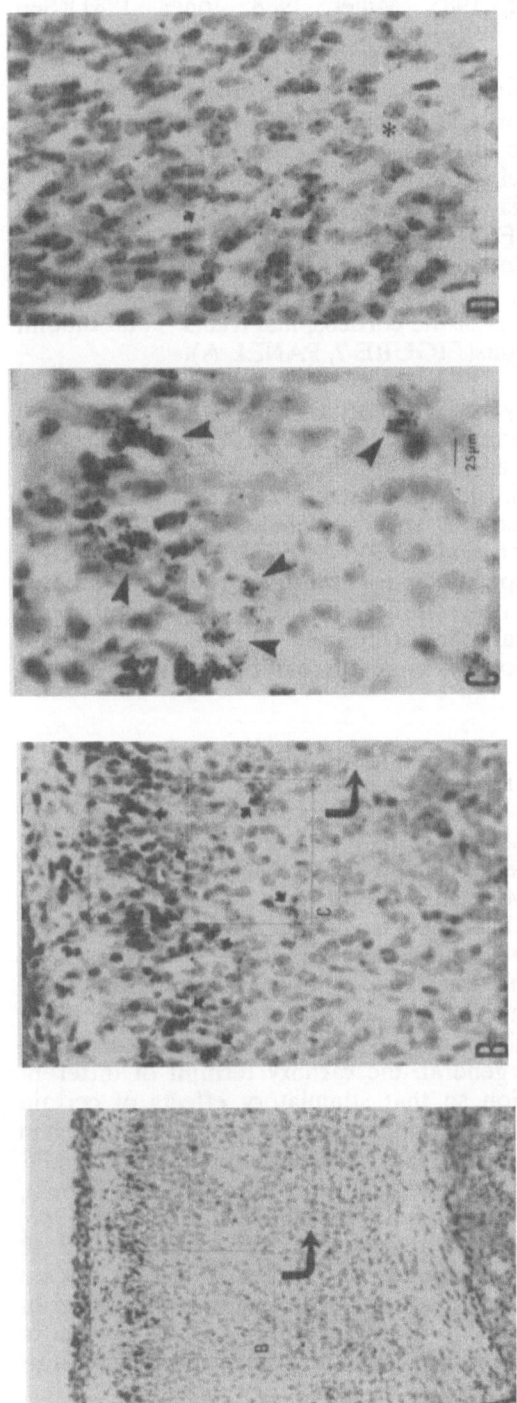

FIGURE 6. IN SITU HYBRIDIZATION OF Mβ5 TO DAY OF BIRTH hyt/+ CEREBRAL CORTEX. Day of birth mice were perfused intracardiac with 4% paraformaldehyde/ 2% glutaraldehyde and 30 micron sections were cut. These sections were hybridized to ^{35}S labeled cDNAs for Mβ5 and Mα1. Sections were autoradiographed and counter-stained with cresyl-violet. PANEL A: Low magnification view taken from a representative field in the sensorimotor cortex of a hyt/+ euthyroid mouse.

PANEL B: Enlargement of the area indicated in A to demonstrate the presence of radiolabeled cells, some of which are indicated by small arrows. The majority of labeled cells are located superficially in the cortex but a few labeled somata can be seen in deep layers.

PANEL C: High magnification view of the area indicated in B. Illustrated here are a large number of neurons (arrowheads) positive for the presence of M 5 as indicated by the dense accumulation of silver grains in the photosensitive emulsion layer overlying the cell bodies.

PANEL D: Representative field of cerebral cortical neurons in a section treated with RNAase. Although a few silver grains are apparent (small arrows), the majority are located in cell body free areas and there is no indication of grain clustering over somata as illustrated in C. Careful inspection reveals that most of the somata exhibit rather large, distinct aggregates of chromatin material (note cells near asterisk). These dark spots should not be confused with the similar appearance of silver grains in the emulsion layer.

occurring(Morreale de Escobar et al, 1983; Miller, 1988; Jones, 1981)(See FIGURES 2 and 4).

Second, by in situ hybridization and immunohistochemistry, Mβ5 mRNA(Hoffman et al, 1988) and its translation product(Burgoyne et al, 1988) are localized to neurons. As an example of this, on in situ hybridization, the Mβ5(FIGURE 6) and Mα1 mRNAs(Stein et al, 1989a) are localized to cell groups in superficial and deep cortex on the day of birth (particularly layer V pyramidal neurons in mouse sensorimotor cortex)(FIGURE 6, Panels A-D); rat Tα1, which is equivalent to mouse Mα1, is localized to similar areas(Miller et al, 1987b). These neurons are the forerunners of layers II, III, IV, and V of the sensorimotor cortex. The layer V neurons represent the origins of the corticospinal tracts. Total tubulin protein is also evident in pyramidal neurons(FIGURE 7, PANEL A).

Third, layer III and layer V pyramidal neurons are: 1) areas of active differentiation in the developing fetal and neonatal cerebral cortex, particularly the motor cortex(Miller, 1988; Jones, 1981); and 2) sites that have shown abnormalities in process growth related to hypothyroidism(Marinesco, 1924, Eayrs, 1955; Ruiz-Marcos, 1989). These regions are also: 1) specific cellular sites for certain developmentally expressed MAPs(Tucker et al, 1989)(See FIGURE 7, PANELS B and C); and 2) sites for localization of specific thyroid hormone receptors(Bradley et al, 1989; Fox and Pfaff, 1987). Certain thyroid hormone receptors, i.e. α1, peak during late gestation and the neonatal period(Strait et al, 1989), and these receptors are required for transduction of some thyroid hormone dependent responses(Samuels et al, 1989).

HOW MIGHT CHANGES IN AN ISOTYPE LIKE Mβ5 AFFECT MICROTUBULES IN MOLECULAR TERMS?

Two general mechanisms that both depend on the sequences of the carboxyterminus sequences of different tubulin isoforms may provide an explanation for the effects of reduced Mβ5. It is known for both α-tubulin and β-tubulin that different isoforms are distinguished from one another primarily by amino acid sequences at their carboxy termini (Sullivan, 1988; Burgoyne et al, 1988). First, certain isotypes may promote microtubule polymerization, because of influences of their distinct carboxyterminal sequences. Mβ2, and perhaps Mβ5, may preferentially promote microtubule subunit polymerization(Joshi and Cleveland cited in Hoffman, 1989). In general, the carboxy termini of different isoforms are inhibitory to polymerization so that stimulatory effects of certain isoforms may have real significance. Second, certain shared or similar sequences in the carboxy terminus of different isoforms may favor the binding of certain MAPs. The binding sites on tubulin for at least some MAPs, i.e. MAP2, tau, and dynein, reside near the carboxy ends of both α-tubulin and β-tubulin (Littauer et al, 1986, Macchioni et al, 1988; Paschal et al, 1989; Serrano et al,1985). A conserved 12 amino acid, consensus carboxy terminus sequence in certain α and β isotype proteins may serve as a preferential binding site for MAP2 and dynein(Paschal et al, 1989). These putative sequences for the binding of certain MAPs are present in Mβ5, but not in Mβ6(Benjamin et al, 1988) or Mβ2. The potential implication of this structural data on Mβ5, is that reductions in Mβ5 may alter the binding characteristics of tubulin for specific MAPs, such as MAP2. In this case, the

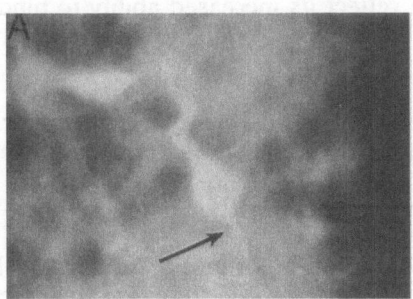

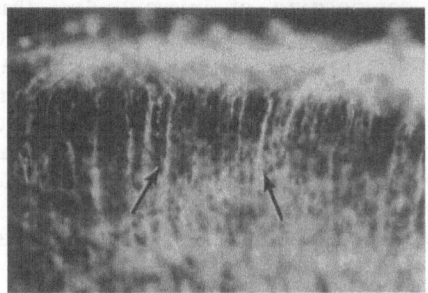
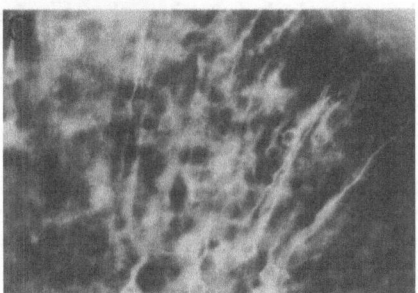

FIGURE 7. IMMUNOCYTOCHEMICAL LOCALIZATION OF TUBULIN
AND MAPS IN DIFFERENTIATING LAYER V PYRAMIDAL NEURONS
IN EUTHYROID MOUSE SENSORIMOTOR CORTEX(Stein and Bloom,
unpublished observations) . Similar to the mRNA localization of Mβ5 and Mα1,
MAP mRNAs and proteins and tubulin protein are found by others(205,210)
and ourselves to be present in differentiating deep and superficial neurons.
Immunoflourescent studies were carried out with 35μ sections and antibodies to
tubulin DM1A(Amersham, Arlington Heights, Illinois), MAP2(Bloom et al,
1984), and MAP1A(Bloom et al, 1984).

PANEL A: Total β Tubulin in pyramidal neurons(mouse sensorimotor cortex)
with axon(arrow) and dendrite on the day of birth.

PANEL B: MAP2 in cerebral cortex pyramidal neuron dendrites(arrows) on the
day of birth.

PANEL C: MAP1A in dendrites of pyramidal neurons on the day of birth

These studies could not distinguish any gross changes between pyramidal
neurons and their processes in day of birth hypothyroid hyt/hyt and euthyroid
cerebral cortex. The lack of distinction on qualitative analysis of hyt/hyt versus
hyt/+ cerebral cortex is consistent with the lack of difference noted on
qualitative golgi analysis(Noguchi, 1988). In a complex structure, such as
cerebral cortex, accurate definition of axons and dendrites and their differences
in the hypothyroid and euthyroid states requires quantitative computerized
analysis of golgi studies(Ruiz-Marcos, 1989).

suitability of Mβ5 might reflect its increased ability to bind one or more proteins, such as MAP2 and tau, that directly stimulate tubulin polymerization and stabilize microtubules.

Additional work is required to isolate and characterize the tubulin isoform and MAP genes and their mRNA and protein products to define the mechanisms of thyroid hormone regulation of gene expression and the effects of thyroid hormone on the interaction of these mRNAs and proteins.

Taken together, the implication of our results with Mβ5 is that diminished levels of Mβ5 in the hyt/hyt mice might alter microtubule composition. One of the mechanisms by which thyroid hormone may affect process growth and microtubules is by selectively altering the biochemical composition of microtubules; this alteration may occur by regulation of specific tubulin isoforms, such as Mβ5, and their translation products, or by alterations in certain MAPs(Nunez et al, 1989) at specific developmental times and in certain areas of the brain. Changes in microtubule composition, in combination with other cytoskeletal effects of thyroid hormone deficiency, may influence the assembly, stability, and perhaps, functional properties and distinctness of neuronal microtubules and may contribute to the process growth abnormalities observed in hypothyroid developing cerebral cortex(MODEL 3) and seen in hippocampal neurons in the hyt/hyt mouse. Changes in microtubule composition must be distinguished as a mechanism affecting microtubule function process growth from reductions in total tubulin(Gonzalez et al, 1978), which is not observed in developing hyt/hyt cerebral cortex(Stein et al, 1991c). Both could lead to the observed diminution in axonal and dendritic microtubule number(Marc and Rabie, 1985; Faivre et al, 1983,1984; Sarafian and Verity, 1986) or alteration in microtubule function that may contribute to the process growth abnormalities in hypothyroid nervous tissues during development and the life cycle(MODELS 3 and 4).

The reduction in Mβ5 mRNA and protein and the localization of the Mβ5 and other tubulin isotypes to neurons of the sensorimotor cortex that give rise to the corticospinal tracts and to neurons that have shown abnormalities in human and rodent serves to link together motor, process growth, and molecular abnormalities in fetal and neonatal hypothyroidism.

THYROID HORMONE DEFICIENCY ALTERS THE DEVELOPING AND MATURING AXON BY REDUCING THE VELOCITY OF AXONAL TRANSPORT AND THE AMOUNT OF AXONAL TUBULIN AND MICROTUBULES AND NEUROFILAMENTS: STUDIES OF AXONAL TRANSPORT WITH HYT/HYT, HYT/+, AND +/+ OPTIC NERVES

Another potential molecular level for regulation by thyroid hormone of process growth is by alteration of axonal transport. Tubulin, neurofilaments, MAPs, and actin are all carried in the axon as part of the slow components of axonal transport, predominantly by one of two different rate components (Hoffman et al 1975; Brady et al, 1982; Lasek et al, 1982, 1984). Microtubules and neurofilaments comprise the bulk of the slowest rate component, Slow Component a (SCa), moving at a rate of 0.1-1.0 mm/day. SCa plays an important role in supporting neuronal shape, process growth and maintenance, and synaptic function (Brady, 1988; Brady

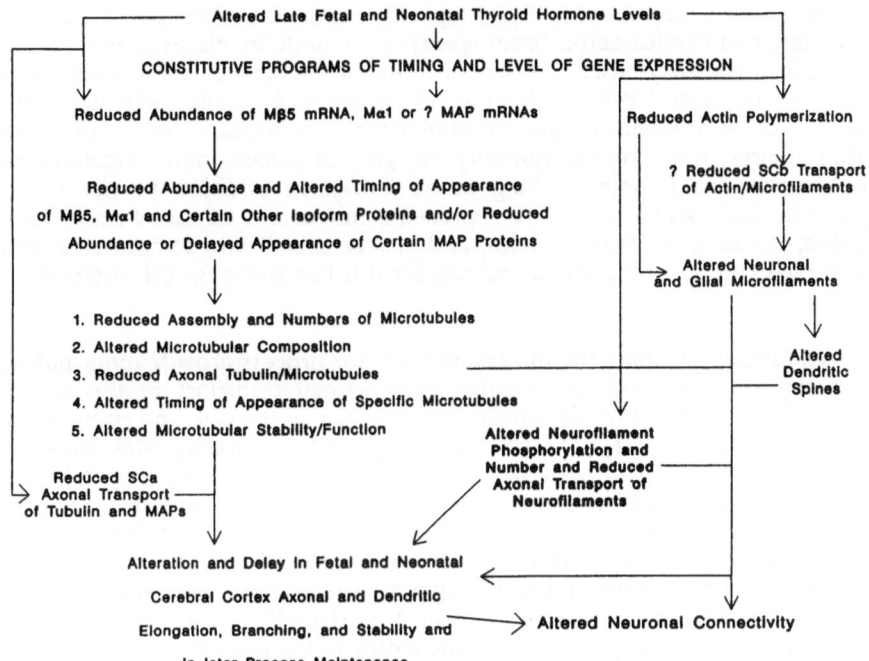

MODEL 4. AN INTEGRATIVE MODEL OF THYROID HORMONE EFFECTS ON THE CYTOSKELETON AND THEIR RELATIONSHIP TO PROCESS GROWTH ABNORMALITIES AND ABNORMAL CONNECTIVITY: The hyt/hyt molecular, neuroanatomical, and behavioral abnormalities suggest a more specific model of thyroid hormone effects related to the potential functions of specific mRNAs and the observed abnormalities in process growth and motoric behavior(MODEL 4). Alterations of thyroid hormone level as seen in the hyt/hyt mouse lead to alteration in the levels of specific developmentally regulated tubulin isotype mRNAs and their translation products, which are precursors of tubulin. This occurs in neurons, such as the pyramidal layer V neurons of the mouse sensorimotor cortex, that are differentiating at this time and whose axons form the corticospinal tracts. Microtubule associated proteins(MAPs)(Matus, 1988) which complex with tubulin isotype proteins to form microtubules may also be altered(Nunez et al, 1989). The delivery of the tubulin isotype proteins to the growing axon is also reduced(see below). These reductions in tubulin isotype levels and transport may lead to alterations in number, composition, assembly, stability, and potential function of the microtubule. The results of these events may be alterations in the extent and timing of process outgrowth, elongation, and branching, in process maintenance, and of connectivity with noted abnormalities in growth of the corticospinal tracts and other tracts. The potential results of these neuroanatomical alterations are delayed, abnormal, and irreversible reflexive, locomotor or adaptive behaviors. These behaviors may be predicated on normal sensorimotor cortex and corticospinal tract development and function and other components of the motor system, as well as, hippocampus and visual and auditory systems.

et al, 1986). Slow Component b (SCb) typically moves faster than SCa, at a rate of 2-4 mm/day, and carries actin, brain spectrin, calmodulin, clathrin and associated proteins, glycolytic enzymes, and other proteins of the cytoplasmic matrix (Black et al, 1979; Garner et al, 1981; Brady et al, 1981; Brady & Lasek, 1982; de Waegh et al, 1989). SCb is thought to play a number of important roles in the neuron, affecting energy metabolism, mobility of growth cones, and regeneration of neuronal processes (Hammerschlag et al, 1989). In addition to the two slow components, fast axonal transport represents movement of membrane bounded organelles, including synaptic vesicles, and plasma membrane components through the axon. Microtubules are also necessary for this fast transport (Hammerschlag et al, 1989).

To further elucidate the molecular basis for process growth abnormalities in the hypothyroid brain, we investigated slow axonal transport in the mouse to determine the effects of thyroid hormone deficiency on the rate and composition of SCa and SCb(SCHEME 1). The pulse-labeling method (Brady, 1985; Brady et al, 1982) was employed in the optic nerve of young(3 month old) age-matched adult CRF hyt/hyt mice and their euthyroid hyt/+ littermates and their progenitor strain, BALB\cBY +/+ mice. SCa transport(See FIGURE 8, TABLE 2) and SCb transport(4,5, and 7 days after radioisotope injection) were evaluated. Comparisons of SCa in the optic nerve of hyt/hyt hypothyroid mouse and euthyroid hyt/+ littermates and euthyroid progenitor strain, BALB/cBY +/+ mice, indicated that the neurofilament protein peaked in a lane closer to the retinal ganglion cells in the hyt/hyt nerves (FIGURE 8). Similar patterns were observed for tubulin(Stein et al, 1991a). These results suggested slower transport of these SCa proteins. Further, the calculated velocity of SCa was significantly reduced in hyt/hyt optic nerve relative to hyt/+ and +/+ nerves. The axonal transport rate for tubulin, which is carried in SCa, was 0.108 +/- .012 mm/day in the hyt/hyt optic nerves. This rate was significantly different(p <.01) from the tubulin rates for the hyt/+ optic nerves(0.127 +/- 0.013 mm/day) and for the +/+ optic nerves(0.138 +/- 0.014 mm/day). Neurofilament proteins, as measured by the 140 KD component, NFM, also were reduced(p < .05) in velocity in the hyt/hyt(0.118 +/- .015) versus the hyt/+(0.134 +/- 0.011) and +/+ (0.143 +/- 0.021) optic nerves(Stein et al, 1991a)(FIGURE 8). These studies revealed that the rates of transport of tubulin, reflective of microtubules, and of neurofilament, reflective of neurofilaments, were reduced in the hyt/hyt optic nerve. This constitutes a slowing of delivery of these components to the distal axon, where process growth is occurring. Further, there was reduction in the bulk rate of SCa. Since certain MAPs, presumably complexed with microtubules, are delivered to the distal axon by SCa, these results also reflect a reduction in MAP SCa transport.

In assessing SCb, clathrin and HSC70 protein(clathrin uncoating ATPase)(de Waegh et al, 1989), demonstrated diminished rates of axonal transport in the hyt/hyt optic nerve. This may have implications for processes dependent on coated vesicles and endocytosis. Specifically, membrane components of the distal axon and the maintenance of synapse formation in the presynaptic axonal terminal may be altered by reduced delivery of clathrin and HSC70(Stein et al, 1991e).

Determinations of the amount of tubulin and neurofilament transported in SCa suggested that the amount of tubulin transported was 30-40% less in the

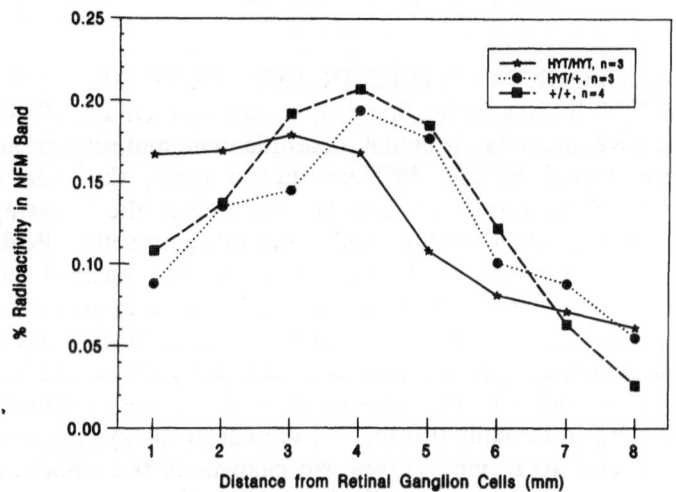

Axonal Transport of NFM (145 kD) In Mouse Optic Nerve
(Transport Time = 4 Weeks)

FIGURE 8.SCa AXONAL TRANSPORT OF CYTOSKELETAL PROTEINS IN hyt/hyt, hyt/+, and +/+ OPTIC NERVE: SCa AXONAL TRANSPORT OF NEUROFILAMENT(NFM, 140,000 DALTONS) ^{35}S-Methionine was injected into the vitreous humor of 3 month old mice. The mice were sacrificed 3 weeks, 4 weeks, and 4.5 weeks after injection. The right optic nerve was removed cut into consecutive 1 mm segments, which were prepared as documented(Stein et al, 1991a). Homogenates were run on SDS-PAGE gels and the neurofilament and other protein, i.e. tubulin, actin, etc., locations on the gels were identified by flurography. These bands were cut out and used to calculate axonal velocity of neurofilament and tubulin and to determine the amount of axonally transported NFM and tubulin. For the plot above, neurofilament protein in each individual hyt/hyt, hyt/+, or +/+ nerve at 4 weeks, the percentage of radioactivity in each 1 mm nerve segment was determined. This percentage of radioactivity was computed by taking the total counts in a single 1 mm segment or lane for a specific protein and dividing this figure by the total number of counts for all lanes for that protein. This percentage was plotted against the distance from the retinal ganglion cells(FIGURE 8). This plot shows that NFM is transported more slowly in the hyt/hyt versus the hyt/+ and +/+ nerves. Similar plots were done for tubulin. Data from these plots were then used to calculate the axonal transport velocity of NFM and tubulin(Stein et al, 1991a).

TABLE 2. THE AMOUNT OF TUBULIN AND NEUROFILAMENT THAT IS TRANSPORTED IN SCa IN hyt/hyt, hyt/+, and +/+ OPTIC NERVES. To compare the relative amounts of tubulin, neurofilament, and actin proteins that were being transported in the different mouse types, the lane of peak radioactivity for each protein in each nerve was used in the following ratios: tubulin:neurofilament, tubulin:actin, and neurofilament:actin. Peaks were defined as the 1 mm segment of the nerve that had the maximal number of counts compared to all other segments. Similar ratios for each protein were also determined in the total radioactivity for each protein in all lanes. Data for actin, which is primarily transported in SCb, was also determined and used as a control. Along with studies of the ratio of total DPM, these studies reveal a reduction of 30-40% in axonally transported tubulin in the hyt/hyt optic nerve versus the hyt/+ and +/+ optic nerves. No changes in the amount of NFM transported were observed in these studies(Stein et al, 1991a).

| | Ratio of DPM in Peak Lane | | |
	Tubulin:NFM	Tubulin:Actin	NFM:Actin
hyt/hyt	10.15 ± 1.50	2.85 ± 0.36	.295 ± .031
hyt/+	15.87 ± 2.17	4.38 ± 0.90	.263 ± .020
+/+	16.79 ± 3.47	5.11 ± 1.47	.340 ± .080
hyt/hyt:hyt/+	0.640	0.651	1.120
hyt/hyt:+/+	0.605	0.558	0.868
hyt/+:+/+	0.945	0.857	0.773

hyt/hyt versus hyt/+ and +/+ nerves(TABLE 2). These results suggested a selective dimunition in the amount of axonally transported tubulin. The amount of neurofilament was unchanged(Stein et al, 1991a)(TABLE 2).

The available data suggest that both reduced assembly of microtubules and reduced synthesis of microtubule proteins may be simultaneously operating in a complex fashion to reduce the amount of axonally transported tubulin in hypothyroid mice(Stein et al, 1991a). This occurs along with the reduction in tubulin transport velocity that is related to reduced thyroid hormone. Such changes in transport rate of cytoskeletal proteins would be expected to affect axonal growth and plasticity. When considered with previously reported changes in specific expression of tubulin isotypes and MAPs (Nunez et al, 1989), changes in the properties and availability of neuronal cytoskeletal elements, particularly tubulin and microtubule proteins, represent another potential mechanism for disturbances of process growth in hypothyroidism(MODEL 4). The importance of slow axonal transport is emphasized by the correlation between process growth rate and regeneration and the rate of axonal transport(Hoffman et al, 1985; Wujek et al, 1985).

Thus, hypothyroidism may influence process growth and maintenance and the microtubules by: 1) alterations in the time, duration, and level of expression of certain tubulin isoforms and MAPs(Nunez et al, 1989); 2) reduction in the assembly of MAPs and tubulin into microtubules(Nunez et al, 1989); and 3)reduction in the velocity and the amount of axonal transport of microtubular components, i.e. tubulin and MAPs composition(Stein et al, 1991a,e). Each of these thyroid hormone effects may affect microtubule composition, assembly, stability and function.

These considerations for thyroid hormone and microtubules must be combined with independent effects of thyroid hormone deficiency that may affect process growth. These occur by: 1) alteration of actin polymerization(Siegrist-Kaiser et al, 1990); 2) alteration of the properties of and axonal transport of neurofilament proteins and other functionally important proteins, i.e., clathrin and HSC70 (Stein et al, 1991a; Stein et al, 1991c; Marc et al, 1986)(MODELS 3 and 4), and 3) dimunitions in the rate of fast axonal transport(Rasool et al, 1985) that is dependent on axonal motor molecules and microtubules(Bloom et al, 1991). Alterations in fast transport may diminish delivery of membrane proteins, organelles, i.e. mitochondria, and synaptic vesicles to the distal axon and presynaptic terminal.

However, the potential importance of the reduced velocity of neurofilament axonal transport is more difficult to understand. Neurofilaments normally function to lead to transverse process(Hoffman et al, 1985b) growth and stabilization of neuronal processes(reviewed in Hammerschlag et al, 1989) by their number(Marc et al, 1985), their ratio with regard to microtubule number(Marc et al, 1985), and their phosphorylation status(Marc et al, 1986). Despite the lack of reduction of neurofilaments in hypothyroid neonatal sciatic axons(Marc et al, 1985), diminished delivery of neurofilament proteins may lead to a reduction in neurofilament number or a change in stability of the mature optic nerve axon. This concept is in line with the regulation of axonal diameter, a function of neurofilaments(Hoffman et al, 1985b), by thyroid hormone(Allpress et al, 1986)(MODEL 4).

AN INTEGRATIVE MODEL OF THYROID HORMONE DEFICIENCY AND THE DEVELOPING AND MATURING BRAIN: NEUROLOGICAL, MOTOR, PROCESS GROWTH, CYTOSKELETAL, AND MOLECULAR ABNORMALITIES

Endemic cretinism, maternal hypothyroidism, maternal hyperthyroidism, and sporadic congenital hypothyroidism represent identifiable causes of cerebral palsy, mental retardation, and learning disability. The hyt/hyt mouse serves as a model of human sporadic congenital hypothyroidism and also demonstrates delayed reflexive behavior, abnormal complex motor behavior, and abnormal learned motor behavior. Complex learned behavior is predicated on normal visual function and the integration of visual and motor cortices and of motor systems and the hippocampus. In the human, disorders of visuomotor integration, visuospatial integration, and cognitive association, that reflect on similar structures, including the parietal lobes, are observed. Abnormalities in process growth and connectivity in pyramdal neurons in visual cortex, motor cortex, and hippocampus in both humans and rodents may contribute to the neurological and behavioral abnormalities that are observed. The human and rodent motor abnormalities may be clinically and neuropathologically explained by abnormalities in the development and maintenance of the corticospinal tracts, which have origin in the rodent sensorimotor cortex and human motor corticies. The abnormalities in process growth of these pyramidal neurons and other neurons may arise in utero and through brain development and maturation by the effects of thyroid hormone on the cytoskeletal structures and their molecular components, particularly tubulins and MAPs. The tubulin isoform mRNAs and their protein products are localized to sensorimotor cortex neurons, that have been altered neuropathologically in hypothyroidism. These molecules are also expressed during a narrow window of large process development in these neurons and regulated by thyroid hormone during this period. The effects of thyroid hormone may not only be at the level of translation and transcription of the tubulin isoforms and different MAPs, but also at the level of assembly of these molecules into microtubules, and the transport of these microtubules to the distal neuron and growth cone. Taken together, these observations provide credence for MODELS 1, 3, and 4. Altered thyroid hormone levels at specific times and in specific cellular locations in the brain may alter functional molecules, i.e. tubulin isoforms or MAPs, which may in turn lead to alterations in number or function of microtubules. The latter may contribute to abnormalities in process growth and together with neuroanatomical abnormalities in other cells in defined locations, these neuropathological abnormalities, involving specific locations, may contribute to specific neurological and behavioral abnormalities.

Multidisciplinary definitions of thyroid hormone action at behavioral, neuroanatomical, and molecular levels in discrete cell populations, including neurons and glia(Gould et al, 1990; Clos et al, 1980; Siegrist-Kaiser et al, 1990), and in animal models are required in the future. The effects of thyroid hormone on additional molecules that have functional roles in brain development needs definition. The molecular and neuroanatomical events influenced by thyroid hormone may occur as a primary effect of thyroid hormone alone or a secondary effect through second messengers that are regulated by thyroid hormone, i.e. NGF,

EGF, and growth hormone. Therefore, the molecular mechanisms of regulation by thyroid hormone in the brain are an important area for future research.

Of the relevant animal models, the fetal and neonatal hyt/hyt cerebral cortex provides a model of the in utero and neonatal insults by thyroid hormone deficiency. This model has relevance for understanding the behavioral and neurological abnormalities in the rodent and human with sporadic congenital hypothyroidism. The hyt/hyt optic nerve is a model for the effects of thyroid hormone on the maturing nervous system, particularly on later axonal process growth and maintenance(consistent with MODEL 1). Clinically, this model has relevance for untreated or late treated sporadic congenital hypothyroidism, as well as for understanding peripheral neuropathy (Swanson et al, 1981; Dyck & Lambert, 1970) and optic neuropathy (Anduze & Merritt, 1980) that occur in adult hypothyroidism, as well as the peripheral neuropathy that occurs in neonates with sporadic congenital hypothyroidism prior to treatment with T4(Giroud et al, 1988). Both the developing cerebral cortex and the maturing optic nerve demonstrate the significant influence of thyroid hormone on the nervous system throughout the life cycle. Both models have direct relevance for defining the pathophysiology, reversibility, and treatment of human sporadic congenital hypothyroidism.

UNDERSTANDING AND TREATING HUMAN SPORADIC CONGENITAL HYPOTHYROIDISM: A CRITICAL SUMMARY AND FUTURE DIRECTIONS

One of the crucial dilemmas in human sporadic congenital hypothyroidism is the efficacy of T4 therapy, starting in the neonatal period, in preventing neurological abnormalities and/or the adequacy of this therapy in reversing neuropathological changes, brought about by prenatal hypothyroidism. These issues can be clarified by dividing the brain effects of hypothyroidism into prenatal versus postnatal in this syndrome. Given the fixed and sequential nature of brain development and the molecular and neuroanatomical sensitivity of the fetal brain to thyroid hormone deficiency, postnatal thyroid hormone therapy is instituted after the occurrence of major prenatal brain developmental events. Therefore, the usual timing of diagnosis and T4 therapy would not be expected to change the structure and wiring of the brain that developed prenatally. However, despite a lack of neuroanatomical reversibility by postnatal T4, a significant number of congenitally hypothyroid children that are treated early may have normal mental and motor function. This may represent evidence of: 1) clinical reversibility of prenatal neuroanatomical abnormalities by thyroid hormone or 2) the primary importance of postnatal development and postnatal T4 effects. The latter would be consistent with the less likely possibility of a lack of or limited sensitivity of the fetal brain to thyroid hormone.

Nevertheless, some children that are treated early will not have normal development and will demonstrate learning disabilities or motor abnormalities that would be classed as mild cerebral palsy. These clinical abnormalities may result from irreversible prenatal thyroid hormone insult and may be related to the type and severity and duration of prenatal hypothyroidism, i.e., athyrosis or levels less than 2 ug/dl(Glorieux et al, personal communication). In these cases of early treatment but persistent clinical abnormality, the targets of the hypothyroidism are prenatal process growth of major neurons and the connection of the processes between these major neurons. These observations along with the molecular and neuroanatomical effects of prenatal thyroid hormone insult suggest that adequate and early postnatal treatment of sporadic congenital hypothyroidism may not be enough to ensure normal neuroanatomical or neurological development.

Unfortunately, we are presently limited diagnostically and therapeutically in the prenatal period for sporadic congenital hypothyroidism. Regarding the other prenatal disorders, positive results with assiduous therapy of maternal hypo and hyperthyroidism must be balanced against improved but still confusing outcomes from early initiation of iodine in endemic cretinism.

In sporadic congenital hypothyroidism, the significant success of early diagnosis and treatment and assiduous maintenance therapy with T4 emphasizes the importance of postnatal T4 and the postnatal period for development and/or prevention of abnormalities in sporadic congenital hypothyroidism. For this disorder, the first postnatal month is also an extremely critical period for installation of therapy. To a lesser degree, but also important, is the maintenance of adequate T4 therapy during the first year and after. These points are emphasized by: 1) rodent studies of hypothyroidism where perinatal insults lead to adult process growth abnormalities(Gottesfeld et al, 1988); and 2) clinical comparisons between early treated(within the first month) and later treated(after the first month) cretinism. In the latter group, striking differences in the severity of mental and motor dysfunction are observed compared to those treated before one month of age. Therefore, inadequate or delayed T4 therapy after birth in combination with the prenatal neuroanatomical abnormalities contribute individually   to increase the occurrence and severity of clinical and neuroanatomical abnormalities. The precedent for the importance of the first postnatal month, the first year, and perhaps, childhood treatment with T4 can be related to the normal cerebral cortex events occurring during this time. The refinement and maintenance of cerebral cortex neurons and their connections and cerebral cortical glia is an ongoing process that remains sensitive to thyroid hormone levels(Ruiz-Marcos et al, 1983) and continues at least to six years of age(Conel, 1967). As with the prenatal effects of thyroid hormone deficiency, this deficiency after birth will interfere with events that are occurring at that time. Therefore, reductions in thyroid hormone level during postnatal and juvenile periods can also cause abnormalities.

These interpretations of the neurological and neuropsychological findings in sporadic congenital hypothyroidism strongly support programs already in place for early neonatal diagnosis and assiduous treatment with adequate levels of T4 in combination with patient compliance. Nevertheless, the therapeutic guidelines, i.e. early replacement doses(Fisher et al, 1989) and maintenance doses and levels(Fisher et al, 1989), and clinical endpoints for evaluation within the first ten years of life still remain empirical. This is related in part to our limited understanding of the normal requirements of the neonatal and juvenile brain for thyroid hormone. Animal studies and further neurological and neuropsychological studies of adolescent patients that had sporadic congenital hypothyroidism will be useful in clarifying issues of treatment, time, dose, and neuroanatomical and clinical reversibility.

CURRENT ISSUES IN HUMAN THYROID HORMONE DISORDERS AND CEREBRAL PALSY: THE RELEVANCE OF HUMAN THYROID HORMONE DISORDERS AND THYROID HORMONE

The clinical and basic research analysis of rodent and human disorders also has a more general relevance for cerebral palsy and mental retardation. The fetal/neonatal nervous system can respond in only a limited number of ways to different effectors/insults, including thyroid hormone. Thus, different effectors may lead to similar neuroanatomical abnormalities and subsequent neurological and behavioral abnormalities. The final common pathways from a variety of prenatal and postnatal insults, i.e. hypothyroidism, include cerebral palsy, learning disability, and mental retardation.

Given the clinical and pathological data, it is evident that thyroid hormone abnormalities represent relatively common and identified prenatal and if treated late, postnatal, causes for cerebral palsy. Animal models and human conditions of thyroid hormone deficiency and excess allow exploration of fetal molecular and neuroanatomical substrates that might underly certain and perhaps, significant numbers of cases of cerebral palsy. Because of these facts, an understanding of the mechanisms of action of thyroid hormone on the fetal and neonatal cerebral cortex may provide us with a means for understanding how cerebral palsy begins prenatally, and its initial neuroanatomical substrates and targets. These studies may also afford a means to diagnose cerebral palsy prenatally using molecular indices, and potential chemical interventions that relate to molecules that are altered in timing or level of expression, such as the tubulin isotypes or second messengers, i.e. EGF. Similarly, approaches of this type may have relevance for mental retardation and learning disabilities.

REFERENCES

Adams, P. M., Stein, S. A., Palnitkar, M., Anthony, A., and Gerrity, L., 1989, Evaluation and Characterization of the hyt/hyt Hypothyroid Mouse I: Somatic and Behavioral Studies, Neuroendo., 49:138-143.

Adams, P. M. and Stein, S. A., Evaluation and characterization of the hyt/hyt mouse III: Abnormalities in Primitive Corticospinal Reflexes and Sensory Behavior, Submitted, 1991.

Adkison, L. R., Taylor, S., and Beamer, W. G., 1989, Mutant gene-induced disorders of structure, function and thyroglobulin synthesis in congenital goitre (cog/cog) in mice, J. Endocrin., 126:51-58.

Allpress, S. J., and Pollock, M., 1986, Morphological and functional effects of triiodothyronine on regenerating peripheral nerve, Exp. Neurol., 91:382-91.

Almazan, G., Honegger, P., and Matthieu, J. M., 1975, Triiodothryoinine stimulation of oligodendroglial differentiation and myelination, Dev. Neuro., 7:45-54.

Anduze, A. L., and Merritt, J. C., 1980, Optic nerve hypoplasia with hyperthyroidism and third nerve palsy, Ann. Opthalmology, 12:1170-1173.

Anthony, A., Adams, P. M., and Stein, S. A., 1990, The effects of congenital hypothyroidism using the hyt/hyt mouse on locomotor activity and learned behavior, Submitted, 1990.

Bakke, J. L., Lawrence, N. L., Robinson, S., and Bennett, J., 1975, Endocrine studies of the untreated progeny of thyroidectomized rats, Pediat. Res., 9:742-748.

Beamer, W. G., Eicher, E. M., Maltais, L. J., and Southard, J. L., 1981, Inherited primary hypothyroidism in mice, Science, 212:61-62.

Beamer, W. G., Maltais, L. J., DeBaets, M. H., and Eicher, E. M., 1987, Inherited congenital goiter in mice, Endocrinol., 120:838-840.

Beamer, W. G. and Cresswell, L. A., 1982, Defective thyroid ontogenesis in fetal hypothyroid (hyt/hyt) mice, Anat. Rec., 202:387-393.

Beierwaltes, W. H., 1959, Instituitionalized cretins in the state of Michigan, Michigan Med., 58:1077-1095.

Benecke, R., 1990, Clumsiness in corticospinal tract lesions, Motor Control, Am. Acad. of Neurol., 47-63.

Benjamin, S., Cambray-Deakin, M.A., and Burgoyne, R.D., 1988, Effect of hypothyroidism on the expression of three microtubule-associated proteins (1A, 1B, and 2) in developing rat cerebellum, Neurosci, 27:931-939.

Benowitz, L. I. and Routtenberg, A., 1987, A membrane phosphoprotein associated with neural development, axonal regeneration, phospholipid metabolism, and synaptic plasticity, TINS, 10:527-532.

Bernal, J. and Pekonen, F., 1984, Ontogenesis of the nuclear 3,5,3'-triiodothyronine receptor in the human fetal brain, Endocrinol., 11:677-679.

Birrell, J., Frost, G. J., and Parkin, J. M., 1987, The development of children with congenital hypothyroidism, Dev. Med. Child Neurol., 25:512-519.

Black, M. M. and Lasek, R. J., 1979, Axonal transport of actin: Slow component b is the principal source of actin for the axon, Brain Res., 171:401-413.

Bloom, G. S., Schoenfeld, T. A., and Vallee, R. B., 1984, Widespread distribution of the major polypeptide component of MAP 1 (microtubule-associated protein 1) in the nervous system, J. Cell. Biol., 98:320-330.

Bloom, G. S. and Vallee, R. B., 1983, Association of microtubule-associated protein 2 (MAP2) with microtubules and intermediate filaments in cultured brain cells, J. Cell Biol., 96:1523-1531.

Bond, J. and Farmer, S., 1983, Regulation of tubulin and actin mRNA production in rat brain: Expression of a new tubulin mRNA with development, Mol. Cell. Biol., 3:1333-1342.

Boyages, S. C., Halpern, J. P., Maberly, G. F., Eastman, C. J., Morris, J., Collins, J., Jupp, J. J., Chen-en, J., Zheng-Hua, W., and Chuan-Yi, Y., 1988, A comparative study of neurological and myxedematous endemic cretinism in Western China, J. Clin. Endocrin. Metabolism, 67:1262-1271.

Boyages, S. C., Collins, J. K., Maberly, G. F., Jupp, J. J., Morris, J., and Eastman, C. J., 1989, Iodine deficiency impairs intellectual and neuromotor development in apparently-normal persons: A study of rural inhabitants of north-central China, Med. J. of Australia, 150:676-77.

Bradley, D. J., Young, W. S., III, Weinberger, C., 1989, Differential expression of and thyroid hormone receptor genes in rat brain and pituitary, Proc. Natl. Acad. Sci., 86:1-6.

Brady, S. T., 1985, Axonal transport: Methods and applications, in:"Neuromethods I: General Methods," Boulton, A., Baker, G., eds., Clifton, NJ, Humana Press.

Brady, S. T., 1988, Cytotypic specialization of the neuronal cytoskeleton and the cytomatrix: Implications for neuronal growth and regeneration, in: "Cellular and Molecular Aspects of Neural Development and Regeneration," A. Goria, et al., eds., Springer-Verlag, New York.

Brady, S. T. and Black, M. M., 1986, Axonal transport of microtubule proteins: Cytotypic variation of tubulin and MAPs in neurons, Ann. NY Acad. Sci., 466:199-217.

Brady, S. T., Lasek, R. J., 1982a, The slow components of axonal transport: Movements, compositions and organization, in: "Axoplasmic Transport," Weiss, D. G., ed., Berlin, Springer-Verlag.

Burgoyne, R. D., Cambray-Deakin, M. A., Lewis, S. A., Sarkar, S., and Cowan, N. J., 1988, Differential distribution of β tubulin isotypes in cerebellum, EMBO. J., 7:2311-2319.

Caviness, V. S., Crandall, J. E., and Edwards, M. A., 1988, The reeler malformation: Implications for neocortical histogenesis, in: "Cerebral Cortex," A. Peters and E. G. Jones, eds., Plenum Press, New York.

Chaouki, M. L., Maoui, R., and Benmiloud, M., 1987, Comparative study of neurological and myxoedematous cretinism associated with severe iodine deficiency, Clin. Endocrinol., 28:399-408.

Chaudhury, S., Chatterjee, D., and Sarkar, P. K., 1985, Induction of brain tubulin by triidothyronine: Dual effect of the hormone on the synthesis and turnover of the protein, Brain Res., 339:191-194.

Christiansen, E. and Melchior, J., 1967, Cerebral palsy: a clinical and neuropathological study, Clin. Dev. Med., 25:1.

Cleveland, D.W., 1989, Autoregulated control of tubulin synthesis in animal cells, Curr. Opinion in Cell Biol., 1:10-14.

Clos, J., Legrand, C., and Legrand, J., 1980, Effects of thyroid state on the formation and early morphological development of Bergmann glia in the developing rat cerebellum, Dev. Neurosci., 3:199-208.

Codaccioni, J. L., Carayon, P., Michel-Bechet, M., Foucault, F., Lefort, G., and Pierron, H., 1980, Congenital hypothyroidism associated with thyrotropin unresponsiveness and thyroid cell membrane alterations, J. Clin. Endocrinol. Metab., 50:932-937.

Comer, C. P. and Norton, S., 1985, Behavioral consequences of perinatal hypothyroidism in postnatal and adult rats, Pharm. Biochem. Behav., 22:605-611.

Conel, J.L., "The postnatal development of the human cerebral cortex," volumes I-VII, 1939, 1941, 1947, 1951, 1955, 1959, 1963, 1967, Harvard Univ. Press, Cambridge, Massachusetts.

Cowan, N. J. and Dudley, L., 1983, Tubulin isotypes and the multigene tubulin families, Intl. Rev. Cytl., 85:147-173.

Crandall, J. E. and Caviness, V. S., 1984, Axon strata of the cerebral wall in embryonic mice, Dev. Brain Res., 14:185-195.

Crandall, J. E. and Caviness, V. S., 1984, Thalamocortical connections in newborn mice, J. Comp. Neurol., 228:542-556.

Davenport, J. W., 1976, Perinatal hypothyroidism in rats: Persistent motivational and metabolic effects, Dev. Psychobiol., 9:67-82.

Davenport, J. and Dorcey, T., 1972, Hypothyroidism: Learning deficits induced in rats by early exposure to thiouracil, Horm. Behav., 3:97-112.

Davenport, J. W., Gonzalez, L. M., Hennies, R. S., and Hagquist, W. W., 1976, Severity and Timing of Early Thyroid Deficiency as Factors in the Induction of Learning Disorders in Rats, Horm. Behav., 7:139.

Davidoff, R. A., 1990, The pyramidal tract, Neurology, 40:332-339.

de Waegh, S. and Brady, S. T., 1990, Axonal transport of a clathrin uncoating ATPase (HSC70): A role for HSC70 in the modulation of coated vesicle assembly in vivo, J. Neurosci. Res., 23:433-440.

de Waegh, S. and Brady, S. T., 1989b, Altered slow axonal transport and regeneration in a myelin-deficient mutant mouse: The trembler as an in vivo model for schwann cell-axon interactions, Neurosci., 10:1855-1865.

Delange F.M., 1989, Endemic cretinism: An overview, in: "Iodine and the Brain," G. R. DeLong, J. Robbins, and P. G. Conliffe, eds., Plenum Press, New York.

Delong, G. R., 1989, Observations on the neurology of endemic cretinism, in: "Iodine and the Brain," G. R. DeLong, J. Robbins, and P. G. Conliffe, eds., Plenum Press, New York.

Delong, G. R., Stanbury, J.B., and Fierro-Benitez, R., 1985, Neurological signs in congenital iodine-deficiency disorder (endemic cretinism), Dev. Med. Child Neurol., 27:317-.

Dememes, D., Dechesne, C., LeGrand, C., and Sans, A., 1986, Effects of hypothyroidism on postnatal development in the peripheral vestibular system, Dev. Brain Res., 25:147-152.

Demeyer, W., 1967, Ontogenesis of the rat corticospinal tract, Arch. Neurol., 16:203-211,

Devries, J.I.P., Visser, G.H.A., and Prechtl, H.F.R., 1982, The emergence of fetal behavior I. Quantitative aspects, Early Hum. Devel., 7:301-322.

Diamond, D.J. and Goodman, H.M., 1985, Regulation of growth hormone messenger RNA synthesis by dexamethasone and triiodothyronine transcriptional rate and mRNA stability changes in pituitary tumor cells, J. Molec. Biol., 181:41-62, 1985.

Donatelle, J. M., 1977, Growth of the corticospinal tract and the development of placing reactions in the postnatal rat, J. Comp. Neur., 175:207-232.

Drubin, D., Kobayashi, S., Kellogg, D., and Kirschner, M., 1988, Regulation of microtubule protein levels during cellular morphogenesis in nerve growth factor-treated PC12 cells, J. Cell. Biol., 106:1583-1591.

Drubin, D. G., Feinstein, S. C., Shooter, E. M., and Kirschner, M. W., 1985, Nerve growth factor-induced neurite outgrowth in PC12 cells involves the coordinate induction of microtubule assembly and assembly-promoting factors, J. Cell Biol., 101:1799-1807.

Dumont, J. E,, Vassart, G., and Refetoff, S., 1989, Thyroid disorders, in: "Metabolic Basis of

Inherited Diseases," 6th ed., Scriver, C. R., Beaudet, A. L , Sly, W. S., Valle, D., eds., McGraw Hill, New York.

Dussault, J. H., Action of thyroid hormones on brain development, in: "Research in Congenital Hypothyroidism," F. Delange, D. A. Fisher, and D. Glinoer, eds., Plenum Press, New York, 95-102.

Dussault, J. H., Glorieux, J., Letarte, J., Guyda, H., and Morissette, J., 1983, The mental development at 3 years of age of hypothyroid infants detected by the Quebec Screening program, in: "Congenital Hypothyroidism," J. H. Dussault and P. Walker, eds., M. Dekker, Inc., New York.

Dyck P.J., Lambert, E.H., 1970, Polyneuropathy associated with hypothyroidism, J Neuropath Exp Neurol, 24:631-658.

Eayrs, J.T., 1968. Developmental Relationships Between Brain and Thyroid, in: "Endocrinol. and Human Behavior," R. P. Michael, ed., Oxford University Press, New York.

Eayrs, J.T., 1955, The cerebral cortex of normal and hypothyroid rats, Acta Anat., 25:160-1832.

Eayrs, J. T. and Lishman, W. A., 1955, The maturation of behavior in hypothyroidism and starvation, Br. J. Animal Behav., 3:17-24.

Faivre C., Legrand C., and Rabie A., 1984, In purkinje cell dendrites of the young rat, thyroid hormone controls the resistance of microtubules to fixation at low temperature, Int. J. Dev. Neurosci., 2:427-436.

Faivre, C., Legrand, C., and Rabie, A., 1983, Effects of thyroid deficiency and corrective effects of thyroxine on microtubules and mitochondria in cerebellar purkinje cell dendrites of developing rats, Dev. Brain. Res., 3:21-30.

Farmer, S., Robinson, G., Mbangkollo, D., Bond, J., Knight, G., Fenton, M., and Berkowitz, E., 1986, Differential expression of the β tubulin multigene family during rat brain development, Ann. NY Acad. of Sci., 466:41-50.

Fernyhough, P., Mill, J. F., Roberts, J. L., and Ishii, D. N., 1989, Stabilization of tubulin mRNAs by insulin and insulin-like growth factor I during neurite formation, Mol. Brain Res., 6:109-120.

Fierro-Benitez, R., Cazar, R., Sandoval, H., Fierro-Renoy F., et al, Early correction of iodine deficiency and late effects on psychomotor capabilities and migration in: "Iodine and the Brain," G. R. DeLong, J. Robbins, and P. G. Conliffe, eds., Plenum Press, New York.

Fisher, D. A. and Foley, B. L., 1989, Early treatment of congenital hypothyroidism, Pediatrics, 83:785-789.

Fisher, D. A. and Klein, A. H., 1981, Thyroid development and disorders of thyroid function in the newborn, NEJM, 304:702-712.

Fisher, A. A., 1989, Development of fetal thyroid system control, in: "Iodine and the Brain," G. R. DeLong, J. Robbins, and P. G. Conliffe, eds., Plenum Press, New York.

Fox, S. R. and Pfaff, D., 1987, Differential expression within neurons and glia of mRNA encoding a putative thyroid hormone receptor(cErbA1), Soc. Neurosci. Abstr., 13(1):376.

Freeman, J. M. and Nelson, K. B., 1988, Intrapartum asphyxia and cerebral palsy, Pediatrics, 82:240-249.

Freund, H. J., 1987, Differential effects of cortical lesions in humans, in: "Motor Areas of the Cerebral Cortex," R. Porter and C. G. Phillips, eds., A. Wiley-Interscience Publication, New York.

Garcia, C. A. and Fleming, R. H., 1977, Reversible corticospinal tract disease due to hyperthyroidism, Arch. Neurol., 34:647-648.

Garner, J. A. and Lasek, R. J., 1981, Clathrin is axonally transported as part of slow component b: The microfilament complex, J. Cell Biol., 88:172-178.

Garza, R., Dussault, J. H., and Puymirat, J., 1988, Influence of triiodothyronine on the morphological and biochemical development of fetal brain acetylcholinesterase-postive neurons cultured in a chemically defined medium, Dev. Br. Res., 43:287-297.

Gerard, C. M., Lefort, A., Christophe, D., Libert, F., Van Sande, J., Dumont, J. E., and Vassart, G., 1989, Control of thyroperoxidase and thyroglobulin transcription by cAMP: Evidence for distinct regulatory mechanisms, Mol. Endocrinol., 3:2110-2118.

Gilman, A. G., 1989, G proteins and regulation of adenylyl cyclase, JAMA, 262:1819-1825.

Giroud, M., Enenbaum, D., D'Athis, P., Dumas, R., and Nivelon, J. L., 1988, Neurophysiological study of peripheral nerves in newborn infants with congenital hypothyroidism. Value in the surveillance of replacement therapy, Arch. Francaises De Pediatrie, 45:175-79.

Glorieux, J., 1989, Mental development of patients with congenital hypothyroidism detected by screening.(Quebec experience), in: "Research in Congenital Hypothyroidism," DeLange, D. A. Fisher, and D. Glinoer, eds., Plenum Press, New York.

Gonzales, L. W. and Geel, S. E., 1978, Quantitation and characterization of brain tubulin (colchicine-binding activity) in developing hypothyroid rats, J. Neurochem., 30:237-245.

Gottesfeld, Z., Garcia, C. J., and Chronister, R. B., 1987, Perinatal, not adult, hypothyroidism suppresses dopaminergic axon sprouting in the deafferented olfactory tubercle of adult rat, J. Neurosci. Res., 18:568-73.

Gould, E., Frankfurth, M., Westlind-Danielsson, A., and McEwen, B. D., 1990, Developing forebrain astrocytes are sensitive to thyroid hormone, Glia, 3(4):283-92.

Gross, H., Jellinger, K., Kaltenback, E., and Rett, E., 1968, Infantile cerebral disorders: clinical-neuropathological correlations to elucidate the aetiological factors, J. Neurol. Sci., 7:551.

Hadjzadeh, M., Sinha, A.K., Pickard, M.R., and Ekins, R.P., 1990, Effect of maternal hypothyroxinaemia in the rat on brain biochemistry in adult progeny, J. Neurochem., In Press.

Hammerschlag, R. and Brady, S. T., 1989, Axonal transport and the neuronal cytoskeleton, in: "Basic Neurochemistry: Molecular, Cellular, and Medical Aspects, 4th Ed.," Siegel, G. J., et al., eds., New York: Raven Press.

Hargreaves, A., Yusta, B., Aranda, A., Avila, J., and Pascual, A., 1988, Triiodothyronine (T3) induces neurite formation and increases synthesis of a protein related to MAP1B in cultured cells of neuronal origin, Dev. Brain. Res., 38:141-148.

Havercroft, J. C. and Cleveland, D. W., 1984, Programmed expression of β-tubulin genes during development and differentiation of the chicken, J. Cell Biol., 99:1927-1935.

Hendrich, T. E., Jackson, W. J., and Porterfield, S. P., 1984, Behavioral testing of progenies of Tx(Hypothyroid) and growth hormone treated Tx rats: An animal model for mental retardation, Neuroendo., 438:429-437.

Hendrich, C. E., Ocasio-Torres, W., Berdecia-Rodriquez, W., and Porterfield, S. P., 1987, Brain and liver ribosomal protein synthesis and profiles in hypothyroid mothers and their progenies, Am. Thyroid Assoc., Abstract #106.

Hoffman, P. N., 1989, Expression of GAP-43, a rapidly transported growth-associated protein, and class II beta tubulin, a slowly transported cytoskeletal protein, are coordinated in regenerating neurons, J. Neurosci., 9(3):893-897.

Hoffman, P. N. and Cleveland, D. W., 1988, Neurofilament and tubulin expression recapitulates the developmental program during axonal regeneration: Induction of a specific β-tubulin isotype, Proc. Natl. Acad. Sci. USA, 85:4530-4533.

Hoffman, P. N. and Lasek, R. J., 1975, The slow component of axonal transport. Identification of major structural polypeptides of the axon and their generality among mammalian neurons, J. Cell Biol., 66:351-366.

Hoffman, P. N., Thompson, G., Griffin, J., and Price, D., 1985, Changes in neurofilament transport coincide temporally with alteration in the caliber of axons in regenerating motor fibers, J. Cell Biol., 101:1332-1340.

Hoskins, S. G. and Grobstein, P., 1983, Induction of the ipsilateral retinothalamic projection in Xenopus laevis by thyroxine, Nature, 307:730-733.

Hoyle, H. D. and Raff, E. C., 1990, Two drosophila beta tubulin isoforms are not functionally equivalent, J. Cell Biol., 111:1009-1026.

Hulse, A., 1987, Congenital hypothyroidism and neurological development, J. Child Psychol. Psychia., 24:629-635.

Jeantet, C. and Gros, F., 1981, One tubulin subunit accumulates during neurite outgrowth in mouse neuroblastoma cells, Biochem. Biophys. Res. Commun., 103:1035-1043.

Jia-Liu, L., Zhong-Jie, Z., Zhon-Fu, S., Jia-Ziu, Z., Yu-Bin, T., and Bin-Zhon, C., 1989, Influence of iodine deficiency of human fetal thyroid gland and brain, in: "Iodine and the Brain," G. R. DeLong, J. Robbins, and P. G. Conliffe, eds., Plenum Press, New York.

Jianquan, L., Xin, W., Yuquin, Y., Kewei, W., Dakai, Q., Zhenfu, X., and Jun, W., 1985, The effects on fetal brain development in the rat of a severely iodine deficient diet derived from an endemic area: observations on the first generation, Neuropath. Appl. Neurobiol., 12:261-276.

Job, J. C., Canlorbe, P., Thomassin, N., and Vassal, J., 1969, L'hypothyroidie infantile a debut precoce avec glande en place, fixation fiable de radioiode et defaut de reponse a la thyrostimuline, Ann. Endocrinol., 30:696-701.

Johanson, I. B., Turkewitz, G., and Hamburgh, M., 1980, Development of home orientation in hypothyroid and hyperthyroid rat pups, Devl. Psychobiol., 13:331-342.

Jones, E. G., Schreyer, D. J., and Wise, S. P., 1982, Growth and maturation of the rat corticospinal tract, Prog. Br. Res., 57:361-379.

Jones, E. C., 1981, Development of connectivity in the cerebral cortex, in: "Studies in Developmental Neurobiology," W.M. Cowan, eds., Oxford University Press, New York.

Ketelbant-Balasse, P., Glinoer, D., and Neve, P., 1975, Ultrastructural aspects of the thyroid in a case of human congenital goitre with cretinism, Path. Europ., 10:155-165.

Klein, R.Z., 1985, Infantile Hypothyroidism then and now: the results of neonatal screening, Curr. Prob. Ped., 15:1-58.

Kristt, D. A., 1978, Neuronal differentiation in somatosensory cortex of the rat. I. Relationship to synaptogenesis in the first postnatal week, Brain Res., 150:467-486.

Kudrjacev, T., 1978, Neurologic complications of thyroid dysfunction, Adv. Neurol., 19:619-636.

Kuypers, H. G. F. M., 1985, The anatomical and functional organization of the motor system, in: "Scientific Basis of Clinical Neurology," M. Swash and C. Kennard, eds., Churchill Livingstone, New York.

Larsen, P. R., 1989, Maternal thyroxine and congenital hypothroidism, NEJM, 321: 44-46.

Lasek, R. J., 1988, Studying the intrinsic determinants of neuronal form and function, in: Intrinsic Determinants of Neuronal Form and Function, R. J. Lasek, ed., A. R. Liss, Inc., New York.

Lasek, R. J., Garner, J. A., and Brady, S. T., 1984, Axonal transport of the cytoplasmic matrix, J. Cell Biol., 99:212s-221s.

Lasek, R. J. and Brady, S. T., 1982, The structural hypothesis of axonal transport: Two classes of moving elements, in: "Axoplasmic Transport," Weiss, D. G., ed., Springer-Verlag, Berlin.

Lauder J.M. and Krebs, H., 1986, Do neurotransmitters, neurohumors, and hormones specify critical periods? in: "Developmental Neuropsychobiology," W. T. Greenough, J. M. Jurask, eds., Academic Press, New York.

Lauder, J.M., 1989. Thyroid influences on the developing cerebellum and hippocampus of the rat, in: "Iodine and the Brain," Plenum Press, New York, G. R. DeLong, J. Robbins, and P. G. Condliffe, eds., New York.

Lee, M. K., Tuttle, J. B., Rebhun, L. K., Cleveland, D. W., and Frankfurter, Anthony, 1990, The expression and posttranslational modification of a neuron-specific β-tubulin isotype during chick embryogenesis, Cell Motility and the Cytoskeleton, 17:118-132.

Lee, V., L. Otvos, M. Carden, M. Hollosi, B. Dietzschold and R. Lazzarini, 1988, Identification of the major multiphosphorylation site in mammalian neurofilaments, Proc Natl. Acad. Sci. USA, 85:1998-2002.

Legrand, J., 1982-1983, Hormones thyroidiennes et maturation du systeme nerveux, J. Physiol., Paris 78:603-652.

Lemire, R. J., Loeser, J. D., Leech, R. W., and Alvord, E. C., 1975, Normal and abnormal development of the human nervous system, Harper & Row, New York.

Letarte, J. and Franchi, S. L., 1983, Clinical features of congenital hypothyroidism, in: "Congenital Hypothyroidism," J. H. Dussault and P. Walker, eds., M. Dekker, New York.

Lewis, S.A., Sherline, P., and Cowan, N.J., 1986, A cloned cDNA encoding MAP1 detects a single copy gene in mouse and a grain-abundant RNA whose level decreases during development, J. Cell Biol., 102:2107-2114.

Lewis, S. A., Lee, M. G., and Cowan, N. J., 1984, Five mouse tubulin isotypes and their regulated expression during development, J. Cell Biol., l0l:852-861.

Lewis, S. A. and Cowan, N. J., 1988, Complex regulation and functional versatility of mammalian a- and β- tubulin isotypes during the differentiation of testis and muscle cells, J. Cell Biol., 106:2023-2033.

Lissitzky, S., Torresani, J., Burrow, G. N., Bouchilloux, S., and Chabaud, O., 1975b, Defective thyroglobulin export as a cause of congenital goitre, Clin. Endocrinol., 4:363-392.

Littauer, U. Z., Giveon, D., Thierauf, M., Ginzburg, I., and Ponsting, l. H., 1986, Common and distinct tubulin binding sites for microtubule-associated proteins, Proc. Natl. Acad. Sci., 83:7162-7166.

Lotmar, F., 1929, Histopathologische befunde in gehirenen von kongenitalem myxodem thyreoaplasie und kachexia thyreopriva, Atschr, Neurol Psychiat., ll9:491-513.

Lotmar, F., 1928, Histopathologische befunde in gehirnen von kongenitalem Myxodem (Thyreoaplasie), Z.f.d.g. Neur. u. Psych., 119:492-513.

Lowe, T. W. and Cunningham, F. G., 1990, Thyroid Disease in Pregnancy, in: "Williams Obstetrics," Supplement #9, 18th ed., Cunningham, F.G., McDonald, P., Gant, N., eds., Appleton-Lange, East Norwalk, Conn., 1-15.

Maccioni, R. B., Rivas, C. I., and Vera, J. C., 1988, Differential interaction of synthetic peptides from the carboxyl-terminal regulatory domain of tubulin with microtubule-associated proteins, EMBO J., 7:1957-1963.

Macfaul, R., Borner, S., Brett, E. M., and Grant, D. B., 1978, Neurological abnormalities in patients treated for hypothyroidism from early life, Arch. Dis. Child., 53:611-619.

Malamud, N., Itabashi, H. H., Castor, J., and Messinger, H. B., 1964, An etiologic and diagnostic study of cerebral palsy, J. Pediatr., 65:270-293.

Man, E. B., Holden, R. H., and Jones, W. S., Thyroid function in human pregnancy, Am. J. Obstet. Gyn., 109:12-18, 1971.

Mandelkow, E. and Mandelkow, E. M., 1989, Microtubular structure and polymerization, Curr. Opin. Cell Biol., 1:5-9.

Marc, C. and Rabie, A., 1985, Microtubules and neurofilaments of the sciatic nerve fibers of the developing rat: Effects of thyroid deficiency, Int. J. Dev. Neurosci., 3:353-358.

Marc, C., Clavel, M., and Rabie, A., 1986, Non-phosphorylated and phosphorylated neurofilaments in the cerebellum of the rat: An immunocytochemical study using monoclonal antibodies, development in normal and thyroid-deficient animals, Dev. Brain Res., 26:249-260.

Marin-Padilla, M., 1970, Prenatal and early postnatal ontogenesis of the human motor cortex: a golgi study. I. The sequential development of the cortical layers, Brain Res., 23:167-183.

Marin-Padilla, M., 1988, Early ontogenesis of the human cerebral cortex, in: "Cerebral Cortex," A. Peters and E. G. Jones, eds, Plenum Press, New York.

Marinesco, M.G., 1924, Contribution a l'etude des lesions du myxoedeme congenital, Encephale, 19:265-293.

Matus, A., 1988, Microtubule-associated proteins: Their potential role in determining neuronal morphology, Ann. Rev. Neurosci., 11:29-44.

Mayerhofer, A., Amador, A. G., Beamer, W. G., and Bartke, A., 1988, Ultrastructural aspects of the goiter in cog/cog mice, J. of Heredity, 79:200-3.

Meller, K., Breipohl, W., and Glees, P., 1968, Synaptic organization of the molecular and outer granular layer in the motor cortex in the white-mouse during postnatal development: A golgi and electron-microscopical study, Z. Zellforsch. Mikrosk. Anat. Abt. Histochem., 92:217-231.

Medeiros-Neto, G. A., Knobel, M., Bronstein, M. D., Simonetti, J., Filho, F. F., and Mattar, E., 1979, Impaired cyclic-AMP response to thyrotropin in congenital hypothyroidism with thyroglobulin deficiency, Acta. Endocrinol., 92:62.

Miller, M. W., 1988, Development of projection and local circuit neurons in neocortex, in: "Development and Maturation of the Cerebral Cortex, Cerebral Cortex," Vol. 7, A. Peters and E. G. Jones, eds., Plenum Press, New York.

Miller, M. W., 1987a, Effect of prenatal exposure to alcohol on the distribution and time of origin of corticospinal neurons in the rat, J. Comp. Neurol., 257:372-382.

Miller, F. D., Naus, C. C. G., Durand, M., Bloom, F. E., and Milner, R. J., 1987b, Isotypes of α-tubulin are differentially regulated during neuronal maturation, J. Cell Biol., 105:3065-3073.

Mills, S. A. and Savage, D. D., 1988, Evidence of hypothyroidism in the genetically epilepsy-prone rat, Epilepsy Research, 2:102-10.

Mitchison, T. and Kirschner, M., 1988, Cytoskeletal dynamics and nerve growth, Neuron, 1:761-772.

Morreale de Escobar, G. M. and Escobar del Rey, F., 1983. Thyroid hormone and the developing brain, in: "Congenital Hypothyroidism," J. H. Dussault and P. Walker, eds., Academic Press, New York.

Morreale de Escobar, G. M., Pastor, R., Obregon, M. J., and Del Ray, F. E., 1985, Effects of maternal hypothyroidism on the weight and thyroid hormone content of rat embryonic tissues, before and after onset of fetal thyroid function, Endocrinol., 117:1890.

Morreale de Escobar, G., Ruiz de Ona, C., Obregon, M.J., and Escobar del Rey, F., 1989, Models of fetal iodine deficiency, in: "Iodine and the Brain," G.R. DeLong, J. Robbins, and P.G. Conliffe, eds., Plenum Press, New York.

Morris, R. G. M. Garrud, P., Rawlines, J. N. P., and O'Keefe J., 1982, Place navigation is impaired in rats with hippocampal lesions, Nature, 297:681-683.

Narayan, P., Towle, H. C., 1985, Stabilization of a specific nuclear mRNA precursor by thyroid hormone, Mol. Cell. Biol., 5:2642-2646.

Narayanan, C. H., Narayanan, Y., and Browne, R. C., 1982, Effects of induced thyroid deficiency on the development of suckling behavior in rats, Physiol. Behav., 29:361-370.

Narayanan, C. H. and Narayanan, Y., 1985, Cell formation in the motor nucleus and mesencephalic nucleus of the trigeminal nerve of rats made hypothyroid by propylthiouracil, Exp. Brain Res., 59:257-266.

Nelson, K. and Ellenberg, J., 1986, Antecedents of cerebral palsy: Multivariate analysis of risk, NEJM, 315:81-86.

New England Congenital Hypothyroidism Collaborative Group, 1990, Elementary school performance of children with congenital hypothyroidism, J. Ped., 116:27-32.

Noguchi, T., 1988, Brain development in dwarf mice, Progr. in Neurobiol., 31:149-170.

Noguchi, T., Kudo, M., Sugisaki, T., and Satoh, I., 1986, An immunocytochemical and electron microscopic study of the hyt mouse anterior pituitary gland, J. Endocrinol., 109:163-168.

Noguchi, T. and Sugisaki, T., 1984, Hypomyelination in the cerebrum of the congenitally hypothyroid mouse (hyt/hyt), J. Neurochem., 42:891-893.

Nunez, J., 1988, Immature and mature variants of MAP2 and Tau proteins and neuronal plasticity, TINS, 11:477-479.

Nunez, J., Couchie, D., and Brion, J. P., 1989, Microtubule assembly: Regulation by thyroid hormones, in: "Iodine and the Brain," G. R. Delong, J. Robbins, P. G. Condliffe, eds., Plenum Press, New York.

Okado, N., 1980, Development of the human cervical spinal cord with reference to synapse formation in the motor nucleus, J. Comp. Neurol., 191:495-513.

Paschal, B. M., Obar, R. A., and Vallee, R. B., 1989, Interaction of brain cytoplasmic dynein and MAP2 with a common sequence at the C terminus of tubulin, Nature, 342:569-572.

Pharoah, P. O. D., Connolly, K. J., Ekins, R. P., and Harding, A. G., 1984, Maternal thyroid hormone levels in pregnancy and the subsequent cognitive and motor performance of the children, Clin. Endocrin., 21:265-270.

Pinto Lord, M. C. and Caviness, V. S., 1979, Determinants of cell shape and orientation: a comparative golgi analysis of cell-axon interrelationships in the developing neocortex of normal and reeler mice, J. Comp. Neuro., 187:49-70.

Porter, R., 1985, The cerebral cortex and control of movement performance, in: "Scientific Basis of Clinical Neurology," M. Swash and C. Kennard, eds., Churchill Livingstone, New York.

Porterfield, S. P. and Hendrich, C. E., 1981, Alterations of serum thyroxine, triiodothyronine, and thyrotropin in the progeny of hypothyroid rats, Endocrinol., 108:1060-1063.

Purpura, D. P., 1975, Dendritic Differentiation in human cerebral cortex: normal and aberrant developmental patterns, in: "Advances in Neurology," G. W. Kreutzberg, ed., Raven Press, New York.

Puymirat, J., Barret, A., Picart, R., Vigny, A., Loudes, C., Faivre-Bauman, A., and Tixier-Vidal, A., 1983, Triiodothyronine enhances the morphological maturation of dopaminergic neurons from fetal mouse hypothalamus cultured in serum-free medium, Neurosci., 10:801-810.

Rabie, A., Patel, A., Clavel, M., and Legrand, J., 1979, Effect of thyroid deficiency on the growth of the hippocampus in the rat, Dev. Neurosci., 2:183-194.

Rami, A., Patel, A., and Rabie, A., 1986a, Thyroid hormone and development of the rat hippocampus: Cell acquisition in the dentate gyrus, Neurosci., 19:1207-1216.

Rami, A., Patel, A. J., and Rabie, A., 1986b, Thyroid hormone and development of the rat hippocampus: Morphological alterations in granule and pyramidal cells, Neurosci., 4:1217-1226.

Rasool CG, Bradley WG, Reichlin S , Reduced axoplasmic somatostatin transport in hypothyroid rats, J Neurochem, 45:973-976.

Rastogi, R. B. and Singhal, R. L., 1976, Influence of neonatal and adult hyperthyroidism on behavior and biosynthetic capacity for norepinephrine, dopamine and 5-hydroxytryptamine in rat brain, J. Pharmacol. Exp. Ther., 198:609-618.

Regard, E., Taurog, A., and Nakashimas, T., 1978, Plasma thyroxine and triiodothyronine levels in spontaneously metamorphosing rana catesbeiana tadpoles and in adult anuran amphibia, Endocrinol., 102:674-683.

Rice, F.L., 1975, The development of the primary somatosensory cortex in the mouse: 1) A nissl study of the ontogenesis of the barrels and the barrel field. 2) A quantitative autoradiographic study of the time of origin and pattern of migration of neuroblasts on area SI. (PH.D. Dissertation) The Johns Hopkins University) University Microfilms, Ann Arbor.

Ricketts, M.H., Simons M.J., Parma J., Mercken, L., Dong O., Vassart G., 1987, A non-sense mutation causes hereditary goiter in the afrikander cattle and unmaks alternative splicing of thyroglobulin transcripts, PNAS, 84:3181-3184.

Rochiccioli, P., Alexandre, F., and Roge, B., 1989, Neurological development in congenital hypothyroidism, in: Research in: "Congenital Hypothyroidism," F. DeLange, D. A. Fisher, and D. Glinoer, eds., Plenum Press, New York.

Roland, P. E., 1987, Metabolic mapping of sensorimotor integration in the human brain, in: "Motor Areas of the Cerebral Cortex," R. Porter and C. G. Phillips, eds., A Wiley-Interscience Publication, New York.

Rosman, N.P., 1975, Neurological and muscular aspects of hypothyroidism in childhood, in: "The Pediatric Clinics of North America," A. L. Prensky, ed., Saunders, Philadelphia.

Ross, E. M., 1989, Signal sorting and amplification through G protein-coupled receptors, Neuron, 3:141-152.

Rovet, J. F., Westbrook, D. L., and Ehrlich, R. M., 1984, Neonatal thyroid deficiency: Early temperamental and cognitive characteristics, J. Am. Acad. of Child Psychi., 23:10-22.

Rovet, J., Glorieus, J., and Heyerdahl, S., 1987, Summary of research findings on the psychological follow-up of CH children identified by newborn screening, "Advances in Neonatal Screening, Proceedings of the Sixth International Newborn Screening Symposium," B. Therrell ed., Elsevier Press, Amsterdam.

Rovet, J., Ehrlich, R., and Sorbara, D., 1987, Intellectual outcome in children with fetal hypothyroidism, J. Ped., 110:700-704.

Rovet, J. F., 1989, Congenital Hypothyroidism: Intellectual and neuropsychological functioning, in: "Psychoneuroendocrinology, Brain, Behavior, and Hormonal Interactions," C. Holmes, ed., Springer-Verlag, New York.

Rudy, J. W. and Stadler-Morris S., Albert P., 1987, Ontogeny of spatial navigation behaviors in the rat: dissociation of "proximal" and "distal"-cue based behaviors, Behav. Neurosci., 101:62-73.

Ruiz-Marcos, A. 1989, Quantitative studies of the effects of hypothyroidism on the development of the cerebral cortex, in: "Iodine and the Brain," G. R. DeLong, J. Robbins, and P. G. Condliffe, eds., Plenum Press, New York.

Ruiz-Marcos, A., Salas, J., Sanchez-Toscano, F., Escobar del Rey, F., and Morreale de Escobar, G., 1983, Effects of neonatal and adult onset hypothyroidism on pyramidal cells of the rat auditory cortex, Dev. Brain Res., 9:205-213.

Samuels, H. H., Forman, B. M., Horowitz, Z. D, and Ye, Z-S, 1989, Regulation of gene expression by thyroid hormone, Ann. Rev. Physiol., 51:623-639.

Sarafian T. and Verity, A.M., 1986, Influence of thyroid hormones on rat cerebellar cell aggregation and survival in culture, Dev. Brain Res., 26:261-270.

Sarlieve, L. L., Bouchon, R., Koehl, C., and Neskovic, N. M., 1983, Cerebroside and sulfatide biosynthesis in the brain of snell dwarf mouse: effects of thyroxine and growth hormone in the early postnatal period, J. Neurochem., 40:1058-1062.

Schapiro, S., Salas, M., and Vukovich, K., 1970, Hormonal effects on ontogeny of swimming ability in the rat: Assessment of central nervous system development, Science, 168:147-150.

Schalock, R. L., Brown, W. J., and Smith, R. L., 1979, Long-term effects of propylthiouracil-induced neonatal hypothyroidism, Exper. Psychobio., 12:187-199.

Schreyer, D. J. and Jones, E. G., 1982, Growth and target finding by axons of the corticospinal tract in prenatal and postnatal rats, Neurosci., 7:1837-53.

Serrano, L., Montejo de Garcini, E., Hernandez, M. A., and Avila, J., 1985, Localization of the tubulin binding site for tau protein, Eur. J. Biochem., 153:595-600.

Shanklin, D. R. and Stein, S. A., 1988, The Ultrastructural Component Phasing of Developing Fetal and Early Neonatal Mouse Thyroid Cells, FASEB J., 2:A394, #571.

Shanklin, D. R., Stein, S. A., et al., 1991, Pathological studies of fetal thyroid development, in: "Advances in Perinatal Thyroidology," B. Bercu, and D. Shulman, eds., Plenum Press, New York.

Shoukimas, G. M. and Hinds, J. W., 1978, The development of the cerebral cortex in the embryonic mouse: an electron microscopic serial section analysis, J. Comp. Neur., 179:795-830.

Sidman, R. L., and Rakic, P., 1982, Development of the human central nervous system, in: "Histology and Histopathology of the Nervous System," W. Haymaker and R. D. Adams, eds., C.C. Thomas, Springfield, Illinois.

Siegrist-Kaiser, Ca. A., Juge-Aubry, C., Tranter, M. P., Ekenbarger, D. M., Leonard, J. L., 1990, Thyroxine-dependent modulation of actin polymerization in cultured astrocytes. A novel, extranuclear action of thyroid hormone, J. of Bio. Chem., 265:5296-302.

Smith, S. J., 1988, Neuronal Cytomechanics: The actin based motility of growth cones, Science, 242:708-715.

Stanbury, J. B., Rochmans, P., Buhler, U. K., Ochi, Y., 1968, Congenital hypothyroidism with impaired thyroid response to thyrotropin, NEJM, 279:1127-1138.

Stein, S. A., 1985, Thyroid hormone control of gene expression in Spraque-Dawley rat brain and liver, Ann. Neuro., 18:385.

Stein, S. A., 1988, 9A6 mRNA, a mouse and rat thyroid regulated brain mRNA: Sequence analysis and in situ hybridization, Soc. for Neuro. Abst., Vol. 14, Part 2.

Stein, S. A., Adams, P. M., Shanklin, D. R., Mihailoff, G. A., Palnitkar, M., 1989a, Thyroid hormone regulation of specific mRNAs in developing brain, in: "Iodine and the Brain," G. R. Delong, J. Robbins, P. G. Condliffe, eds., New York, Plenum Press.

Stein, S. A., Shanklin, D. R., Krulich, L., Roth, M. G., Chubb, C. M., Adams, P. M., 1989b, Evaluation and characterization of the hyt/hyt hypothyroid mouse II. Abnormalities of TSH and the thyroid gland, Neuroendocrin., 49:509-519.

Stein, S. A., Bloom, G. S., Mihailoff, G. A., Adams, P. M., and Shanklin, D. R., 1989c, Thyroid hormone effects on microtubular composition in developing cerebral cortex, Soc. Neurosci. Abst., 15(1):95.

Stein, S. A., Kirkpatrick, L., Shanklin, D. R., Adams, P. M., and Brady, S., 1991a, Hypothyroidism reduces the rate of slow component A(SCa) axonal transport and of total tubulin protein in the hyt/hyt mouse optic nerve, J. Neurosci. Res., 28:121-133.

Stein, S. A., Zakarija, M., MacKenzie, J. M., and Shanklin, D. R., 1991b, The site of the molecular defect in the thyroid gland of the hyt/hyt mouse: Abnormalities in the TSH receptor-G protein adenylyl cyclase complex, THYROID, In Press.

Stein, S. A., Bloom, G. S., Shanklin, D. R., and Adams, P. M., 1991c, The effect of thyroid hormone on microtubular composition in developing mouse cerebral cortex, Submitted for publication.

Stein, S. A., et al., 1991d, The role of thyroid hormone in adult and developing brain, in: "Molecular Genetics of Neurological Disease," R. N. Rosenberg and S. Prusiner, eds., Churchill-Livingstone.

Stein, S.A., Kirkpatrick, L., Adams, P.M., Shanklin, D.R., and Brady, S.T., 1991e, Specific proteins of slow component b(SCb) axonal transport are slowed in the hypothyroid hyt/hyt mouse optic nerve, Submitted.

Strait, K.A., , Schwartz, H.L., Perez-Castillo, A.M.,and Oppenheimer, J.H., 1990, Relationship of c-erbA mRNA content to tissue triiodothyronine nuclear binding capacity and function in developing and adult rats, J. Biol. Chem., 265:10514-10521.

Strupp, B. J. and Levitsky, D. A., 1983, Early brain insult and cognition: A comparison of malnutrition and hypothyroidism, Dev. Psych., 16:535-40.

Sturrock, R. R., 1974, Histogenesis of the anterior limb of the anterior commissure of the mouse brain, I. A quantitative study of changes in the glial population with age, II. A quantitative study of pre and postnatal mitosis, J. Anat., 117:17-35.

Sullivan, K. F., 1988, Structure and utilization of tubulin isotypes, Ann. Rev. Cell Biol., 4:687-716.

Sutherland, R. J. and Rudy, J. W., 1988, Place learning in the Morris place navigation task is impaired by damage to the hippocampal formation even if the temporal demands are reduced, Psychobiol., 16:157-163.

Takahashi, T., 1983, Transplacental effects of 3,5-dimethyl-3'-isopropyl-l-thyronine on tubulin content in fetal brains in rats, Jap. J. Physiol., 34:365-368.

Taylor, B. A. and Rowe, L., 1987, The congenital goiter mutation is linked to the thyroglobulin gene in the mouse, Proc. Natl. Acad. Sci. USA, 84:1986-90.

Tucker, R.P., Garner, C.C., and Matus, A., 1989, In situ localization of microtubule-associated protein mRNA in the developing and adult rat brain, Neuron, 2:1245-1256.

Uziel, A., 1986, Periods of sensitivity to thyroid hormone during the development of the organ of Corti, Acta Otolaryngol. Suppl., 429:23-27.

Vallee, R. B. and Bloom, G. S., 1991, Mechanisms of fast and slow axonal transport, Ann. Rev. Neurosci., 14:59-92.

Van Middlesworth, L. and Norris, C. H., 1980, Audiogenic seizures and cochlear damage in rats after perinatal antithyroid treatment, Endocrin., 106:1686.

Vulsma, T., Gons, M. H., and de Vijlder, J. J. M., 1989, Maternal-fetal transfer of thyroxine in congenital hypothyroidism due to a total organification defect of thyroid agenesis, NEJM, 321:13-16.

Weinstein, S. L. and Tharp, B. R., 1989, Etiology and timing of static encephalopathies of childhood (cerebral palsy), in: "Fetal and Neonatal Brain Injury," D. K. Stevenson and P. Sunshine, eds., B.C. Decker, Inc., Philadelphia.

Wise, S. P., Fleshman, J. W., and Jones, E. G., 1979, Maturation of pyramidal cell form in relation to developing afferent and efferent connections of rat somatic sensory cortex, Neurosci., 4:1275-1297.

Wolter, R., Noel, P., de Cock, P., Craen, M., Ernould, C., Malvaux, P., Verstraeten, F., Simons, J., Mertens, S., Van Broeck, N., and Vanderschueren-Lodeweyck, M., 1979, Neuropsychological study in treated thyroid dysgenesis, Acta Paediatr Scand. Suppl., 277:41-45.

Wujek, J. and Lasek, R. J., 1983, Correlation of axonal regeneration and slow component b in two branches of a single axon, J. Neurosci., 3:243-257.

Yamada, K.M., Spooner B.S., Wessells N.K., 1971, Ultrastructure and function of growth cones and axons of cultured nerve cells, J. Cell Bio., 49:614-635.

THE THYROIDECTOMIZED PREGNANT RAT - AN ANIMAL MODEL TO

STUDY FETAL EFFECTS OF MATERNAL HYPOTHYROIDISM

Susan P. Porterfield and Chester E. Hendrich

Department of Physiology and Endocrinology
Medical College of Georgia
Augusta, GA

INTRODUCTION

If women are hypothyroxinemic during pregnancy, their children show an abnormally high incidence of behavioral and neurological disorders. The most common problems seen in the children of hypothyroxinemic mothers are: visual disturbances, poor fine-motor coordination, signs of cerebral palsy and lower I.Q. scores (Man et al., 1971; Man and Jones, 1969; Jones and Man, 1969; Nelson and Ellenburg, 1986). Clinically this is a concern because the hypothyroxinemia may occur only during pregnancy and need not be severe to produce these effects; hence, the pregnancy could proceed to parturition with the thyroid disturbance remaining undiagnosed (Jones and Man, 1969). Support for the proposal that normal maternal thyroid function is essential for normal fetal development is provided by the well-established data which show that the developmental deficiencies seen in children suffering from congenital iodine deficiency-in which both the mother and the fetus have a thyroid hormone deficiency-are greater than those resulting only from a fetal thyroid hormone deficiency (Hetzel, 1983; Escobar del Rey et al., 1986; Van Middlesworth and Norris, 1980; Fierro-Benitez et al., 1974; Pharoah et al., 1981). Furthermore, if supplemental iodine is given to iodine deficient pregnant women after the fifth month of gestation (Fierro-Benitez et al., 1972), the iodine does not prevent significant neurological impairment. These data emphasize the importance of normal maternal thyroid hormone production, particularly during the first one-half of gestation.

The impairment of fetal development resulting from maternal hypothyroidism during pregnancy is multifaceted. Abnormal maternal metabolism can alter the availability of nutrients to the fetus. Placental development, metabolism, and/or

Advances in Perinatal Thyroidology, Edited by B.B. Bercu and
D.I. Shulman, Plenum Press, New York, 1991

blood flow could be abnormal, thereby impairing placental function; uterine development could also be abnormal. In addition, recent evidence indicates that, in both humans and rodents, maternal thyroid hormones reach the fetus and, therefore, could play an important direct role in early fetal brain development (Vulsma et al., 1989; Obregon et al., 1984, Porterfield and Hendrich, 1990).

Rodents (rats and mice) have been the most commonly used animal models for the study of human endocrine disorders, including thyroid disorders. The rat fetus at 16-18 days post-conception (d pc) corresponds developmentally to the human fetus at about 12 weeks of gestation and the 10 day postnatal (d pn) rat is at the neurologically equivalent stage of development as the newborn human (Morreale de Escobar and Escobar del Rey, 1983; Fisher et al., 1977). Humans and rats have similar relative development of the fetal thyroid gland, the appearance of T_3 receptors and maturation of the hypothalamo-pituitary-thyroid axis (Morreale de Escobar and Escobar del Rey, 1983; Fisher et al., 1977). The effects of thyroid hormones on the developing rat brain have been studied extensively (Dussault and Ruel, 1987; Bass et al., 1977). Similar pathological changes are seen in the brains of hypothyroid humans and rats (Eayrs, 1955; Bass et al., 1977; Legrand 1982; Morreale de Escobar et al., 1983), and behavioral and learning abnormalities are similar (Morreale de Escobar et al., 1983).

The thyroidectomized (Tx) rat appears to be a suitable model to study the effects of maternal hypothyroidism on fetal development. Radiothyroidectomy was utilized to alleviate the need for parathormone replacement therapy. With this procedure for thyroidectomy, when the pregnant rats were treated with replacement doses of thyroxine (T_4) (1.0 µg/100g BW/day), they were indistinguishable from controls (Porterfield et al., 1975).

PROCEDURE FOR RADIOTHYROIDECTOMY

Rats were maintained on a low iodine diet (Remington diet, U.S. Biochemical Corp, Cleveland, OH) for 7 days and were then given a single i.p. injection of 300 µ Ci [131] I followed 4 hours later with a s.c. injection of 10 µg L-T_4 and thereafter with 1.0 µg L-T_4/100 g BW/day for three weeks. At this time, they were placed with fertile males and the day spermatozoa were detected in the vaginal tract was considered as day 1 of the pregnancy. T_4 replacement therapy was discontinued on day 1 of pregnancy. As the prolonged $t_{1/2}$ of thyroid hormones in hypothyroidism delays the loss of hormone from the rat, the animals never became athyroxinemic in these studies. Following parturition, Tx rats were given T_4 s.c. during lactation (1.0 µg/100g BW/day). Cross-fostering studies showed this treatment provided for normal maternal lactation.

REPRODUCTIVE PERFORMANCE OF TX RATS.

Maternal hypothyroidism seriously impairs fetal development. Fetal and perinatal mortality are increased significantly with nearly a 50% fetal-neonatal wastage (fig.1) (Porterfield et al., 1975; Hendrich et al., 1984). Fetal body, and brain growth are stunted (fig.2). Catch-up growth occurs in the progenies of Tx mothers by 30-60 days postpartum (Hendrich et al., 1984).

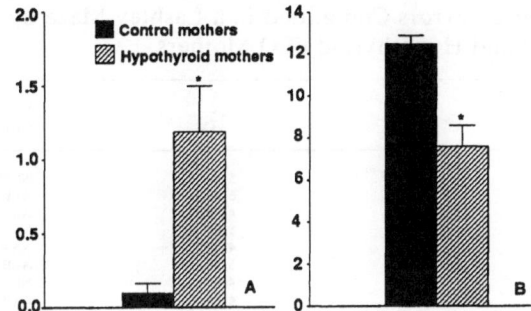

Fig. 1   A. Number of fetal resorptions at 22 days
B. Number of live neonates at 1 day
* Significantly different from controls, P<0.05

BEHAVIORAL STUDIES OF PROGENIES OF TX RATS

At between 40 and 60 days of age, the progenies of control, and Tx mothers were tested for differences in behavior (Hendrich et al., 1984) in a Lashley type 3 enclosed maze (Munn, 1950) and in stabilimeters (Strong, 1957). The progenies from both groups were tested for their ability to learn the maze and to retain that knowledge by placing them in the maze every other day for 10 trials. The number of errors per trial and the time to traverse the maze were recorded. An animal was considered to have learned the maze when it completed two or more consecutive trials free of errors.   On alternate days, the progenies of all groups were placed in stabilimeters and their spontaneous activity recorded. These devices are sensitive to fine movements such as grooming, scratching or eating as well as general movements around the cage area. Animals were tested for 30 minutes every other day over a 20 day period. The control animals were able to learn the maze fairly rapidly so that the number of errors/trial went from 5.3±1.4 on trial one to 0.8±.03 on trial 10 and the time to traverse the maze went from 1.72±.36 min on trial 1 to 0.24±0.4 min on trial 10 (Table I). Ten out of 24 animals were able to traverse the maze without error on

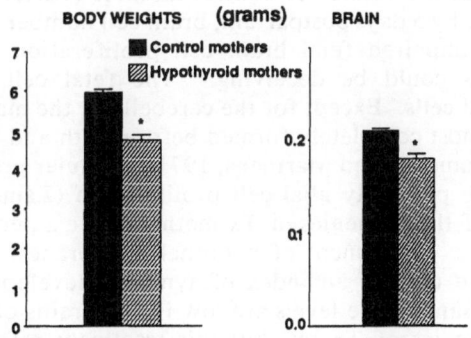

Fig. 2   Fetal weights (g) at 22 days gestation
* Significantly different from controls, P<0.05

Table I   Number of Errors Committed in a Lashley Maze by Progeny of
Control and Hypothyroid (Tx) Mothers.

	A Control	B Tx	Significant Differences
Trial 1	5.3±1.4*	5.5±0.6	NS**
Trial 2	3.1±0.5	6.1±1.1	A<B
Trial 3	2.7±0.4	4.5±1.7	NS
Trial 4	2.3±0.3	5.1±1.9	NS
Trial 5	3.4±0.5	4.6±0.8	NS
Trial 6	2.2±0.5	5.2±0.6	A<B
Trial 7	3.0±0.5	4.4±0.6	NS
Trial 8	1.5±0.5	4.3±0.5	A<B
Trial 9	1.3±0.4	4.6±0.7	A<B
Trial 10	0.8±0.3	5.3±0.6	A<B

Errors for trials 4,6,8,9,10 are < trial 1	No decrease in error number occurred from trial one to trial ten	* Mean ± S.E. ** P<0.05

The number of progenies to traverse the maze without error on 2 or more consecutive trials was
10/24 for group A and 0/24 for group B.

2 consecutive trials during the 10 trial series. However, the performance of the progenies of Tx mothers was poor. The number of errors per trial did not improve from 5.5±.06 on trial 1 to 5.3±.06 on trial 10. No animals (out of 24) were able to traverse the maze error-free in 2 consecutive trials. The data from the Lashley Maze study show that the progenies of Tx mothers have learning deficits involving memory. Results of the studies using the stabilimeters show that the progenies of the Tx mothers are hyperactive. Hyperactivity is a trait commonly found in retarded children and it indicates deficits in control mechanisms regulating one's ability for selective and sustained attention to a given task.

BRAIN DEVELOPMENT

In assessing various conventional parameters used as indices of brain development, it appeared that fetal/perinatal brain development is delayed in progenies of Tx mothers. At 22 d pc the fetuses of Tx mothers have only 70% of the number of brain cells as the fetuses of control mothers (Porterfield and Hendrich, 1982) (fig.3). However, by 5 days postpartum, brain cell number is normal. While the data suggest that the impaired fetal brain cell proliferation is compensated for postnatally, the results could be deceiving. The fetal cell number represents predominantly neuronal cells. Except for the cerebellum, the mature complement of rat brain neurons is almost completely formed before birth and is definitely formed before 5 days of age (Zamenof and Marthens, 1971). Cellular proliferation after this period is thought to be primarily glial cell proliferation (Zamenof and Marthens, 1971). It isn't known if the progenies of Tx mothers have a persistent neuronal cell deficit. Gangliosides, a component of neuronal membranes, are concentrated in nerve terminals and are used as an index of synaptic development (Karlsson and Svennerholm, 1978). Ganglioside levels are low in the brains of the 22-day fetuses and the 5 day-old neonates of Tx rats but this treatment effect is limited to the perinatal period. These data suggest delayed brain maturation in the progenies of Tx

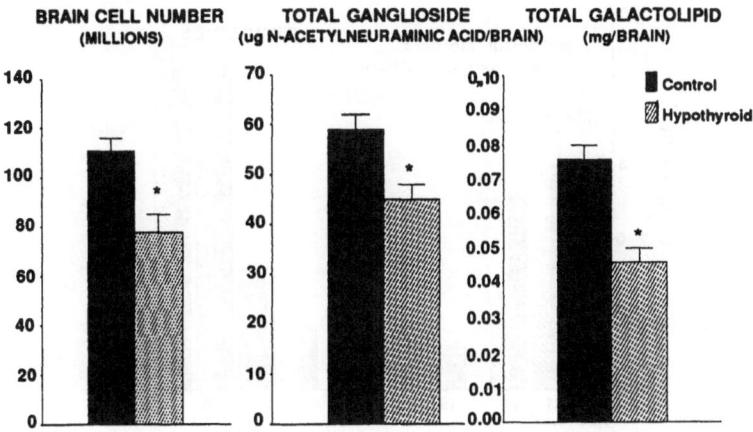

Fig 3   Parameters of brain development in 22 day fetuses
*Significantly different from controls, P<0.05

mothers. As thyroid hormones are known to be important for normal neuronal myelination in the neonatal rat, total galactolipid (cerebrosides plus sulfatides) was measured as an index of myelination. During the period of rapid myelin formation (birth-30 days) the progenies of Tx mothers show development comparable to controls. At 22 d pc, the fetal brain galactolipid concentrations are low in fetuses of Tx mothers relative to controls (fig.3). As there is little myelin present at this time, the treatment effects might not be very significant.

SERUM THYROID HORMONE LEVELS IN FETUSES AND PROGENIES OF TX MOTHERS

When replacement thyroxine therapy is withdrawn on day one of pregnancy, there is a progressive decline in serum thyroid hormone levels during the pregnancy such that over the course of 22 days, the mother goes from euthyroidism to having serum T_4 levels that are 39% of normal (fig.4), T_3 levels that are 44% of normal (fig.5) and TSH levels that are 9.3x greater than normal (fig.6). As is readily apparent, these mothers demonstrate moderate hypothyroidism, not athyroidism, during gestation. Fetal serum hormone levels were measured at 19, 21, and 22 d pc. On day 21 pc, fetuses of Tx mothers have serum T_4 levels comparable to those of the control fetuses (fig.4). By this stage in gestation, the fetus is actively synthesizing and secreting thyroid hormones and serum T_4 levels are progressively increasing from day 19 to 22. However, by day 22 (the time of parturition in control rats), the fetuses of Tx mothers are hyperthyroxinemic. The presence of a hyperfunctioning fetal thyroid is supported by elevated TSH levels. The high serum T_4 levels just prior to parturition have been confirmed repeatedly and the data reported represent an excess of 50 litters per treatment group. Obviously, the fetuses of Tx mothers are not hypothyroxinemic late in gestation. However, the situation appears to be entirely different earlier in gestation when the fetal thyroid is either non-functional or minimally functional. The

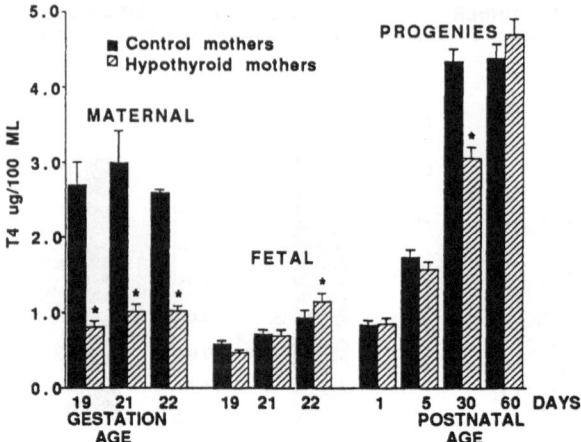

Fig. 4   Total serum thyroxine levels in pregnant rats, their fetuses and postnatal
progenies
* Significantly different from controls, P<0.05

development of a highly sensitive technique for extraction of iodothyronines from
small quantities of tissues and measurement by HPLC of the dansyl chloride derivatives of the iodothyronines has allowed for the measurement of extremely low levels of
thyroid hormones in the mid-gestational fetus.   These results show that thyroid
hormones of maternal origin are present in the mid-gestational rat fetus and that, if
the mother is hypothyroid, the fetal tissue T_4 and T_3 levels are lower than normal.
These data are presented later in the paper.

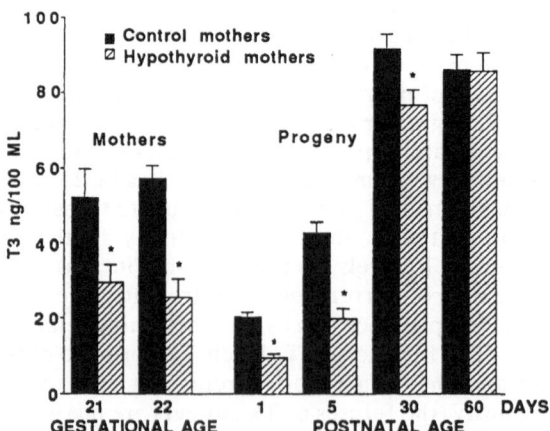

Fig. 5   Total serum T_3 in pregnant rats, their fetuses and postnatal progenies
* Significantly different from controls, P<0.05

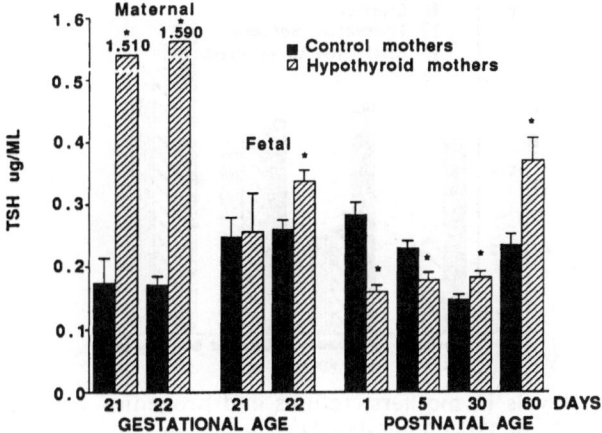

Fig. 6   Serum TSH in pregnant rats, their fetuses and postnatal progenies
* Significantly different from controls, P<0.05

Disorders in perinatal thyroid function are present in the progenies of Tx mothers. While serum T_4 levels are abnormally low only at 30 days postpartum, serum T_3 levels are low throughout the perinatal period and this treatment effect persists through 30 days postpartum. The perinatal period is a critical period for the influence of thyroid hormones on brain development and the serum levels of T_4 and T_3 suggest that the developing brains of the progenies of Tx mothers could be exposed to inappropriately low levels of thyroid hormones during this time.

As can be seen in figs. 4-6, at least minor alterations in thyroid function persist through adulthood in the progenies of Tx mothers. This has been confirmed by others (Bakke et al., 1975). Bakke et al., (1977) have shown that if a neonatal rat is given a single large dose of T_4, a permanent decrease in the "set point" for hypothalamic-pituitary regulation occurs. They used the term "neo- T_4 syndrome" to refer to this condition. To ascertain if the progenies of Tx rats could be showing persistent thyroid disturbances because of the brief exposure to hyperthyroxinemia at 22 d pc, pregnant dams were treated either with the synthetic thyromimetic agent 3,5-dimethyl-3'-isopropyl-L-thyronine (DIMIT) (2.5 µg/100 g BW) at 0900 hrs or with L-T_4 (10 µg/100 g BW at 0900 and 1600 hrs on day 21 pc. DIMIT is known to cross the placenta readily and a significant depression of fetal serum TSH was noted on day 22 pc (fig 7). One day after the T_4 administration to the mother, maternal serum T_4 was elevated 8-fold and fetal serum T_4 was elevated 4-fold (fig 8) (Porterfield, 1985). The only possible mechanism for the elevation of fetal T_4 is transplacental passage of maternal T_4. Subtle, but persistent, changes in hypothalamic-pituitary-thyroid function as well as growth retardation were seen as a result of late-gestational fetal exposure to high serum T_4 or DIMIT levels, suggesting that the critical developmental period for establishment of normal regulation of the hypothalamic-pituitary-thyroidal control system includes not only the neonatal but also the late-gestational period. Furthermore, slightly different results were obtained

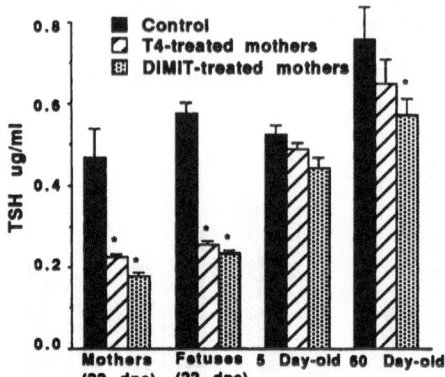

Fig. 7 Serum TSH levels in mothers, fetuses and progenies when the mothers are treated with T_4 or DIMIT on day 21 of gestation
* Significantly different from controls, P<0.05

when excessive T_4 exposure occurred <u>in utero</u> rather than postnatally. Therefore, it is possible that exposure of the developing fetus to abnormal thyroid hormone levels could result in permanent, albeit mild, thyroid dysfunction.

THYROID HORMONE DEIODINATION

Approximately 80% of the serum T_3 comes from peripheral 5'monodeiodination of thyroidally produced T_4. The primary tissues contributing to serum T_3 production are the liver and kidney. Silva and Matthews (1984) have shown that brain, unlike many other tissues, obtains less than 20% of its T_3 from serum while greater than 80% comes from intracellular deiodination of T_4. Because T_3 has a greater affinity than T_4 for the thyroid hormone nuclear receptor, tissue 5'monodeio-

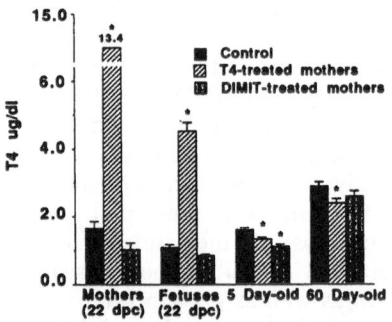

Fig. 8 Serum T_4 levels in mothers, fetuses and progenies when the mothers are treated with T_4 or DIMIT on day 21 of gestation.
* Significantly different from controls, P<0.05

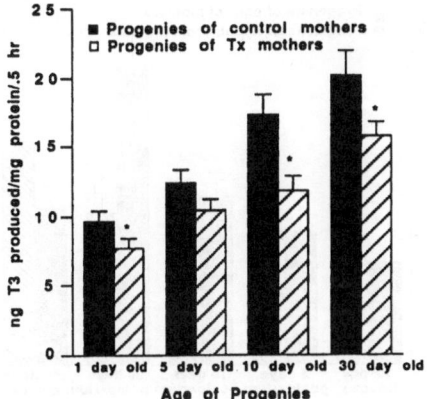

Fig. 9   Liver in vitro T_3 production
* Significantly different from controls, P<0.05

dinase (5'DI) activity can be an important regulator of the biological action of thyroid hormones. As serum T_3 is low in the perinates of the Tx mothers, even at ages when serum T_4 is normal, liver 5'DI activity could be deficient in these pups. Indeed, measurements of liver 5'DI activity show that activity is low in the progenies of Tx mothers through 30 days of age (Porterfield, 1985) (fig.9). The decrease in enzyme activity could result in a decrease in the availability of adequate T_3 to some tissues, even when T_4 is normal. The effect of the treatment on cerebral cortical 5'DI is particularly notable (fig 10). At all ages studied (22 d pc-60 d pn) enzyme activity was lower in the tissues of the progenies of Tx mothers than in the controls. The differences were statistically significant at all postnatal ages except 10 d pn. These data would suggest that brain T_3 availability would be low even when T_4 levels were normal.

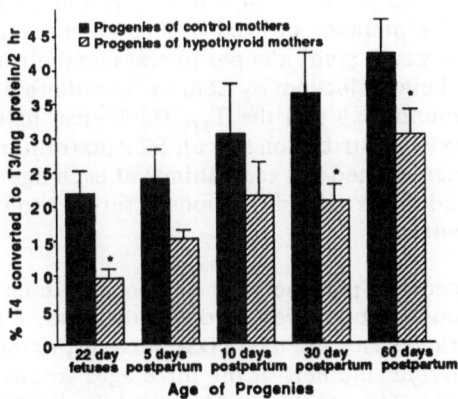

Fig. 10 Cerebral cortex 5' monodeiodinase activity
* Significantly different from controls, P<0.05

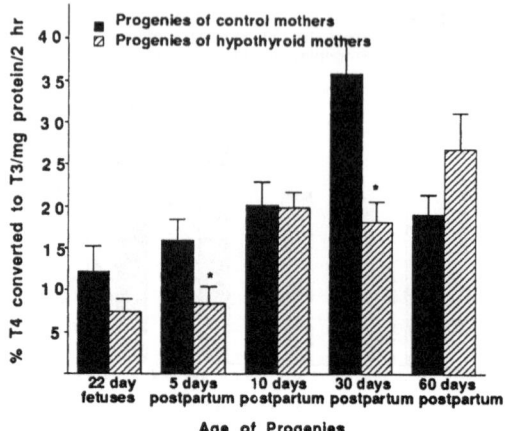

Fig. 11 Cerebellar 5' monodeiodinase activity
* Significantly different from controls, P<0.05

Such a persistent deficiency in T_3 availability could cause a functional brain thyroid deficient state and could explain the reported problems with amino acid uptake and with protein deficiency. Similar results were obtained in the cerebellum except that a statistically significant depression of 5'DI activity was seen only at 5 and 10 d pn (fig. 11). The cause of the low 5'DI activity is not known. However, poor nutrition is thought to suppress enzyme activity, and, hence, the problems shown with perinatal nutrient availability could be factors in enzyme suppression.

PERIPHERAL METABOLISM OF THYROXINE

As the progenies of hypothyroid mothers showed multiple alterations from normal in their developmental patterns of serum T_4 and T_3 levels and in their brain 5'DI activities, it became of interest to determine other metabolic parameters for T_4. These studies were conducted on the 5, 30, and 60 day-old progenies of control and hypothyroid mothers. The methods utilized by Ingenbleek and Malvaux (1980) on children were adapted to rats to study the peripheral metabolism of ^{125}I-T_4 with the thyroidal iodide pump being blocked by sodium perchlorate. The total T_4 was determined by radioimmunoassay and the $T_{1/2}$ (biological half life), K (fractional turnover rate of T_4), TDS (T_4 distribution space), ETP (extrathyroidal pool), TDR (T_4 degradation rate) were determined for each animal at each age interval. The purity of the injected ^{125}I-T_4 and the extracted fractions of serum and tissue labeled T_4 were determined by HPLC methods.

Only those parameters of peripheral T_4 metabolism which differed significantly between the two groups are being reported, i.e. the TDS, ETP, TDR, and MCR. The thyroxine distribution space in ml (TDS) was surprisingly increased in the progenies of the hypothyroid mothers at the three ages studied as compared to the control progenies (fig.12). The mean extrathyroidal pools (ETPs) in the 5 and 60 day-old progenies of hypothyroid mothers were also much greater than those of the

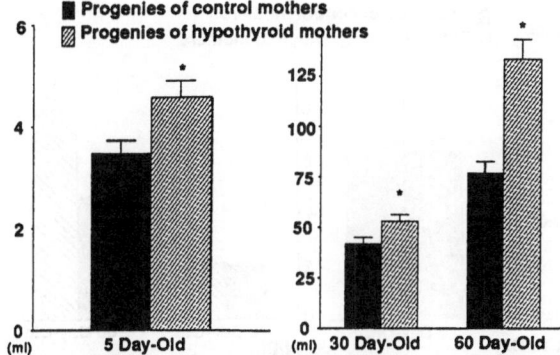

Fig. 12 Thyroxine distribution space
* Significantly different from controls, P<0.05

controls (fig.13). A similar pattern was seen also for the thyroxine degradation rate (TDR) with the TDRs of the 5 and 60 day-old progenies of hypothyroid mothers being greatly increased over those of the control progenies (fig.14).

The metabolic clearance rates (MCR), like the TDS, were increased significantly in the progenies of the hypothyroid mothers over those of the progenies of the control mothers at all of the three ages that were studied (fig.15). The most surprising aspect of these data is the magnitude of the increased rate of peripheral T_4 metabolism which was observed in the progenies of the hypothyroid mothers. These data help to explain the low tissue levels of T_4 and T_3 and the elevated serum TSH levels required to maintain normal serum T_4 conconcentrations in the 60 day-old progenies of hypothyroid mothers.

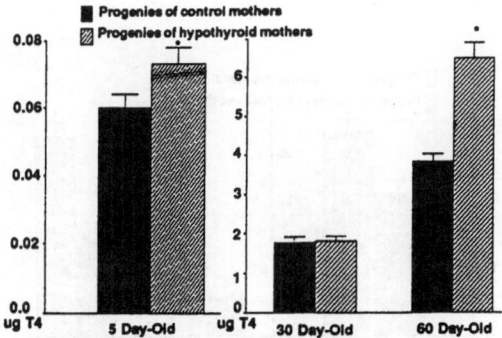

Fig. 13 Thyroxine extrathyroidal pool
* Significantly different from controls, P<0.05

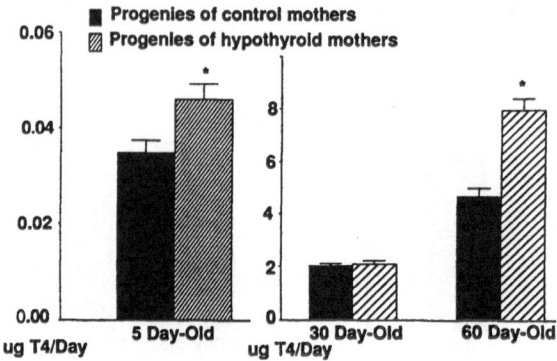

Fig. 14 Thyroxine degradation rate
* Significantly different from controls, P<0.05

HPLC MEASUREMENT OF TISSUE IODOTHYRONINES

Serum thyroid hormone levels are important but they are not the only determinant of tissue levels. It now appears that thyroid hormone transport into many, or even all cells, may involve a carrier-mediated system (Centanni et al., 1988; Docter et al., 1987; Blondeau et al., 1988; Francon et al. 1989). Furthermore, in some tissues, intracellular T_3 is primarily of extracellular origin, while in other tissues, like brain, it is predominantly synthesized in situ (Silva and Matthews, 1984). In the latter tissues, T_4 transport becomes more important than T_3 transport. It is now possible to extract thyroid hormones from very small quantities of tissues and measure minute quantities of hormones using an ultra-sensitive HPLC method. Previously, a very sensitive HPLC method had been developed for the determination of amine-containing compounds in body tissues and fluids (Wiedmeier et al., 1982). This method involves the precolumn derivatization of the amine-containing compound with dansyl chloride (5-Dimethyl-aminonaphthalene-1-sulfonyl chloride) which results in a highly fluorescent derivative. Fluorescence detection is many times more sensitive

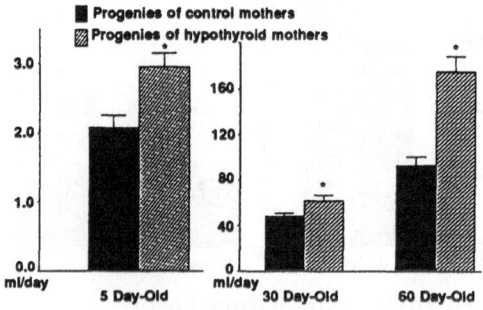

Fig. 15 Thyroxine metabolic clearance
*Significantly different from controls, P<0.05

than absorbance detection. This method was modified to derivatize iodothyronines extracted from body tissues and fluids. Resolution of the peaks is accomplished using gradient elution over a 60 minute period. Serum T_4 and T_3 concentrations measured by this method agree with those obtained by radioimmunoassay.

Iodothyronines were extracted from tissues with three washes of 80% ethanol; 100 μM phloretin; 0.02 N NaOH, pH 11.5. The supernatants were dried, taken up in 100 μM phloretin in HPLC water, pH 6.0, and ultracentrifuged. The precipitate was dissolved in 100 μM phloretin; 0.02 N NaOH, pH 11.5, ultracentrifuged and the supernatants used for HPLC analyses of T_4, T_3, rT_3 and T_2. An aliquot of the supernatant was reacted overnight with dansyl chloride (6 mg/ml in acetone) in 0.5 M $NaHCO_3$, pH 9.5. An aliquot of the dansylated sample was eluted from a reverse phase column (ultrasphere ODS, 5 μm, 25 cm x 4.6 mm) using gradient elution with a mobile phase in pump A of phosphate buffer and in pump B of acetonitrile-tetrahydrofuran. Extractions provide from 90-100% recovery, depending upon the tissue. Brain iodothyronines were measured in 13, 16, and 22 d pc fetuses as well as in postnatal progenies.

For the experiments on 13 and 16 d pc fetuses, the Tx dams were without T_4 treatment for an average of 24 days preconception so the maternal hypothyroidism was more severe than in any of the other experiments reported in this paper. At 13 d pc, the fetal brains were removed and the remainder of the body was classified as the carcass. At this age the carcass included the liver. At 16 d pc, the liver was isolated from the carcass. Brain iodothyronine levels were well within the limits of detection by 13 d pc. T_4, T_3, rT_3, and T_2 were all measurable in these fetuses. Obregon, et al., (1984) have shown by RIA that T_4 is present in whole fetuses at 14 d pc (13 d pc using their system of calculating gestational age). As this period precedes the onset of function of the fetal thyroid, these hormones must be of maternal origin. As is apparent in figure 16, tissue T_3 levels are almost as high as T_4 levels.

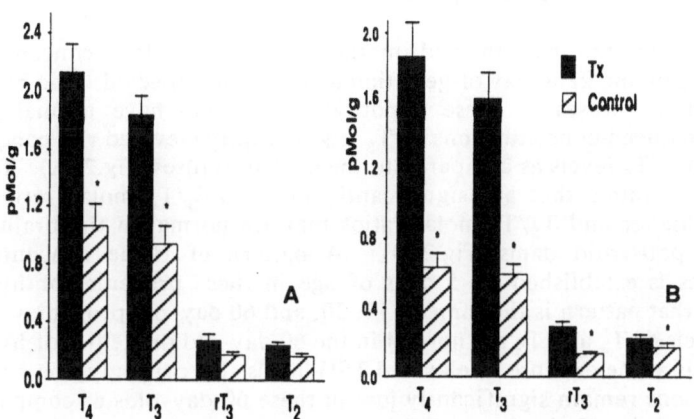

Fig. 16 Brain (A) and carcass (B) iodothyronine levels of 13 day
post-conception fetuses of control and hypothyroid (Tx) rats.
* Significantly different from controls, P<0.05

Morreale de Escobar et al., (1985) also noted high tissue T_3 levels in late-gestational fetuses. T_4 and T_3 concentrations are lower in the carcass than in the brain. These data suggest that at this age the fetal brain is selectively accumulating thyroid hormones and apparently monodeiodinating T_4 to T_3. At 13 d pc, if the mother is hypothyroid, the fetal brain T_4 concentrations are 50% and T_3 concentrations are 51% of the levels measured in control fetuses. Carcass T_4 levels are 34% and T_3 levels are 36% of the levels in controls (fig.16). These data show that maternal thyroid hormones do reach the fetus and might even be selectively accumulated in critical tissues such as the brain. At this stage in gestation, when the mother is thyroid hormone deficient, the developing fetus becomes thyroid hormone deficient. The fetuses of the Tx mothers actually have proportionately more tissue hormone than would be predicted from the maternal levels suggesting that perhaps placental hormone transport improves when maternal hormone levels drop. The ratios of T_4 to T_3 in brain and carcass are not altered by the treatment, suggesting that fetal deiodinase activity is probably not altered by the treatment at this age. At 16 d pc, brain T_4 levels were comparable to those seen at 13 d pc (fig.17). However, T_3 levels were higher. Again, fetuses of Tx mothers have lower brain T_4 and T_3 levels than fetuses of control mothers. It is feasible to separate the liver from the remaining carcass at 16 d pc, and the liver levels of T_4 and T_3 are quite high (fig.18) relative to the carcass (fig.19) and even the brain concentrations of hormone (fig.17). The brain T_4 levels of fetuses of control mothers are 1.6x greater than the T_4 levels in the carcass and the liver T_4 concentrations are 3x greater than the carcass levels. Again, fetal brain thyroid hormone levels are depressed in maternal hypothyroidism, and brain and liver appear to selectively accumulate thyroid hormones relative to the rest of the body. The T_3/T_4 molar ratios are not changed by the treatment suggesting that tissue deiodinases are probably not altered at this age as they are later in development.

In another experiment, the brain iodothyronines were extracted and measured in control and hypothyroid mothers, and their fetuses at 22 d pc and in 5, 10, 30, and 60 day-old progenies. For this experiment, the Tx rats were maintained on T_4 until day 1 of pregnancy; hence, maternal serum T_4 and T_3 levels were higher during pregnancy than in the experiment with mid-gestational fetuses.

The brains of the hypothyroid mothers have very low concentrations of iodothyronines by the 22nd day of gestation as would be expected. The brains of the 22 day-gestation fetuses of these hypothyroid mothers have normal T_4 levels, significantly reduced concentrations of T_3, significantly elevated rT_3 concentrations and reduced 3, 5 T_2 levels as compared to the control fetuses (fig.20A). This results in T_3/T_4 molar ratios that are significantly lower, rT_3/T_4 molar ratios that are significantly higher and T_2/T_3 molar ratios that are normal in the brains of these fetuses of hypothyroid dams (fig.20B). A pattern of tissue concentrations of iodothyronines is established by 5 days of age in these progenies of hypothyroid mothers, and that pattern is the same in 10, 30, and 60 day-old progenies. Although the serum levels of T_4 and T_3 are normal in the 60 day-old progenies of hypothyroid mothers, this is at the expense of elevated TSH levels. Surprisingly, the brain T_4 and T_3 concentrations remain significantly low in these 60 day-olds as compared to the controls. The brain rT_3 is elevated above the controls, and the T_2 is significantly reduced from control levels. This results in a brain T_3/T_4 molar ratio that is relatively normal, a T_2/T_3 molar ratio that is decreased from normal and a rT_3/T_4 molar ratio that is increased significantly above that of the 60 day-old control progenies (fig.21).

The 5, 10, and 30 day-old progenies of the hypothyroid mothers have brain iodothyronine profiles that are quite similar to these shown for the 60 day-old offspring. Only the absolute concentrations change with age.

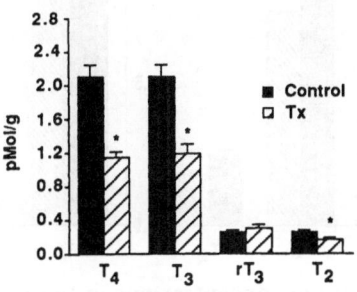

Fig. 17 Brain iodothyronine levels of 16 day post-conception fetuses of control and
      hypothyroid (Tx) rats
      * Significantly different from controls, P<0.05

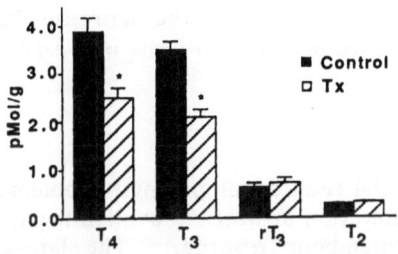

Fig. 18 Liver iodothyronine levels of 16 day post-conception fetuses of control and
      hypothyroid (Tx) rats
      * Significantly different from controls, P<0.05

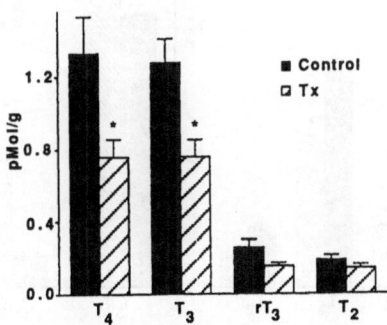

Fig. 19 Carcass iodothyronine levels of 16 day post-conception fetuses of control and
      hypothyroid (Tx) rats
      * Significantly different from controls, P<0.05

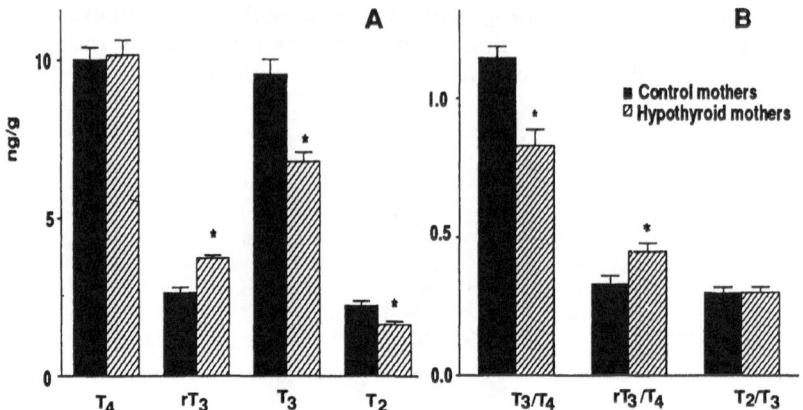

Fig. 20 A. Iodothyronine concentrations in brains of 22 day fetuses
        B. Iodothyronine molar ratios
        * Significantly different from controls, P<0.05

The low concentrations of T_3 in the brains of these offspring of hypothyroid dams may represent a primary reason for the depressed brain protein synthesis in these animals. Reduced brain protein synthesis is recognized as a cause of mental retardation.

METABOLIC STUDIES

A likely cause of the fetal developmental problems in the progenies of Tx mothers is the potential alteration of nutrient availability to the fetus resulting from the maternal endocrine/metabolic disorder. The late-gestational fetus of the hypothyroid mother has a 22% lower serum glucose level than control fetuses, even

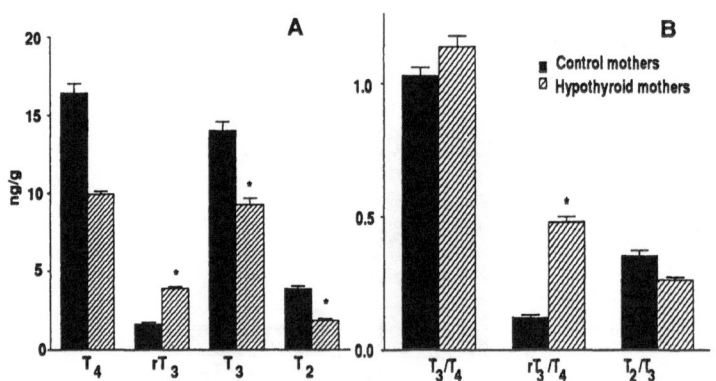

Fig. 21 A. Brain iodothyronine concentrations in 60 day-old progenies
        B. Iodothyronine molar ratios
        * Significantly different from controls, P<0.05

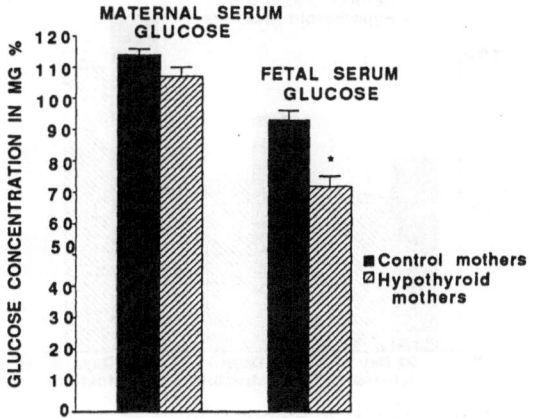

Fig. 22 Maternal and fetal serum glucose
    * Significantly different from controls, P<0.05

though maternal serum glucose levels do not differ significantly (fig.22) (Porterfield et al., 1975). Liver glycogen concentrations in the fetuses of hypothyroid dams are 52% lower than normal and brain concentrations are 46% lower than normal (fig.23). Tissue glycogen levels remain low in the perinatal progenies of Tx mothers (Porterfield et al., 1975). This deficit, coupled with the decreased liver gluconeogenic capacity (Hendrich et al., 1982) could impair the ability of the neonate to maintain serum glucose homeostasis in the period immediately post-partum.

While tissue glucose and glycogen levels are normal by 30 days post-partum in the progenies of Tx mothers, protein concentrations remain low. The inability to maintain normal protein concentrations is apparent in the fetal brain, liver, and serum at 22 d pc and the treatment effect persists to adulthood, thereby suggesting a major

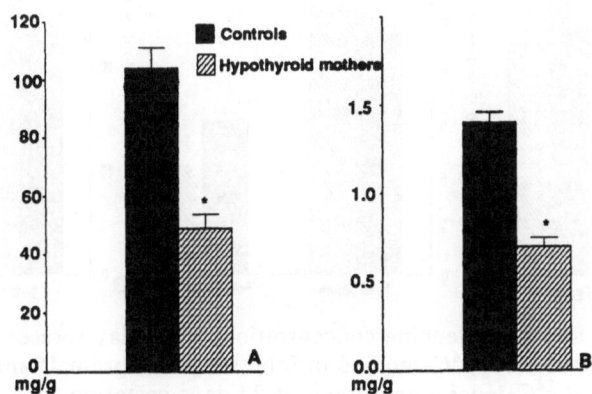

Fig. 23 A. Fetal liver glycogen
        B. Fetal brain glycogen
        * Significantly different from controls, P<0.05

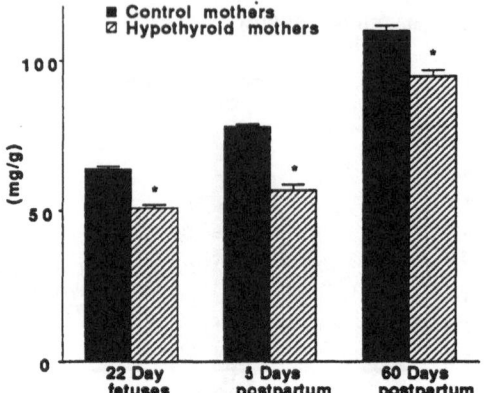

Fig. 24 Brain protein concentration
* Significantly different from controls, P<0.05

disorder in protein metabolism in these animals (Hendrich et al., 1982) (fig.24). Even though fetal serum leucine levels are normal in the 22 day fetus of Tx mothers (fig.25A), fetal brain levels are low, suggesting either impaired uptake or increased turnover. When ^{14}C-leucine is used as a tracer, both serum free ^{14}C-leucine (fig.25B) and serum leucine specific activity (fig.25C) are high in the fetuses of Tx mothers 15 minutes after maternal injection. However, brain free ^{14}C-leucine is low (fig.25B). Brain ^{14}C-leucine specific activity (fig.25C) is the same in fetuses of Tx mothers as

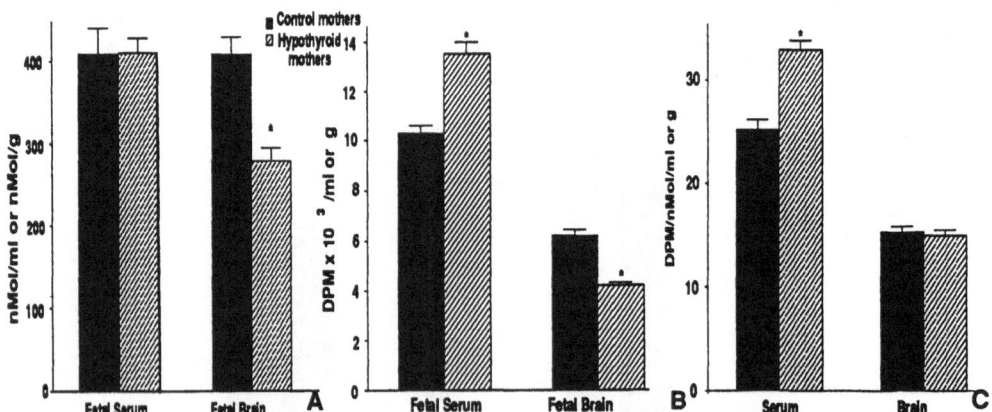

Fig. 25 A. Serum and brain leucine concentrations in 22 day fetuses
B. Extractable free ^{14}C-leucine in fetal serum or brain 15 minutes after maternal ^{14}C-leucine injections at 22 days gestation
C. Specific activity of free leucine pools in serum and brain of 22 day fetuses 15 minutes after ^{14}C-leucine injection
* Significantly different from controls, P<0.05

in control fetuses, reflecting the very low brain leucine concentration in these fetuses (fig.25A). These data suggest that there are problems with the uptake of leucine into the brain of these fetuses. Similar results were seen with ^{14}C-alanine (Hendrich, et al., 1982) and ^{14}C-glutamate. When brain and serum amino acid levels were measured in these fetuses by HPLC (Wiedmeier et al., 1982), it was found that while there were not treatment effects on any of the fetal serum amino acids (Table II), all measured amino acids were lower in the brains of fetuses of Tx mothers than in the brains of control fetuses (Table III). While the low brain amino acid levels were apparent through 5 days postpartum, most brain amino acid levels were normal by 30 days. However, deficiencies in amino acid incorporation into protein persisted.

RIBOSOMAL PROTEIN SYNTHESIS:

Shashoua and co-workers (1972, 1976, 1977, 1979, 1981, 1982) have demonstrated in a somewhat unique animal model, the goldfish, that there are specific changes in brain protein synthesis associated with learning and memory. Continual brain protein synthesis and turnover are essential for these behavioral processes of learning and memory.

The initial metabolic studies conducted on this hypothyroid maternal rat model indicated that brain protein concentrations per gram of tissue are decreased significantly in the fetuses and progenies of these hypothyroid rats as compared to the offspring of control mothers. In vivo studies utilizing this animal model showed that there were deficiencies in the rate of transport of amino acids into brain and liver of

Table II   Serum Amino Acid Levels of 22 Day-old Fetuses:
Treatments Refer to Mothers (nMoles/ml)

Amino Acid	A Control	B Tx	Significant Differences
Taurine	409±20	456±20	N.S.
Asparagine	105±8	114±12	N.S.
Glutamine	433±27	423±20	N.S.
Serine	452±41	432±37	N.S.
Aspartic Acid	92±10	86±9	N.S.
Hydroxyproline	142±11	109±7	N.S.
Glutamic Acid	620±41	576±30	N.S.
Threonine	305±20	299±10	N.S.
Glycine	351±13	320±15	N.S.
Alanine	1100±63	997±47	N.S.
Arginine	285±20	280±15	N.S.
Methionine	128±9	126±15	N.S.
Proline	477±18	440±21	N.S.
Valine	524±39	443±20	N.S.
Phenylalanine	258±14	315±19	N.S.
Tryptophan	184±14	166±5	N.S.
Leucine	409±31	410±18	N.S.
Isoleucine	247±13	261±12	N.S.
Lysine	1101±85	1012±85	N.S.
Tyrosine	331±13	306±15	N.S.
Histidine	144±12	134±7	N.S.

Mean ± S.E.

Table III   Brain Amino Acid Levels of 22 Day-old Fetuses:
Treatments Refer to Mothers (nMoles/g)

Amino Acid	A Control	B Tx	Significant Differences
Taurine	17.26±.65	11.64±.39	B<A
Asparagine	.45±.038	.312±.020	B<A
Glutamine	3.55±.097	2.81±.123	B<A
Serine	2.19±.042	1.55±.062	B<A
Aspartic Acid	1.72±.092	1.20±.071	B<A
Glutamic Acid	4.93±.173	3.07±.151	B<A
Threonine	.858±.050	.583±.055	B<A
Glycine	1.65±.050	.862±.048	B<A
Alanine	3.08±.141	2.15±.103	B<A
Arginine	.776±.035	.582±.031	B<A
Methionine	.120±.006	.073±.007	B<A
Proline	.522±.037	.305±.026	B<A
Valine	.503±.025	.295±.020	B<A
Phenylalanine	.259±.015	.178±.011	B<A
Tryptophan	.148±.013	.085±.008	B<A
Leucine	.408±.021	.278±.016	B<A
Isoleucine	.220±.015	.151±.010	B<A
Lysine	.754±.051	.420±.017	B<A
Tyrosine	.521±.028	.335±.022	B<A
Histidine	.415±.022	.277±.013	B<A

Mean ± S.E. $P<.05$

the fetuses and progenies of hypothyroid mothers (Hendrich and Porterfield 1980; Hendrich et al, 1982). In order to determine if the deficient protein synthesis in the offspring of hypothyroid mothers was due primarily to the decrease in amino acid transport or if other mechanisms were involved, studies were conducted to determine the rates of ribosomal amino acid incorporation into protein in a cell-free system. Brain ribosomes from control and hypothyroid rats and their fetuses and progenies were isolated, the ribosomal profiles were compared, the rRNA was determined and the ribosomes were incubated with ^{14}C-leucine. The methods utilized were based on those of Dreskin and Kostyo (1980) but were modified for small amounts of tissue and the incubation media contained amino acids in the concentrations that had been measured previously in the brains of control mothers and their offspring. These data are presented in figures 26 and 27. The brain rRNA, and the ^{14}C-leucine incorporation into brain ribosomal protein are depressed significantly in the 22 day-gestation fetuses of hypothyroid mothers as compared to controls. The polyribosomal fraction of the brain ribosomal profiles of these animals is likewise reduced significantly below that of the controls. Other dams were allowed to go through parturition and representative progenies were utilized for similar studies at 5, 30, and 60 days of age. The data obtained on brain ribosomes from 5 and 30 day-old progenies were similar to that of the 22 day-gestation fetuses and are therefore not shown. The adult 60 day-old progenies, from hypothyroid mothers had normal levels of brain rRNA. However, the brain ribosomal profiles of these offspring contained only 44% of the amount of polyribosomes as did those of the 60 day-old control progenies. Whether the ribosomal incorporation of ^{14}C-leucine is expressed as DPM per gram of brain or DPM per μg of rRNA per gram of brain, the brain ribosomal incorporation of ^{14}C-leucine into protein continues to be decreased significantly in the 60 day-old progenies of hypothyroid mothers (fig.27). It must therefore be concluded, that the derangement of protein synthesis in the progenies of hypothyroid mothers that is initiated

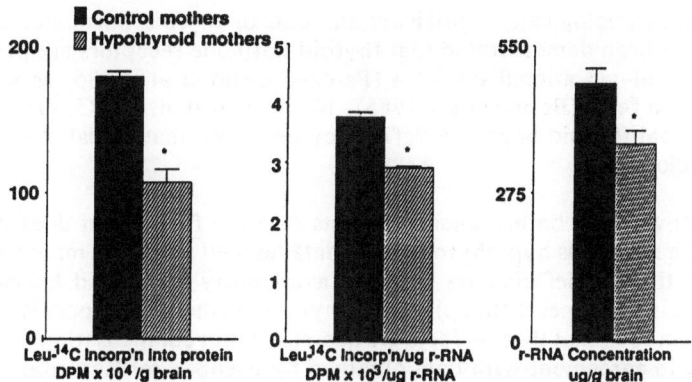

Fig. 26 Brain ribosomal protein synthesis in 22 day fetuses
* Significantly different from controls, P<0.05

during fetal life is permanent. As protein synthesis is, in part, thyroid hormone dependent and as it has been demonstrated that brain T_4 and T_3 levels in these progenies of hypothyroid dams are reduced significantly below normal, this intracellular hormone deficiency represents a probable cause for the alterations of protein synthesis observed in these progenies.

CONCLUSIONS

For some time, placental transport of thyroid hormones was considered to be so minimal that it was not physiologically significant. Hence, fetal thyroid function was considered to occur autonomously of maternal thyroid function. In addition, fetal development, prior to the age of fetal thyroid hormone synthesis, was thought to proceed independently of any need for thyroid hormones. However, the works of Vulsma, et al. (1989), Narayanan, et al. (1985), Morrealle de Escobar et al., (1985), Obregon, et al., (1984) and Porterfield (1985) have shown that thyroid hormones do cross the placenta in both humans and rats. The data presented in the present paper

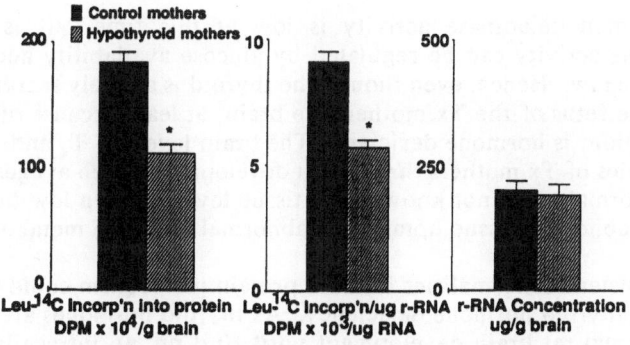

Fig. 27 Brain ribosomal protein synthesis in 60 day-old progenies
* Significantly different from controls, P<0.05

also show that physiologically significant amounts of thyroid hormones cross the rat placenta. It has been demonstrated that thyroid hormone receptors are present in the tissues of the mid-gestational rat fetus (Perez-Castillo et al., 1985) as well as the 7 week-old human fetus (Bernal et al., 1985). Narayanan et al., (1985) have shown that a mid-gestational thyroid hormone deficiency produces anatomical abnormalities in rat brain development.

It is obvious in both humans and rats that the fetal brain does not develop normally if the mother is hypothyroid. Our data, as well as that of many others, show that maternal thyroid deficiencies alter the availability of thyroid hormones to the developing brain. It appears that placental thyroid hormone transport is improved in hypothyroidism such that these effects are decreased, but compensation is only partial. Furthermore, in agreement with data obtained by Escobar del Rey et al., (1986), the fetal brain appears to selectively accumulate thyroid hormones. The period in mid-gestation in which the fetal thyroid hormones are entirely of maternal origin corresponds with the period of neurogenesis and cellular migration in the cerebral cortex.

Cerebellar neurogenesis and cellular migration occur later in development and are known to be dependent upon normal thyroid hormone levels. Perhaps cerebral development is also thyroid hormone dependent. Potter has shown that cerebral neurogenesis is impaired in iodine-deficient sheep fetuses (Potter et al., 1982, 1981). Many of the neurological problems reported for the inhabitants of severe endemic goiter areas are deficits associated with early and mid-gestational brain development. He does note that after fetal thyroid function begins, cerebral catch-up growth occurs.

Similar results were also obtained in the fetuses of hypothyroid sheep where brain growth was stunted until the age when autonomous fetal thyroid function began and then catch-up growth occurred (Potter et al., 1986). If the fetuses were also Tx, catch-up growth never occurred.

Our results show that once fetal thyroid function begins, it is not normal in the fetuses of Tx rats. Serum RIA's demonstrated that the fetuses become hyperthyroxinemic just prior to birth. However, brain T_4 levels are normal and T_3 levels are low at 22 d pc.

Brain 5'monodeiodinase activity is low at this time. It is known that 5'monodeiodinase activity can be regulated by glucose availability and fetal serum glucose levels are low. Hence, even though the thyroid is actively secreting hormone by 22 d pc in the fetus of the Tx mother, the brain, at least because of impaired T_4 5'monodeiodination, is hormone deficient. The brain levels of T_4 and T_3 are low in postnatal progenies of Tx mothers throughout development-even at ages when serum T_4 and T_3 are normal. It is not known why tissue levels remain low but it could be a result of both abnormal tissue uptake and abnormal hormone metabolism.

The persistent abnormalities in brain protein metabolism could be a result of the intracellular thyroid hormone deficiency. As thyroid hormones are known to be essential for normal rat brain development until 10 d pn, an intracellular hormone deficiency could prevent normal postnatal development. Furthermore, normal protein metabolism is necessary for normal neurological development and function and the progenies of Tx mothers never show normal brain amino acid uptake and protein synthesis. Both of these processes are thyroid hormone regulated in the brain (Pickard et al., 1987). While all of the causes of impaired neurological development in the progenies of Tx mothers are not known, it does appear that the lack of normal availability of thyroid hormones and the lack of normal nutrient availability in utero are probable causes. Furthermore, the subsequent postnatal persistence of defects in thyroid hormone metabolism, tissue thyroid hormone availability and tissue amino acid uptake and utilization would likely produce the observed behavioral deficiencies seen in the progenies of hypothyroid mothers.

ACKNOWLEDGEMENTS

The authors would like to thank Drs. V.T. Weidmeier, W.J. Jackson and R.C. Little for their valuable contributions to this project.

REFERENCES

Bakke, J.L., Lawrence, N.L., Robinson, S., Bennett, J., 1975. Endocrine studies of the untreated progeny of thyroidectomized rats, Pediat. Res., 9:742-748.

Bakke, J.L., Lawrence, N.L., Robinson, S., Bennett, J., 1977. Endocrine studies in the untreated F_1 and F_2 progeny of rats treated neonatally with thyroxine, Biol. Neonate, 31:71-83.

Bass, N.H., Pelton, E.W., Young, E., 1977. Defective maturation of cerebral cortex: an inevitable consequence of dysthyroid states during early postnatal life, in: "Thyroid Hormones and Brain Development," G.D. Grave, ed., pp. 199-214.

Bernal, J., Liewendahl, K., Lamberg, B.A., 1985. Thyroid hormone receptors in fetal and hormone resistant tissues, Scand. of Clin Lab. invest., 45:577-583.

Blondeau, J.P., Osty, J., Francon, J., 1988. Characterization of the thyroid hormone transport system of isolated hepatocytes, J. Biol. Chem., 263:2685-2692.

Centanni, M., Pontecorvi, A., Robbins, J., 1988. Insulin effect on thyroid hormone uptake in skeletal muscle, Metabolism, 37:626-630.

Docter, R., Krenning, E.P., Bernard, H.F., Hennemann, G., 1987. Active transport of iodothyronines into human cultured fibroblasts, J. Clin. Endocrinol. Metab., 65:624-628.

Dreskin, S.C., Kostyo, J.L., 1980. Acute effects of growth hormone on the function of ribosomes of rat skeletal muscle, Horm. Metab. Res., 12:60-66.

Dussault, J.H., Ruel, J., 1987. Thyroid hormones and brain development, Ann. Rev. Physiol., 49:321-334.

Eayrs, J.T., 1955. The cerebral cortex of normal and hypothyroid rats, Acta Anat., 25:160-183.

Escobar del Rey, F., Pastor, R., Mallot, J., Morreale de Escobar, G., 1986. Effects of maternal iodine deficiency on the L-thyroxine and 3,5,3'-triiodo-L-thyronine contents of rat embryonic tissues before and after onset of fetal thyroid function, Endocrinology, 118:1259-1265.

Fierro-Benitez, R., Ramirez, I., Garces, J., Jaramillo, C., 1974. The clinical pattern of cretinism as seen in highland Ecuador, Amer. J. Clin. Nutr., 27:531-543.

Fierro-Benitez, R., Ramirez, I., Suarez, J., 1972. Effect of iodine correction early in fetal life on intelligence quotient. A preliminary report, in: "Human Development and the Thyroid Gland," J.B. Stanbury, R.L. Kroc, eds., Plenum Press, N.Y., pp. 239-247.

Fisher, D.A., Dussault, J.H., Sack, J., Chopra, I.J., 1977. Ontogenesis of hypothalam ic-pituitary-thyroid function and metabolism in man, sheep and rat, Rec. Prog. Horm. Res, 33:59-116.

Francon, J., Chantoux, F., Blondeau, J.P., 1989. Carrier-mediated transport of thyroid hormones into rat glial cells in primary culture, J. Neurochem., 53:1456-1 463.

Hendrich, C.E., Jackson, W.J., Porterfield, S.P., 1984. Behavioral testing of progenies of Tx (hypothyroid) and growth hormone-treated Tx rats: An animal model for mental retardation, Neuroendocrinology, 38:429-437.

Hendrich, C.E., Wiedmeier, V.T., Porterfield, S.P., 1982. Utilization of alanine by hypothyroid and growth hormone treated hypothyroid rats, their fetuses and progeny, Horm. Metabol. Res., 14:658-666.

Hetzel, B.S., 1983. Iodine deficiency disorders (IDD) and their eradication, Lancet, 2.3:1126-1129.

Ingenbleek, Y., Malvaux, P., 1980. Peripheral turnover of thyroxine and related parameters in infant protein-caloric malnutrition, Am. J. Clin. Nutri., 33:606-616.

Jones, W.S., Man, E.B., 1969. Thyroid function in human pregnancy. IV. Premature deliveries and reproductive failures of pregnant women with low serum butanol-ex tractable iodines, Maternal serum TBG and TBPA capacities, Amer J. Obstet. Gynec., 104:909-914.

Karlsson, I., Svennerholm, L., 1978. Biochemical development of rat forebrains in severe protein and essential fatty acid deficiencies, J. Neurochem., 31:657-662.

Legrand, J., 1982-1983. Hormones thyroidiennes et maturation du systeme nerveux, J. Physiol. Paris, 78:603-652.

Man, E.B., Holden, R.H., Jones, W.S., 1971. Thyroid function in human pregnancy. VII. Development and retardation of 4-year-old progeny of euthyroid and of hypothyroxinemic women, Amer. J. Obstet. Gynec., 109:12-18.

Man, E.B., Jones, W.S., 1969. Thyroid function in human pregnancy. V. Incidence of maternal serum low butanol-extractable iodines and of normal gestational TBG and TBPA capacities; retardation of 8-month-old infants, Amer. J. Obstet. Gynec., 104:898-908.

Morreale de Escobar, G., Escobar del Rey, F., Ruiz-Marcos, A., 1983. Thyroid hormone and the developing brain, in: "Congenital Hypothyroidism", J.H. Dussault, P. Walker, eds. Marcel Dekker, Inc., N.Y., pp. 85-125.

Morreale de Escobar, G., Pastor, R., Obregon, M.J., Escobar del Rey, F., 1985. Effects of maternal hypothyroidism on the weight and thyroid hormone content of rat embryonic tissues, before and after onset of fetal thyroid function, Endocrinolo gy, 117:1890-1900.

Munn, N.W., 1950. The role of sensory processes in maze behavior, in: "Handbook of Psychological Research on the Rat, pp. 213 ff, Houghton-Miffin, Boston.

Narayanan, C.H., Narayanan, 1985. Cell formation in the motor nucleus and mesencephalic nucleus of the trigeminal nerve of rats made hypothyroid by propylthiouracil, Exp. Brain Res., 59:257-266.

Nelson, K.B., Ellenburg, J.H., 1986. Anticedents of cerebral palsy, New Eng. J. Med., 315:81-86.

Obregon, M.J., Mallol, J., Pastor, R., Morreale de Escobar, G., Escobar del Rey, F., 1984. L-thyroxine and 3,5,3'-triiodo-L-thyronine in rat embryos before onset of fetal thyroid function, Endocrinology, 114:305-307.

Perez-Castillo, A., Bernal, J., Ferreiro, B., Pans, T., 1985. The early ontogenesis of thyroid hormone receptor in the rat fetus, Endocrinology, 117:2457-2461.

Pharoah, P., Connolly, K., Hetzel, B., Ekins, R., 1981. Maternal thyroid function and motor competence in the child, Develop. Med. Child. Neurol., 23:76-82.

Pickard, M.R., Sinha, A.K., Gullo, D., Patel, N., Hubank, M., Ekins, R.P., 1987. The effect of 3,5,3'-triiodothyronine on leucine uptake and incorporation into protein in cultured neurons and subcellular fractions of rat central nervous system, Endocrinology, 121:2018-2026.

Porterfield, S.P., 1985. Prenatal exposure of the fetal rat to excessive L-thyroxine or 3,5-dimethyl-3'-isopropyl-thyronine produces persistent changes in the thyroid control system, Horm. Metab. Res., 17:655-659.

Porterfield, S.P., Hendrich, C.E., 1982. Brain and liver deoxyribonucleic acid and ribonucleic acid in the progeny of hypothyroid and growth hormone-treated hypothyroid rats, Endocrinology, 111:406-411.

Porterfield, S.P., Whittle, E., Hendrich, C.E., 1975. Hypoglycemia and glycogen deficits in fetuses of hypothyroid pregnant rats, Proc. Soc. Expt'l Biol. Med. 149:748-753.

Potter, B.J., Mano, M.T., Belling, G.B., McIntosh, G.H., Hua, C., Cragg, B.G., Marshall, J., Wellby, M.L., Hetzel, B.S., 1982. Retarded fetal brain development resulting from severe dietary iodine deficiency in sheep, Neuropathol. Appl. Neurobiol., 8:303-313.

Potter, B.J., McIntosh, G.H., Hetzel, B.S., 1981. The effect of iodine deficiency on fetal brain development in the sheep. In: B.S. Hetzel, R.M. Smith (eds), "Fetal Brain Disorders-Recent Approaches to the Problem of Mental Deficiency, pp. 119-148, Elsevier North Holland, Amsterdam.

Potter, B.J., McIntosh, G.H., Mano, M.T., Baghurst, P.A., Chavadej, J., Hua, C.H., Cragg, B.G., Hetzel, B.S., 1986. The effect of maternal thyroidectomy prior to conception on foetal brain development in sheep, Acta Endocrinol., 112:93-99.

Shashoua, V.E., 1972. A multistage transduction model for informati on processing in the nervous system, Intl. J. Neurosci., 3:299-304.

Shashoua, V.E., 1976. Brain metabolism and the acquisition of new behaviors. I. Evidence for specific changes in the pattern of protein synthesis, Brain Res., 111:347-367.

Shashoua, V.E., 1977. Brain metabolism and the acquisition of new behaviors. II. Immunological studies of the α, β, γ proteins of gold fish brain, Brain Res., 122:113-124.

Shashoua, V.E., 1979. Brain metabolism and the acquisition of new behaviors. III. Evidence for secretion of two proteins into the brain extracellular fluid after training, Brain Res., 166:349-358.

Shashoua, V.E., 1981. Extracellular fluid proteins of goldfish brain: Studies of concentration and labeling patterns, Neurochem. Res., 6: (10) 1129-1147.

Shashoua, V.E., 1982. Molecular and cell biological aspects of learning: Towards a theory of memory, Adv. Cellular Neurobiol., 3:97-141.

Shashoua, V.E., 1982. The role of extracellular proteins in learning and memory, Am. Sci., 73:364-369.

Shashoua, V.E., Moore, M.E., 1978. Effects of antisera to β and α goldfish brain proteins on the retention of a newly acquired behavior, Brain Res., 154:441-449.

Silva, J.E., Matthews, P.S., 1984. Production rates and turnover of triiodothyionine in rat-developing cerebral cortex and cerebellum. Responses to hypothyroidism. J. Clin. Invest., 74:1035-1049.

Strong, P.N., Jr., 1957. Activity in the white rat as a function of apparatus and hunger, J. Comp. Physiol. Psychol., 50:596-600.

Van Middlesworth, L., Norris, C.H., 1980. Audiogenic seizures and cochlear damage in rats after perinatal antithyroid treatment, Endocrinology, 106:1680.

Vulsma, T., Gons, M.H., DeVijlder, J.J.M., 1989. Maternal-fetal transfer of thyroxine in congenital hypothyroidism due to a total organification defect or thyroid agenesis, N. Engl. J. Med., 321:13-16.

Wiedmeier, V.T., Porterfield, S.P., Hendrich, C.E., 1982. Quantitation of Dns-amino acids from body tissues and fluids using high-performance liquid chromatography, J. Chromatography, 231:410-417.

Zamenof, S., Van Marthens, E., 1971. Hormonal and nutritional aspects of prenatal brain development, In: "Cellular Aspects of Neural Growth and Differentiation," D.C. Pease, ed., U. of California Press, Berkley, p.329, 1971.

MATERNAL THYROID HORMONES DURING PREGNANCY: EFFECTS ON THE FETUS IN CONGENITAL HYPOTHYROIDISM AND IN IODINE DEFICIENCY

Gabriella Morreale de Escobar, María Jesus Obregón, Rosa Calvo and Francisco Escobar del Rey

Unidad de Endocrinología Molecular, Instituto de Investigaciones Biomédicas, Consejo Superior de Investigaciones Científicas y Facultad de Medicina UAM, Arzobispo Morcillo 4, 28029-Madrid (Spain)

Introduction

For years it has been generally accepted that the mammalian placenta is virtually impermeable to the natural iodothyronines, thyroxine (T4) and 3,5,3' triiodothyronine (T3). Although some placental transfer might occur, it would be physiologically irrelevant, with the pituitary-thyroid system being already functional and autonomous before birth [1-3]. In most babies with congenital hypothyroidism, prompt onset of treatment with thyroid hormones soon after birth prevents most manifestations of brain damage. From this observation, and the idea that no thyroid hormone had been available to the fetus with an impaired thyroid, the conclusion has been drawn that thyroid hormones are not necessary for normal human brain development until birth [3, 4].

On the other hand, such notions were not easily reconciled with the observation that in areas of severe iodine deficiency neurological cretins are born. Their central nervous system (CNS) damage is irreversible by birth and much more severe than observed in congenitally hypothyroid babies [5-7]. The ideas summarized above also posed serious difficulties to the understanding of the developmental and behavioral defects described for the progeny of women who were hypothyroxinemic during pregnancy [8], or of thyroidec-tomized [9] or hypothyroxinemic [10] rats.

A survey of earlier literature (appearing in refs [11-17]) did not, in our opinion, support the idea that maternal thyroid hormones do not reach the mammalian fetus. This prompted us more than seven years ago, to re-investigate many aspects of maternal-fetal thyroid hormone interrelationships in several experimental rat models, using highly sensitive RIAs for the determination of T4 and T3 in tissue extracts as the basic methodology. The different studies we have performed have tried to answer several questions:

Advances in Perinatal Thyroidology, Edited by B.B. Bercu and D.I. Shulman, Plenum Press, New York, 1991

1) Are thyroid hormones present in embryonic tissues before onset of
fetal thyroid function (FTF)?
2) If so, are they of maternal origin?
3) Does maternal transfer continue after onset of FTF? If so, what happens
    i) when maternal thyroid function is impaired,
    ii) when FTF is impaired, and
    iii) when both are impaired, as is the case of severe iodine deficiency?
4) Are the findings in the rat relevant for human maternal-fetal thyroid
hormone relationships.

We will mostly use data from our laboratory. Much excellent work has been
carried out in other species, especially in sheep, and has been reviewed [1-3,18,19]
by others

Thyroid hormone economy in normal embryos and fetuses, before and after onset of FTF.

Normal Wistar female rats were mated with normal males, and the day of
appearance of the vaginal plug was taken as =0 days of gestation (dg). With this
timing, onset of FTF (considering as such the onset of *secretion* of thyroid
hormones by the fetal thyroid) would take place at 17.5-18 dg [20], with birth at
22.5 dg. Embryo-trophoblasts, embryos, placentas and amniotic fluid were
obtained before and after 17 dg, extracted, purified by paper chromatography and
submitted to specific RIAs. Typical results[11] are illustrated in Figure 1: serial
dilutions of all extracts contained compounds which had co-chromatographed with
labeled T4 and T3 tracers and which behaved as the T4 and T3 standards in the
respective RIAs. Such results were confirmed repeatedly [12, 22] with a different
extraction and purification procedure which we later developed and used in
further studies [14, 15, 23-29]. The concentrations of both T4 and T3 were
relatively constant from the earliest age studied, namely from 9 dg, until 18 dg.
After this they increased suddenly, confirming onset of FTF at 17.5-18 dg.

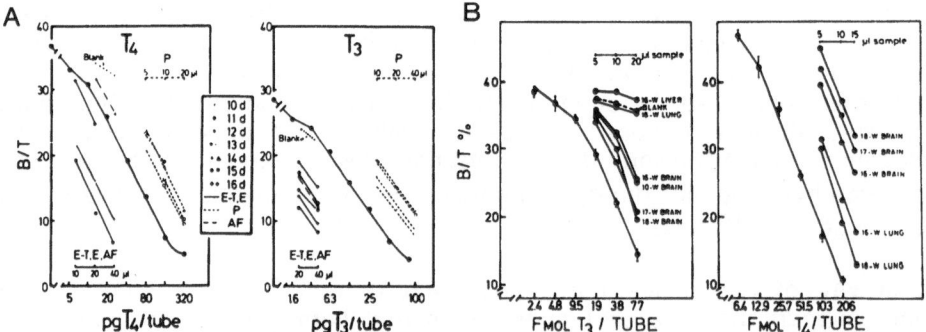

Figure 1 *A and B. A compares the T3 and T4 standard curves with serial
dilutions of eluates obtained from embryo-trophoblasts (E-T), embryonic (E),
placental (P) and amniotic fluid (AF) samples obtained from rat dams at different
gestational ages, shown in the inset. B shows similar curves with extracts from
human fetal brain, between 10 to 16 wks of gestation. From Obregón et al.*[11]*,and
Bernal and Pekonen* [21]*; reproduced with permission .*

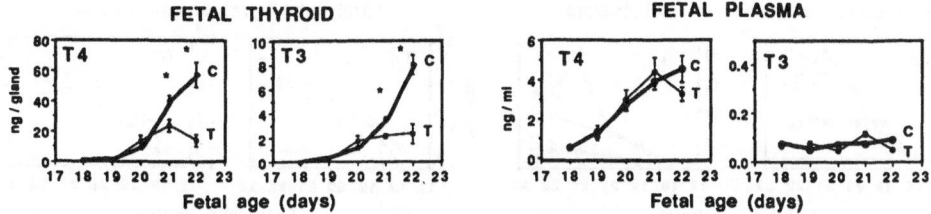

Figure 2 *Total T4 and T3 contents of proteolytic digests of thyroid glands from 18-22 dg-old fetuses from normal (C) and from thyroidectomized (T) dams (left-hand panels), as well as circulating concentrations of both iodothyronines in fetal plasma (right-hand panels). Asterisks identify statistically significant differences between values for fetuses from T versus C dams. Data are from Ruiz de Oña et al. [29].Reproduced with permission of the publishers.*

The changes in thyroid hormone economy of rat fetuses after onset of FTF have been studied in detail up to 22 dg and have been reported elsewhere [29]. The total amounts of T4 and T3 in the fetal thyroid gland (after proteolytic digestion) increase about 100-fold and 400-fold, respectively, from almost undetectable levels (0.66 ng T4 / gland and 0.021 ng T3 / gland) to 57.7 ng T4 / gland and 8.07 ng T3 / gland, as illustrated in Figure 2. During this period there is a steady increase (approximately 9-fold) in circulating T4 levels (from 0.53 ng / ml to 4.55 ng / ml). Although 10-fold lower than the plasma T4 of non-pregnant adult rats, this concentration is only 3-fold lower than that of their mothers, as rat dams late in gestation have very low circulating and tissue levels of T4 and T3, brain T3 excepted [27]. T3 in fetal plasma was very low between 18 and 22 dg. Levels of T3 fluctuated betweeen 0.05 and 0.09 ng / ml, with the increase being less than two-fold (Figure 2).

As illustrated in Figure 3, the concentration of T4 increased in all fetal tissues studied, namely brain, liver, lung, and carcass # showing patterns similar to fetal plasma T4. In contrast, the patterns of changes of T3 in fetal tissues were different from the pattern for plasma T3, and for different tissues, as also illustrated in Figure 3: The concentration of T3 increased almost 18-fold [24] in the fetal brain , despite the very small increase in the circulating T3. Brain T3 levels actually followed a pattern resembling that of fetal plasma T4 and fetal brain T4 much more than that of plasma T3. The increase in the concentration of T3 in other tissues [29] is also greater than might be expected from the changes in plasma T3, except for the liver, where the increase was only three-fold. The concentration of T4 and T3 are also very high in fetal BAT [25]. During this developmental period 5' iodothyronine deiodinase (5' D) activities increase in the brain (type II, 6-fold), lung (type I, 10-fold) and liver (type I, 5-fold), and are also very high in BAT [24, 25, 29]. It is likely that the role of these activities as determinants of intracellular T3 concentrations is greater during fetal than during adult life, especially in the fetal brain.

#*Carcass corresponds to the total fetus minus the thyroid, blood, liver, brain, lung, brown adipose tissue (BAT) interscapular pads.*

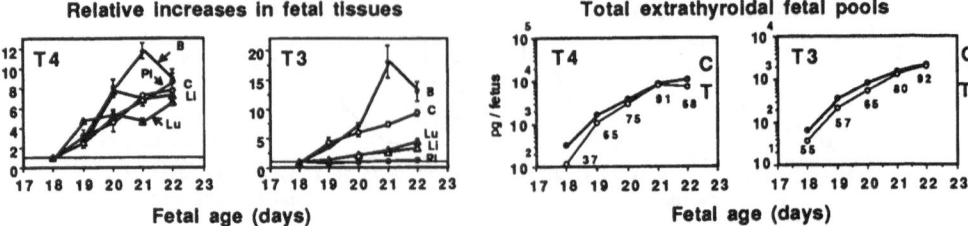

Figure 3 Left-hand panels: Relative increase of T4 and T3 concentrations in plasma and tissues of normal fetuses between 18 and 22 dg. Values were normalized for each tissue taking the concenration at 18 dg as = 1 (horizontal dotted line). Pl=plasma; C=carcass; Lu=lung; Li=liver and B=brain. From Ruiz de Oña et al. [29], with permission of the publishers.

Figure 4 Right-hand panels: Total extrathyroidal T4 and T3 pools in fetuses from normal (C) and thyroidectomized (T) dams, represented on a logarithmic scale versus fetal age. Numbers below the symbols represent the percentage of T4 or T3 in fetuses from T dams with respect to those from C mothers. From Ruiz de Oña et al. [29], with permission of the publishers.

The total fetal extrathyroidal T4 and T3 pools increase approximately 34-fold between 18 and 22 dg: from 0.32 to 11.08 ng T4 and from 0.067 to 2.260 ng T3 (Figure 4). The concentrations reached in different fetal tissues near term, namely at 21 dg, are shown in Table 1, and compared to those in their mothers and in non-pregnant adult rats. The data show that fetuses near term are not as thyroid hormone-deficient as was previously thought from the circulating levels. The concentration of T4 in the fetal brain, for instance, is as high as that of their mothers, and the concentration of T3 is 74 %. Part of this T3 is in the nucleus, where receptor occupancy is almost 40 %, as high as 30 days after birth [37].

Maternal versus fetal contributions to fetal thyroid hormone economy.

How much of the T4 and T3 found in the conceptus is derived from the mother, how much from the fetal thyroid?

<u>Before onset of FTF</u>

Contribution of T4 and T3 by the fetal thyroid before onset of FTF is obviously negligible. Thus, T4 and T3 found in embryonic tissues before 17.5-18 dg are likely to be of maternal (or placental) origin. This has been investigated in several laboratories using different methodological approaches. We have studied the effects of maternal thyroidectomy on the concentrations of T4 and T3 in embryonic and fetal tissues, both before and after onset of FTF. Main results are illustrated in Figure 5: when the embryo-trophoblast, embryos or placentas were

Table 1 Comparison of the thyroidal T4 and T3 contents (ng / gland), and of thyroid hormone concentrations (ng / g or ml) in different tissues from fetuses and dams at 21 dg, and from non-pregnant adult rats.

	FETUS§	DAM§	ADULT§§	FETAL value as % of: DAM	ADULT
T4:					
Thyroid†	37.93 ± 1.05 (4)	- - -	2982 ± 142 (2)	- - -	1.2 %
Plasma	4.00 ± 0.23 (6)	11.11 ± 1.43 (8)	33.10 ± 5.75 (5)	36 %	12 %
Liver	2.87 ± 0.20 (8)	19.59 ± 0.69 (8)	29.90 ± 7.00 (5)	15 %	9 %
Lung	1.78 ± 0.21 (7)	4.10 ± 0.21 (4)	7.66 ± 0.07 (3)	43 %	23 %
Brain	1.19 ± 0.13 (7)	1.16 ± 0.11 (8)	2.04 ± 0.28 (6)	102 %	58 %
BAT	1.74 ± 0.08 (4)	- - -	3.44 ± 0.95 (2)	- - -	50 %
EXT.¥	2.13 ± 0.10 (8)	2.74	5.47	78 %	40 %
T3:					
Thyroid†	2.37 ± 0.39 (4)	- - -	258 ± 11 (2)	- - -	0.9 %
Plasma	0.08 ± 0.01 (4)	0.31 ± 0.05 (6)	0.62 ± 0.07 (6)	25 %	13 %
Liver	0.35 ± 0.03 (6)	2.80 ± 0.30 (6)	3.27 ± 0.39 (6)	13 %	11 %
Lung	0.44 ± 0.03 (6)	1.61 ± 0.45 (3)	1.75 ± 0.22 (4)	28 %	25 %
Brain	0.94 ± 0.14 (5)	1.27 ± 0.10 (6)	1.61 ± 0.20 (7)	74 %	58 %
BAT	2.45 ± 0.45 (4)	- - -	1.72 ± 0.70 (2)	- - -	142 %
EXT.¥	0.45 ± 0.04 (5)	0.40	0.75	113 %	60 %

§ Data shown for fetuses and dams at 21 dg are means of data obtained over the last 8 years in studies reported in Calvo et al., ms in preparation and in 12, 14, 15, 22-26, 28, 29. The SEM was calculated from the mean values obtained in each experiment. The number in brackets corresponds to the number of experimental means involved.

§§ Data for adults have been calculated from results reported in references 27, 30-33 and in Obregón et al., ms in preparation.

† Thyroid data correspond to proteolytic digests of the gland, representing the hormonal contents of thyroglobulin.

¥ EXT.= Total extrathyroidal concentration of T4 or T3, that is, the total pool referred to 1 g. For the fetus: The total extrathyroidal pools of T4 and T3 were obtained by direct determinations as described 28 and divided by the mean body weight of the fetus. For the dams: Calculated from the T4 and T3 distribution volumes reported for rats in late pregnancy by Dussault and Coulombe 34 and by Dubois et al 35, and the T4 and T3 plasma levels given above. For adults: Calculated from data reported by DiStefano 36. Both for these animals and for the dams the values correspond only to plasma-exchangeable pools of T4 and T3, and thus the value for T3 is likely to be underestimated, inasmuch as not all T3 generated in tissues from T4 is in equilibrium with plasma T3.

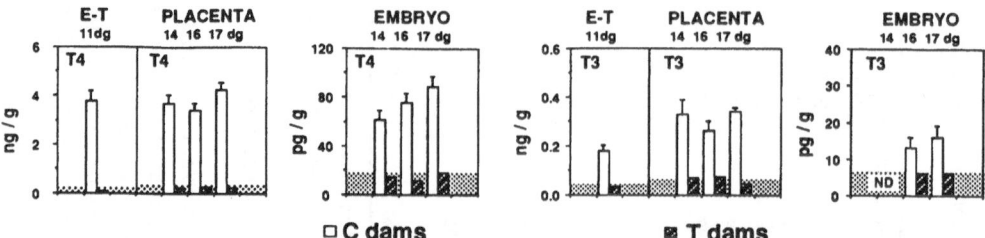

Figure 5 *T4 and T3 concentrations in embryo-trophoblasts (E-T) at 11 dg, and in placentas and embryos before (14, 16 and 17 dg) onset of FTF, from normal (C) and from thyroidectomized (T) dams. Shaded areas correspond to the limits of detection in the assay. Drawn from data by Morreale de Escobar et al.* [12].

derived from thyroidectomized (T) dams before 17.5-18 dg. These findings are compatible with transfer of both iodothyronines from the mother to the developing embryo and fetus; the conceptus of a thyroid hormone -deficient mother would be deprived of this source of hormone early in development.

It might be argued that the iodothyronines found in the embryos before 17.5-18 dg merely represent residual molecules which had been taken up from the maternal blood supply during the implantation stage, before development of the placenta. Such a supply would obviously have been lower in the case of embryos from T dams. This possibility, however, is highly unlikely, considering that, although concentrations are relatively constant, the total amounts of T4 and T3 found in developing embryos are increasing steadily. It is well known that the placenta actively synthesizes many hormones, but there is no evidence either in favour of placental synthesis of T4 or T3, or of its eventual impairment in T dams. Thus, the most likely explanation for our findings is transfer of the iodothyronines from the mother to the embryo, a possibility which is clearly supported by the findings of others. Thus, Sweeney and Shapiro [38] and later Woods et al. [39] showed that after injection of labelled iodothyronines into the dams early in gestation (9-14 dg), these were identified in embryonic tissues.

Is this transfer of any biological importance? The possibility that thyroid hormones play a role in early embryonic development had been previously excluded because the idea prevailed that before onset of FTF normal development takes place in the absence of any thyroid hormone. The evidence summarized here shows this concept is no longer tenable. Although direct evidence of a role of thyroid hormone early in development has not yet been obtained, there is abundant indirect evidence of developmental anomalies in the progeny of women who have been hypothyroxinemic during pregnancy [8], and in the progeny of T rats [9]. We have previously [12] discussed the possibility that the thyroid hormone-deficiency of the developing embryo plays *per se* a causal role in the CNS damage which has been detected in such progeny. The point in discussion is whether it is maternal hypothyroidism *per se* or the deficiency of thyroid hormones in the embryo which cause the developmental anomalies. Bonet and Herrera [40] have presented data showing that thyroid hormone deprivation of the dam during the initial phases of pregnancy interferes with the early anabolic phase of pregnancy needed to face the maternal and fetal metabolic requirements of late gestation, and attributed the developmental defects of the progeny mostly to the poor condition of the T dams Others propose [10, 12] that it is the hypothyroxinemia of the mother which is damaging, inasmuch as it deprives the early embryo of T4: developmental anomalies are very severe in iodine deficiency, although normal maternal T3 levels prevent overt clinical hypothyroidism despite hypothyroxinemia. Moreover, Ekins et al. [10] have shown CNS damage in the adult progeny of rats which were submitted to subtotal T: these dams had very low T4, but normal T3 levels during pregnancy. Although we do not know how the early anabolic phase of pregnancy proceeds in such dams, severe maternal hypothyroidism is not likely; their early embryos would, however, be T4 deficient. Both explanations are not mutually exclusive, as both maternal hypothyroidism and fetal thyroid hormone deficiency may well play complementary causative roles. We will not expand on this point any further, because it is fully discussed in another chapter of this book .("The thyroidectomized pregnant rat-an animal model to study fetal affects on maternal hypothyroidism" by Hendrich and Porterfield).

In conclusion, thyroid hormones are available during early embryonic development in the rat. They are also available to early developing embryos of non-mammalian species: T4 and T3 have been found in the eggs of chicken [41] and salmon [42], in concentrations comparable to those we find in rat placenta. The nuclear receptor for T3 is also present before onset of FTF in the brain of the rat [43, 44] and chick [45] embryos. The message for a thyroid hormone receptor is expressed very early in the brain of the chick embryo, namely at 4 days of incubation [46], several days before onset of thyroid secretion. Thyroid hormone receptor mRNA and the capacity to respond biologically to T3 have been reported for Xenopus laevis very early in development [47]. Thus, the elements necessary for biological effects to result from the hormone-receptor interaction are present in early embryos of different types of animals, and it is reasonable to assume that a deficiency of thyroid hormone might interfere with their development.

After onset of FTF

The maternal transfer of T4 and T3 might either continue after onset of FTF, or stop completely once the fetal thyroid is able to meet fetal thyroid hormone requirements. Previous studies injecting labeled T4 and T3 into late pregnant dams, showed transfer of labeled compounds into the fetal circulation [34, 35] (for a review of older literature, ref. [48]). However, attempts to quantify the amounts transferred in normal conditions were fraught with difficulties. Early findings by Geloso and Barnard [49] suggested that only 60 % of fetal plasma BEI (butanol extractable iodine) was of fetal origin. The experiment, however, involved both T of the dam and / or the fetus (by decapitation), and findings would not necessarily represent the degree of maternal contribution under normal conditions. Attempts using a kinetic approach after injection of labeled T4 or T3 late in pregnancy involved precise determinations of T4 and T3 in fetal plasma and the knowledge of the fetal metabolic clearance rate of both iodothyronines. T3 was not detected in the fetal plasma, and the assumption was made that the metabolic clearance rates

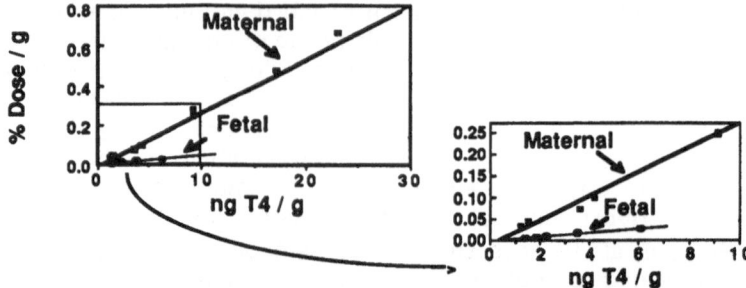

Figure 6 ^{125}I-T4 *concentrations in maternal and fetal samples, plotted against the corresponding T4 concentrations. Maternal samples comprise plasma, liver, kidney, brain, lung, muscle; fetal samples comprise plasma, liver, brain, carcass and total fetuses (minus thyroid), obtained near term (21 dg). Mean values were used.for each tissue The data for the maternal side of the placenta and for the membranes + amniotic fluid fell on the same regression line as the data for maternal tissues. The inset shows the lower values at higher magnification. The slopes of the two lines are equal to the mean specific activities of the* ^{125}I-T4 *in the maternal and fetal T4 pool. Drawn from data by Morreale de Escobar et al.* [26].

would be the same in the fetus as in new-born pups. It was concluded that maternal T4 represented < 1 % of fetal T4 [34], and maternal T3 even less [35].

We have recently investigated this problem using a steady-state approach [26] which does not require determination of fetal metabolic clearance rates. Normal dams were infused continuously with trace amounts of ^{125}I labeled T4 (^{125}I-T4) starting at 11 dg, together with KI in amounts which avoided radioiodide recycling without inhibiting thyroid function. At 21 dg the concentrations both of ^{125}I-T4 and of stable T4 were determined in many maternal and fetal tissues. Figure 6 shows that both the maternal and fetal T4 pools were in isotopic equilibrium: the plots of ^{125}I-T4 versus T4 were closely fit to linear functions. The specific activity (s.a.: ratio of ^{125}I-T4 to T4, derived from the slopes of the linear functions) was the same in all maternal tissues, and also the same for fetal samples, but clearly different in the fetal as compared to the maternal compartment. The s.a. decreased in fetal tissues as a consequence of the contribution by the fetal thyroid to the extrathyroidal fetal T4 pool. From such data it was found that the mean net contribution of maternal T4 to the total fetal extrathyroidal T4 was 17.5 ± 0.9 %. The net maternal contribution of different dams to their litter was proportional to the maternal ^{125}I-T4, as illustrated in Figure 7.§

---

§ *Similar results have been obtained in dams on MMI, infused with different doses of T4 ranging from 2.4 to 7.2 μg / 100 g BW per day : the total T4 and T3 pools of these fetuses with a blocked gland were derived from the mother, and were proportional to maternal thyroxinemia* [50].

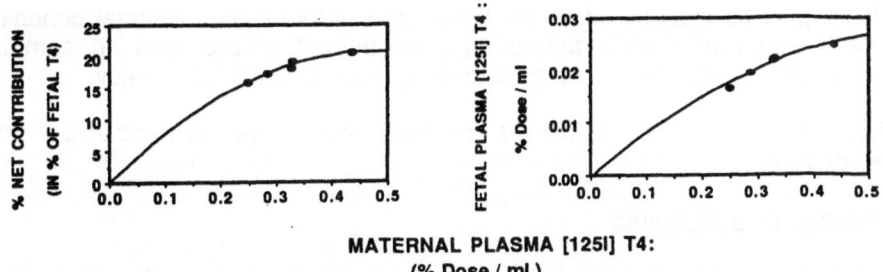

Figure 7 *The left-hand panel shows the net contribution of maternal T4 to the extrathyroidal fetal pool of T4, as % of the T4 available to the fetus, plotted against the maternal ^{125}I-T4 level. The right-hand panel shows the concentration of ^{125}I-T4 in fetal versus maternal plasma. Data points correspond to litters from 5 different dams, infused from 11 to 21 dg with ^{125}I-T4. From experiments reported by Morreale de Escobar et al. [26].*

The net maternal T4 contribution was 100 % for the T4 in the maternal (basal) side of the placenta and in fetal membranes plus amniotic fluid, and 71.5 % of the T4 in the fetal (labyrinthine) side of the placenta, which receives both maternal and fetal blood. Although some amniotic fluid was lost, and metabolites of ^{125}I-T4 were not quantified, several assumptions were made to estimate the amount of T4 actually transferred to the total uterine contents: for a litter of 12 fetuses it might well exceed 20 % of the total maternal extrathyroidal T4 pool.¶

Thus, in normal rat pregnancies maternal to fetal transfer of T4 continues after onset of FTF. Before onset of FTF all of the T4 available to the embryo is derived from the mother, but near term the maternal contribution represents less than one fifth of the total extrathyroidal T4 pool of the fetus. This does not mean that the *amounts* actually transferred decrease in late pregnancy. They may be constant, or even increase, but appear as a decreasing *percentage* of the total extrathyroidal T4 available to the fetus, because the T4 secreted by the fetal thyroid is increasing. This point is now being studied.

Is the T4 transferred from the mother to the fetus of any relevance late in gestation, once the fetal thyroid has become active? It has often been stated that the fetal pituitary-thyroid system functions with complete autonomy from maternal

---

¶ *The net maternal contribution of T3 has not been determined with the same approach, because the fetal extrathyroidal T3 pool, as determined by RIA, not only contains T3 secreted by the fetal thyroid plus T3 transferred from the mother, but also T3 generated from T4 in the fetal tissues. The latter not only includes T3 generated from T4 secreted by the fetal thyroid, but also T3 generated from maternal T4 reaching the fetus. This complicates interpetation of the s.a. data with respect to the net maternal contribution of T3.*

thyroid status. To test this idea, we have used different experimental conditions, namely i) maternal thyroid failure, ii) fetal thyroid failure, and iii) combined maternal-fetal thyroid failure, as produced by severe iodine deficiency.

Relevance of the maternal thyroid hormone contribution after onset of FTF.

i) Maternal thyroid failure

Our findings regarding the effects of maternal T on fetal thyroid hormone economy after onset of FTF, namely between 18 and 22 dg [29] are summarized in Figures 2 and 4. As already indicated, during this period of development there is a 100-fold increase in the T4 content of the thyroid gland of fetuses from normal mothers, a 400-fold increase in the T3 content. Initially, T4 and T3 also increased in the glands of fetuses obtained from T dams, but this increase stops entirely after 20 dg (Figure 2, left-hand panels). Just before birth the thyroid glands of fetuses from T dams contain only 24 and 28 %, respectively, of the T4 and T3 which are accumulating in normal fetal thyroids. T4 and T3 concentrations at 17 and 18 dg were found [29] to be lower in all fetal tissues obtained from T dams, as expected considering that these fetuses were thyroid hormone-deficient before onset of FTF. The total extrathyroidal T4 and T3 pools were accordingly lower at 18 dg (Figure 4), but the difference with respect to fetuses from C dams became smaller after 20 dg, so that near birth fetal extrathyroidal pools have reached values which are practically normal. We believe this catch-up is likely to result from an increased thyroidal secretion of both iodothyronines.

This interpetation is supported by the increased TSH levels found by others in fetuses from T dams [40, 51]. We have also found that 5' D activities are decreased in liver and lung [29] and that the response of fetal BAT 5' D-II to changes in T4 and T3 availability are also affected by maternal T [25]. Thus, despite the fact that the fetal thyroid becomes active at the same gestational age as the gland of fetuses from normal mothers, the fetal pituitary-thyroid system somehow senses the lack of contribution of thyroid hormone by the mother. It responds to this situation with a typical compensatory reaction . All these observations strongly suggest that fetal thyroid hormone economy is not entirely independent of maternal thyroid status even after onset of FTF. They also suggest that the maternal contribution is not irrelevant for the fetus, considering that mechanism(s) are operative to compensate for its absence.

ii) Fetal thyroid failure

In order to sudy this situation, we have blocked FTF by treating the pregnant dams with 0.02 % MMI (methimazole) in the drinking water from 14 dg until 21 dg. This treatment avoids the increase in fetal thyroidal T4 and T3 contents usually observed after onset of FTF. Thus, the thyroidal T4 and T3 in fetuses from MMI-treated dams were 0.6 % and 5 %, respectively, of normal values at 21 dg. But the treatment is also blocking maternal thyroid function, and by 21 dg T4 and T3 in maternal plasma and tissues are decreased as compared to C dams. As a result the concentrations of T4 and T3 in fetal plasma and tissues were markedly

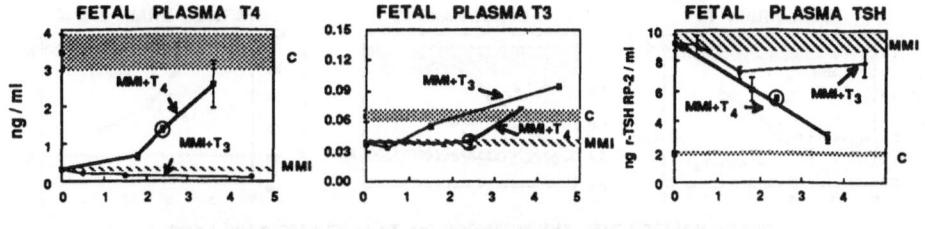

Figure 8 *Changes in the concentrations of T4, T3 and TSH in plasma obtained from fetuses of normal (C) dams and MMI-treated dams (MMI), and from MMI-treated dams infused with T4 (MMI+T4) or T3 (MMI+T3), shown as dose-response curves versus the infusion dose. Data are mean values ± SEM. Values for the C and MMI fetuses are shown as shaded areas. Those corresponding to dams infused with 2.4 μg T4 / 100 g BW per day are highlighted by enclosing them in circles. Drawn from data by Calvo et al.* [28] *, with permission.*

decreased [14, 15, 28]. To avoid the combined effects of maternal and fetal thyroid failure, maternal hypothyroidism was prevented by infusing the dams continuously from 15 to 21 dg with either T4 or T3, at different doses. These hormonal infusions into the mothers did not increase the T4 and T3 contents of the fetal thyroid. They did, however, clearly mitigate the T4 and T3 deficiency of the fetal tissues.

Results obtained in four different studies [14, 15, 28] (and Calvo et al., manuscript in preparation) using this methodology and infusion doses ranging from 1.8 to 7.2 μg T4 / 100 g BW per day and from 0.5 to 4.5 μg T3 / 100 g BW per day have disclosed several points we believe are important for our understanding of maternal-fetal thyroid hormone interrelationships.

When T4 is infused into the MMI-treated dams, both T4 and T3 increase in all fetal tissues, as compared to the levels found in the fetuses from MMI-treated dams infused only with the solvent. On the contrary, when the MMI-treated dams are infused with T3, the concentrations of T3 increase in some, but not all, fetal tissues, with T4 concentrations remaining at MMI levels, or lower. These results show that both T4 and T3 are transferred from the mother to the hypothyroid fetus near term, but affect fetal thyroid hormone concentrations differently.

As the T4 infusion doses increase, so do the concentrations of both T4 and T3 in the fetal plasma, and plasma TSH levels decrease. The daily infusion dose of 2.4 μg per 100 g BW restored to normal values the concentration of T3 in maternal plasma as well as in maternal tissues, and several end-points of thyroid hormone action. This dose, however, did not ensure normal T4, T3 and TSH concentrations in the fetal plasma: T4 and T3 were still lower than in C fetuses, and TSH was high (see Figure 8). Thus, fetuses from such dams would have been identified at birth as a congenitally hypothyroid pups. The same conclusion would have been drawn from the determination of the total extrathyroidal T4 and T3 pools, which were lower (23 % for T4 and 33 % for T3) than in normal fetuses, and signs of fetal

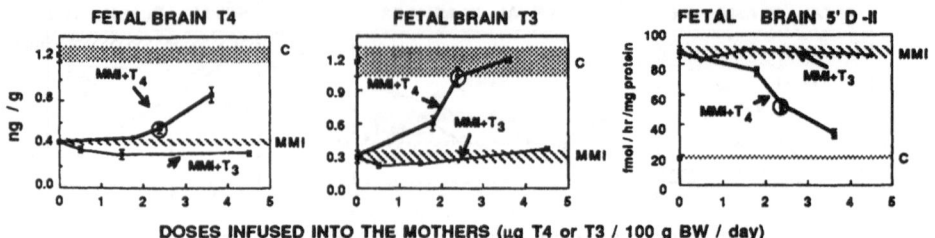

Figure 9 *Changes in brain T4 and T3 concentrations and 5'D-II activity corresponding to the same fetuses described in Figure 8. Drawn from data by Calvo et al. [28], with permission.*

hypothyroidism could ensue. Despite this, normal cerebral T3 concentrations were attained in the otherwise thyroid hormone-deficient fetus, with increased cerebral 5'D-II playing an important role, as illustrated in Figure 9. These results show that a normal maternal thyroxinemia and a compensatory increase in 5'D-II activity are essential for the preferential protection of the fetal brain from T3 deficiency. Although the transfer of maternal T4 is not enough to prevent T4 and T3 deficiency of the fetus it would be wrong to conclude that its brain had likewise been deprived of T3 during late fetal life. Provided the maternal T4 supply is adequate and nothing interferes with 5'D-II activity, the fetal brain is preferentially protected from T3 deficiency until birth.

A further increase of the daily T4 dose infused into the dams to 7.2 µg per 100 g BW (Calvo et al., ms in preparation) resulted in normal T4 and TSH levels in fetal plasma, normal T4 and T3 concentrations in most fetal tissues, normal 5' D-II activity in fetal brain. This dose of T4 raised T4 and T3 levels in maternal plasma above the low concentrations which are normal for the 21 dg pregnant rat [27], but not above those of non-pregnant age and sex-paired rats. Thus, increasing maternal thyroxinemia can lead to complete protection of the hypothyroid fetus within doses which appear to be well tolerated by the dam. Throughout the range of infusion doses used (0 to 7.2 µg T4 / 100 g BW) the total fetal extrathyroidal T4 and T3 pools were proportional to maternal thyroxinemia [50], despite active deiodinations in the placenta [2].

When the MMI-treated dams were infused with increasing doses of T3, the mothers became less T3-deficient, but their T4-deficiency was either unmitigated, or actually increased. Thus the mothers were severely hypothyroxinemic, although their plasma and tissue T3 were raised to normal values, or were higher than normal. The T3 deficiency of most (but not all) fetal tissues was mitigated, but their T4 deficiency was the same, or greater. With the infusion of T3 instead of T4, the mother and her fetuses are deprived of mechanism(s) for the generation of T3 from T4, which are an important source of intracellular T3. This appears to be especially dramatic for the fetal brain, which is totally dependent on local generation of T3 from T4 for its supply of T3 [24, 28]. Even the infusion of a T3 dose as high as 4.5 µg / 100 g BW, which increased

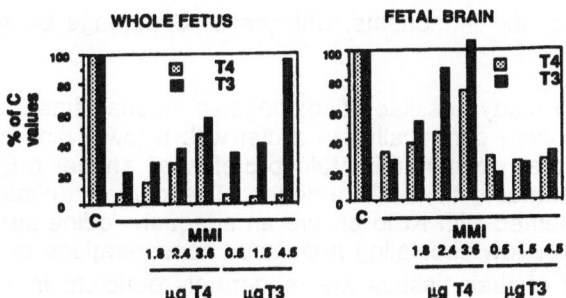

Figure 10 *The total extrathyroidal T4 and T3 pools and the total cerebral T4 and T3 contents were determined for fetuses from the same experimental groups decribed in Figures 8 and 9. The values for C fetuses (11.2 ng T4 and 2.42 ng T3 for the extrathyroidal pools, 0.224 ng T4 and 0.208 ng T3 for the cerebral contents) were taken as the corresponding 100 % value. Drawn from data by Calvo et al. [28], with permission.*

maternal T3 to about twice that of non-pregnant rats, did not increase fetal cerebral T3 above the values found in the MMI-treated fetuses (Figure 9), despite the fact that T3 in the fetal plasma and other fetal tissues was higher than in C fetuses (Figure 8). Likewise, the high TSH level in fetal plasma was not decreased by the infusion of T3 into the mothers, strongly suggesting that the fetal thyrotroph is also totally dependent on the intracellular generation of T3 from T4 for feed-back control of TSH secretion. Pups born from a T3-infused dam would have low plasma T4 and high plasma TSH levels and would be identified as congenitally hypothyroid, unless plasma T3 were also determined. This would be true with regard to the concentration of T3 in the brain (and, possibly in the anterior pituitary), but the T3 deficiency of most other fetal tissues might be mitigated to a point where clinical evidence of hypothyroidism would no longer be found. This is the opposite situation of the one found in fetuses from T4-infused dams, which might have generalized tissue hypothyroidism, but normal brain T3.

These results (summarized in Figure 10) support the idea that the maternal transfer of thyroid hormones is relevant, and especially so in cases of fetal thyroid failure. Although the transfer of T4 and T3 from the normal mother to its hypothyroid fetus are not sufficient to prevent thyroid hormone deficiency of the fetus as a whole, normal maternal levels of T4 (but not of T3) play a crucial role for the protection of the fetal brain from T3 deficiency until birth.

iii) Maternal and fetal thyroid failure

Combined congenital hypothyroidism with severe maternal thyroid failure is not frequent. But it is at present estimated that more than 600 million people may be living in areas where the iodine intake is too scarce to ensure adequate thyroidal production of T4. The dramatic effects of severe iodine deficiency on the development of the CNS are well-documented [5-7]: neurological cretinism might affect up to 10 % of the inhabitants, with the CNS damage being irreversible by birth.

We have tried to study possible etiopathogenic mechanisms in an experimental model: female rats were chronically fed a diet with a low iodine content (LID, low iodine diet). This diet does not contain proteins of animal origin to avoid the possibility that it contains thyroid hormones. The control animals were fed the same diet, supplemented with KI to ensure an adequate iodine intake (LID+I). Rats on this LID have very low circulating and tissue concentrations of T4 and elevated plasma TSH. Most of their tissues are moderately deficient in T3 [22, 23, 33, 52, 53], despite normal plasma T3 levels. Females on LID and on LID+I were mated, and their embryo-trophobalsts were obtained at 11dg, embryos and placentas at 17 dg, and fetal plasma and tissues at 21 dg [22, 23]. Pups born from such LID and LID+I dams, and from dams on the normal laboratory diet, were also studied at different ages after birth [23]. It was found that the concentrations of T4 were very low in all embryonic and fetal tissues from concepta of LID rats, at all gestational ages studied, as illustrated in Figure 11. The concentration of T3 was not lower in embryo-trophoblasts from LID as compared to LID+I dams at 11 dg, but by 17 dg and thereafter it was lower in placentas and embryos. At 21 dg, near term, both T4 and T3 were decreased in all extrathyroidal tissues [22, 23]. Their body and organ weights were smaller, their thyroid glands were hyperemic and clearly enlarged. The thyroidal iodine content was only 4.7 % of that of LID+I fetuses, a decrease similar to that found in the maternal thyroid.

A later study comprised the whole period between 18 and 21 dg (ref [54] and Obregón et al, ms in preparation). The thyroidal T4 and T3 contents increased rapidly in fetuses from LID+I dams (Figure 12), the pattern being comparable to that found for fetuses from C dams (Figure 2). On the contrary, neither T4 or T3 increased in LID fetuses, with values as low as in fetuses from MMI-treated dams, or even lower. The increases in T4 and T3 concentrations in liver, lung and carcass from LID+I fetuses were also comparable to those described previously for fetuses from C dams, but in extrathyroidal tissues of LID fetuses the concentrations of T4 and T3 did not increase above the 18 dg values. This is illustrated for the fetal

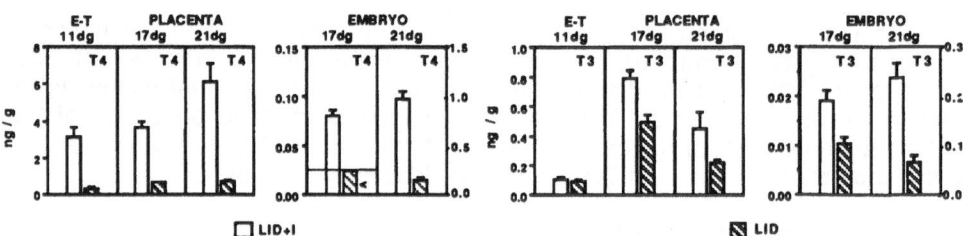

Figure 11 *Concentrations of T4 and T3 in embryo-trophoblasts (E-T) at 11 dg, and placentas and embryos before (17 dg) and after (21 dg) onset of FTF, obtained from dams on low iodine intake (LID) and from dams on iodine-supplemented LID (LID+I). All values for LID samples were different from those obtained from LID+I dams, except for T3 in 11 dg E-T. Drawn from data by Escobar del Rey et al.* [22].

brain in Figure 13. By birth the concentrations of both iodothyronines were comparable, or actually lower, than in the brain of fetuses from MMI-treated dams, and certainly lower than in MMI-treated fetuses from euthyroid mothers (infused with T4). At 21 dg the total extrathyroidal T4 and T3 contents of fetuses from LID+I dams were 8.35 ng T4 and 2.5 ng T3, values comparable to those found for fetuses from C dams (Figure 4). In contrast, the corresponding pools in fetuses from LID dams were 0.43 ng T4 and 0.44 ng T3, which are 5.2 % and 17.7 %, respectively, of the LID+I values. This contrasts with the practically normal values attained by fetuses from T dams. 5' D activities were decreased in liver and lung of LID fetuses at 20 and 21 dg. On the contrary, 5'D-II increased markedly in the brain of LID fetuses, four times more than the normal developmental increase occurring in the brain of LID+I fetuses (Figure 13). Thus, the low cerebral T3 concentration was not caused by an inability of 5'D-II activity to respond, but to the very low supply of substrate.

It is evident that in situations of severe iodine deficiency, the fetal thyroid is unable to compensate for the lack of maternal T4 and, in contrast to what happens in fetuses from T dams, there is no catch-up in the extrathyroidal T4 and T3 pools by birth. Maternal hypothyroxinemia is severe throughout pregnancy [22, 23, 54] and, as a consequence, the fetal brain is much more deficient in T3 than in a congenitally hypothyroid fetus from a normal mother.

Findings after birth were unexpected [23]: During the suckling period the LID pups receive 4-5 times more iodine through the milk than was available to them *in utero* , or will be available to them later in life, when eating the LID. As a consequence, there is a small postnatal increase in their plasma T4 which, coupled to a 7-fold increase in cerebral 5'D-II activity [54] results in practically normal brain T3 concentrations. This contrasts with the very low T3 concentrations in LID fetuses, and the lower than normal cerebral T3 after weaning [23, 53]. It is difficult to obtain cretin rats with iodine deficiency alone, and we believe this is related to the fact that the marked cerebral T3 deficiency of the fetus is later mitigated througout the early postnatal period, when important phases of brain maturation take place.

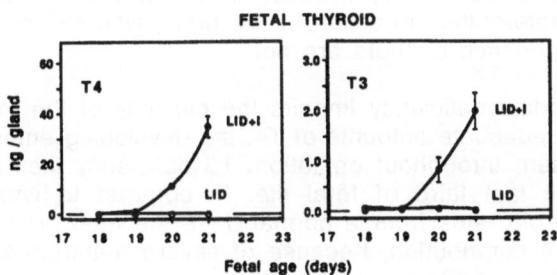

Figure 12 *Changes in the content of T4 and T3 in digests of thyroid glands obtained from LID and LID+I dams, between 18 and 21 dg. From Obregón et al., ms in preparation, and* [54].

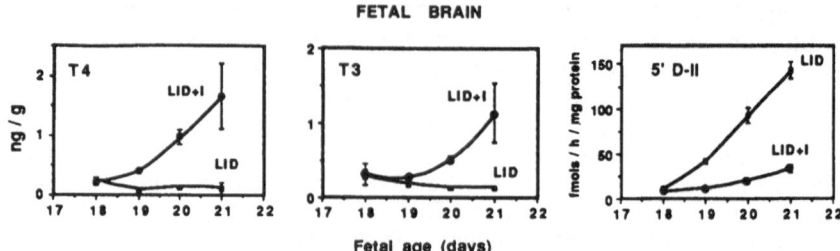

Figure 13 *Changes in brain T4 and T3 and 5' D-II activity of fetuses obtained from LID and LID+I dams. From Obregón et al., ms in preparation, and ref. [54].*

Summary of findings in rats.

As regards the questions posed at the beginning, we may state now that T4 and T3 are present in embryonic tissues before the fetus has an active thyroid secretion, and that they are of maternal origin. Transfer of thyroid hormones continues until term, and fetal thyroid thyroid hormone economy is not completely autonomous nor totally independent from maternal thyroid status.

If the maternal contribution is missing, embryonic tissues are thyroid hormone-deficient until the fetal thyroid is able to compensate for this by increasing secretion, at the expense of depleted thyroidal stores of both iodothyronines by birth.

If the mother is normal and the fetus congenitally hypothyroid, early development takes place under normal conditions, and later in gestation the thyroid hormone deficiency of most fetal tissues is mitigated by the maternal contribution of T4 and T3. The fetal brain is totally protected from T3 deficiency throughout fetal development, although other tissues are not. Normal maternal thyroxinemia and a compensatory increase in cerebral 5' D-II activity are crucial elements in this preferential protection of the brain, whereas normal maternal T3 levels in hypothyroxinemic mothers are not.

When severe iodine deficiency impairs the capacity of the maternal and fetal thyroids to secrete adequate amounts of T4, the developing embryo and fetus are markedly T4 deficient throughout gestation. T3 deficiency increases and is very marked during the last third of fetal life. In contrast to what occurs in the congenital hypothyroid fetus from a normal dam, the fetal brain is not protected by the maternal T4 contribution, because of severe maternal hypothyroxinemia. Thus, thyroid hormone deficiency of the fetus as a whole, included the brain, is more prolonged and severe than in the cases of maternal or of fetal thyroid failure. It seems reasonable to conclude that the damaging effects of thyroid hormone deficiency on CNS development would also be more severe.

Are findings in rats relevant to human maternal-fetal thyroid hormone interrelationships?

In man (for reviews, refs. [1,17-19], and the chapter on "Thyroid system ontogeny in the sheep: a model for precocial mammalian species" by D. A. Fisher, in this book) the critical components ensuring pituitary-portal vascular continuity by linking the sinusoids with the median eminence (primary plexus) is not established until 18-22 weeks (wks) of gestation, near the end of the second trimester. TSH is often undetectable in fetal plasma until mid-gestation, with a sudden increase in both plasma and pituitary TSH at 22-24 wks. Thyroidal uptake of ^{131}I is very low until 18-22 wks. Fetal plasma T4 is very low. although detectable, up to 24 wks, when it suddently increases; plasma T3 is undetectable almost until 30 wks, when it suddenly increases. Thus, available data suggest that FTF (as regards T4 and T3 secretion) does not start until mid-gestation.

T3 has been clearly detected and quantified before onset of FTF: in purified extracts from human fetal brain as early as 9-10 wks by Bernal & Pekonen [21] and by Ferreiro et al [55]. T3 concentration in the brain increased from 0.27 ng / g at 10 wks (brain weight 0.4 g) to 0.48 ng / g at 18 wks (brain weight 24.7 g) despite the fact that during this period plasma T3 is undetectable and less than 10 % of adult values. The T3 concentration reported for adult brain cortex is 1.4 ng / g [56]. By mid-gestation the concentration of T3 in the brain is much higher (34 % of adult values) than would be inferred from comparison of the circulating T3 levels, a finding similar to that described for fetal rats at the end of pregnancy, when the rat brain is undergoing a phase of neurogenesis comparable to that of the human fetus by mid-gestation. It is also likely that cerebral 5'D-II is playing an important role, as reported by Karmarkar et al. [57].

The nuclear T3 receptor is also present in the brain of the 10 wk-old fetuses, although at a low concentration (46 fmols / mg DNA), and increases more than 6-fold by 12 wks (to 264 fmol / mg DNA) and 10-fold by 16 wks (to 479 fmol / mg DNA) (21). Later experimental findings suggest the presence of a small gradient between nuclear and cytosolic free T3, which would contribute to further increase availability of T3 for receptor occupancy [55]. Receptor occupancy was 25 % throughout this period, so that the amount of receptor-bound T3 per mg DNA would also increase 10-fold between 10 and 16 wks. Similar quantitative results have been reported by Su et al. [58] in another series of human fetal cerebra. In the study by Ferreiro et al. [55] T4 was found in the nuclei of the fetal brains in concentrations comparable to those found for T3. They did not clarify whether, or not, this nuclear T4 was also bound to the nuclear receptor, in which case it would double the number of hormone-receptor units capable of eliciting a biological effect.

The iodothyronines and the nuclear T3 receptor were also found very early in gestation in other fetal tissues, such as lung [21, 55, 59] Receptor concentration is quite high, increasing from 200-300 fmol / mg DNA at 12-13 wks to 350-500 fmol / mg DNA at 16-19 wks [59]. T4 concentrations in lung by 16 wks were as high as in the brain, but T3 could not be detected up to, and including, 16-18 wks. Despite this, T3 could be measured in lung nuclei as early as 10.5 wks : it increased from 11 fmol / mg DNA to 38 fmol / mg DNA by 14 wks, also suggesting a gradient of free T3 between the nucleus and the cytosol [21, 55].

Thus, the few available data for human fetuses show that the hormones and T3-occupied nuclear receptors are found several weeks before the end of the 1st trimester, clearly preceding onset of active FTF near the end of the second trimester. As in rats, this constitutes strong indirect evidence that T3 might influence early brain development.

It is likely that the T4 and T3 found in early human fetuses is of maternal origin. Early work using ^{131}I labeled T4 or T3, injected into the mother before elective abortion at 11-25 wks of gestation resulted in levels of TCA-precipitable radioiodine ("organic iodine") in fetal plasma which were 5.8 % of those in maternal plasma when T4 was injected, and double this amount when T3 was used [60]. Although radioactivities were usually 10 %, or less, those in the maternal circulation, considerable amounts might reach the fetus over longer periods of time, as pointed out by Osorio and Myant [61], if the rate of transfer were slow. In this respect we should like to point out that in the steady-state experiment illustrated in Figure 7, the labeled T4 found in rat fetal plasma was approximately 5-6 % of that in the maternal circulation, despite which the net maternal T4 contribution amounted to 17.5 % of the fetal extrathyroidal T4 pool.

There have been relatively few experiments investigating transfer of T4 and T3 after onset of FTF, namely after the end of the 2nd trimester. For obvious reasons the scant available information was obtained at term. All the results clearly show that, despite active deiodinations by the human placenta, some maternal thyroid hormone reaches the fetus.

Grumbach and Werner [62], Costa and Ravera [63], and Kerns and Hutson [64] injected the labeled hormones into normal pregnant women receiving iodide to block radioiodine recycling, and found that significant amounts were transferred, although at a slow rate. Fisher et al. [65] injected a single large T4 dose in women at term, and found that cord-blood BEI (butanol-extractable iodine) increased with dose and diffusion time. In uncomplicated pregnancy as much as 36 μg T4 might be transferred to the fetus per day, an amount which might mitigate the thyroid hormone deficiency of a congenitally hypothyroid fetus [66]. T3 was injected into normal women at term by Raiti et al. [67] and by Dussault et al [68] at different doses. An increase in T3 could be detected even with the lowest dose (50 μg T3 per day), but a decrease in cord blood TSH could only be shown with higher doses. The latter finding suggests that, as we have shown for rats, sensitivity of the pituitary to T3 might still have not matured in the human fetus near term, and that changes in circulating TSH would not be the adequate end-point to assess transfer of T3.

Vulsma et al. [69] have recently shown transfer of T4 in women at term: cord blood T4 levels in neonates with total organification defect are 20-50 % of normal values, then decreasing rapidly with a half-life of disappearance of 3.6 days. This finding is clearly reminiscent of the data shown for rat fetuses from MMI-treated dams on the T4 dose which maintained the mothers euthyroid (2.4 μg T4 / 100 g BW per day) which had plasma T4 levels 40 % of normal (Figure 8).

In conclusion, we believe that there is more evidence in favour of maternal to fetal transfer of thyroid hormones in women throughout pregnancy, than evidence of virtually complete placental impermeability. To define the amounts transferred are "small" or "large" is rather subjective, and influenced by the circulating thyroid hormone levels and requirements determined mostly for adults. It would appear more objective to assess whether the amounts contributed by the mother are needed for normal development of the human fetus, or not.

In this respect, most of the information we have is indirect, and based on the adverse consequences of maternal hypothyroxinemia, congenital hypothyroidism and severe iodine deficiency.

During the 1st and possibly during part of the 2nd trimester, the human fetus would be entirely dependent on the mother for its supply of thyroid hormone. Maternal hypothyroidism or hypothyroxinemia would result in decreased availability of the hormone during gross brain development and during the spurt in forebrain neuroblast proliferation. The outcome of pregnancies in untreated hypothyroid (or hypothyroxinemic) women is poor[8, 70, 71], with an increased proportion of aborted fetuses, stillbirths and perinatal deaths, and decreased I.Q. of the progeny, detected when infants are followed at least a year after birth.

In most cases of congenital hypothyroidism the mother has adequate circulating T4, and early fetal development ought to proceed normally. But during the third trimester the fetus, as a whole, would be thyroid hormone-deficient, considering that maternal transfer of T4 does not compensate totally for fetal thyroid failure. Indeed, the newborns in the study of Vulsma et al. [69] presented signs of bone retardation, a finding which indicates that the amount of T4 transferred from the mother was not sufficient to avoid clinical signs of fetal hypothyroidism. However, it is quite tempting to conclude that their brain T3 deficiency had been mitigated, or even totally avoided. No major brain damage would ensue until after birth, when maternal T4 is no longer available. In most cases of congenital hypothyroidism early onset of treatment with T4 results in normal brain development. This has been interpreted as proof that the human fetal brain does not need thyroid hormone for normal development until after birth. Present results support an alternative explanation, namely, that the brain has not been T3-deficient until birth and early treatment prevents this deficiency postnatally, so that there is no insult to the developing brain. The study by Vulsma et al. [69] shows individual variability in the maternal transfer of T4 and, therefore, the degree of mitigation of the fetal thyroid hormone deficiency and of the protection of the fetal brain might also be variable. Several studies [72-75] have indeed shown that the babies with the most severe retardation of bone maturation and the lowest T4 values at screening are those with the poorest prognosis as regards CNS development, despite prompt postnatal treatment.

In areas of marked iodine deficiency there is a correlation between the degree of maternal hypothyroxinemia and the risk of CNS damage of the progeny [76, 77]. The CNS damage of the neurological endemic cretin is irreversible by birth, and much more severe than in congenital hypothyroidism. It almost invariably includes hearing loss, and the insult is likely to have started by mid-gestation [7]. It is quite interesting that a case has been reported [7] in the United States with the full clinical picture of the neurological cretin, but without iodine deficiency: he was the congenitally hypothyroid son of a hypothyroid mother. The severe neurological damage persisted despite early postnatal treatment.

Our results in rats could afford an explanation for these observations. Early development would be affected by the low levels of maternal T4, the only source of this hormone for the fetus before onset of FTF. But, contrary to what occurs in the case of maternal thyroid failure alone, the fetal thyroid is later unable to compensate for the lack of maternal T4, and the fetal brain would be left unprotected until birth. The normal T3 levels of the iodine-deficient mother would not protect the fetal brain. A possible mitigation of the cerebral T3 deficiency by a postnatal increase in the availability of iodine through milk might occur when it is too late to avoid permanent damage to the CNS.

Very important phases of forebrain development take place in man between onset of FTF and birth, which in the rat and mouse are occurring in the early postnatal period (for reviews and references to previous reviews see [75, 78-80], and the chapter in this book "Thyroid hormone control of cerebral cortex development: Molecular, neuroanatomical and behavioral studies", by S.A. Stein). These phases of development are affected by fetal thyroid hormone deficiency: thus, for instance, Stein et al [80] have recently found decreased abundance of EGF and Mβ5 tubulin isomer mRNAs in the cerebral cortex of congenitally hypothyroid mice on the day of birth, a period which would correspond to mid-gestation in man, and Muñoz et al [81] have shown decreased abundance of a neuron-specific mRNA (RC3) by 5-10 days after the birth of pups from MMI-treated rat dams.

Whichever the mechanism(s) involved, present evidence strongly suggests an important role for a normal maternal thyroxine for fetal development, especially in cases of fetal thyroid failure (congenital hypothyroidism, iodine deficiency). We wish to point out that maintainance of a "euthyroid state" during pregnancy is not enough. The degree of protection of the fetal brain is related to maternal thyroxinemia ("free" T4 in women), and not to maternal "euthyroidism". Therefore it is a normal availability of maternal T4 which has to be ensured throughout pregnancy, whether or not normal levels of T3 maintain clinical euthyroidism.

References

1. D. A. Fisher, and A. K. Klein, Thyroid development and disorders of thyroid function in newborn, N Engl J Med 304:702 (1981).

2. E. Roti, A. Gnudi, and L. E. Braverman, The placental transport, synthesis and metabolism of hormones and drugs which affect thyroid function, Endocr Rev 4:131 (1983).

3. D. A. Fisher, and D. H. Polk, Maturation on thyroid hormone actions, in: "Research in Congenital Hypothyroidism," F. Delange, D. A. Fisher, and D. Glinoer, eds., Plenum Press, New York (1989).

4. M. Hamburgh, The role of thyroid and growth hormone in neurogenesis, in: "Current Topics in Developmental Biology," A. A. Moscona, and A. Monroy, eds., Academic Press, New York (1969).

5. M. Choufoer, M. van Rhijn, and A. Querido, Endemic goiter in Western New Guinea. II. Clinical picture, incidence and pathogenesis of endemic cretinism, J Clin Endocrinol Metab 25:385 (1965).

6.   E. Delange, A. Costa, A. M. Ermans, H. K. Ibbertson, A. Querido, and J. B. Stanbury, A survey of clinical and metabolic patterns of endemic cretinism, in: "Human Development and the Thyroid Gland. Relation to endemic cretinism," J. B. Stanbury, and R. L. Krock, eds., Plenum Press, New York (1972).

7.   G. R. DeLong, Observations on the neurology of endemic cretinism, in: "Iodine and the Brain," G. R. DeLong, J. Robbins, and P. G. Condliffe, eds., Plenum Press, New York (1989).

8.   E. B. Man, and S. A. Serunian, Thyroid function in human pregnancy. IX. Development or retardation of 7-year-old progeny of hypothyroxinemic women, Am J Obst Gynecol 125:949 (1976).

9.   C. E. Hendrich, W. J. Jackson, and S. P. Porterfield, Behavioral testing of progenies of Tx (hypothyroid) and growth hormone-treated Tx rats: an animal model for mental retardation, Neuroendocrinol 38:429 (1984).

10. R. Ekins, A. Sinha, M. Ballabio, M. Pickard, M. Hubank, Z. al Mazidi, and M. Khaled, Role of the maternal carrier proteins in the supply of thyroid hormones to the feto-placental unit: Evidence of a feto-placental requirement for thyroxine, in: "Research in Congenital Hypothyroidism," F. Delange, D. A. Fisher, and D. Glinoer, eds., Plenum Press, New York (1989).

11. M. J. Obregón, J. Mallol, R. M. Pastor, G. Morreale de Escobar, and F. Escobar del Rey, Thyroxine and triiodothyronine in rat embyos before onset of fetal thyroid function, Endocrinology 114:305 (1984).

12. G. Morreale de Escobar, M. R. Pastor, M. J. Obregón, and F. Escobar del Rey, Effects of maternal hypothyroidism on the weight and thyroid hormone content of rat embryonic tissues, before and after onset of fetal thyroid function, Endocrinology 117:1890 (1985).

13. G. Morreale de Escobar, M. J. Obregón, and F. Escobar del Rey, Fetal and maternal thyroid hormones, Hormone Res 26:12 (1987).

14. G. Morreale de Escobar, M. J. Obregón, C. Ruiz de Oña, and F. Escobar del Rey, Transfer of thyroxine from the mother to the rat fetus near term: Effects on brain 3,5,3'-triiodothyronine deficiency, Endocrinology 122:1521 (1988).

15. G. Morreale de Escobar, M. J. Obregón, C. Ruiz de Oña, and F. Escobar del Rey, Comparison of maternal to fetal transfer of 3,5,3'-triiodothyronine versus thyroxine in rats, as assessed from the 3,5, 3'-triiodothyronine levels in fetal tissues, Acta Endocrinol (Copenh.) 120:490 (1989).

16. G. Morreale de Escobar, M. J. Obregón, and F. Escobar del Rey, Transfer of thyroid hormone from the mother to the fetus, in: "Research in Congenital Hypothyroidism," F. Delange, D. A. Fisher, and D. Glinoer, eds., NATO ASI Series. Plenum Press, New York (1989).

17. G. Morreale de Escobar, and F. Escobar del Rey, Thyroid physiology in utero and neonatally, in: "Iodine Prophylaxis following Nuclear Accidents," E. Rubery, and E. Smales, eds., Proceedings of a Joint WHO/CEC Workshop, Pergamon Press, U.K. (1990).

18. D. A. Fisher, J. H. Dussault, J. Sack, and I. J. Chopra, Ontogenesis of hypothalamic-pituitary-thyroid function in man, sheep and rat, Rec Progr Horm Res 33:59 (1977).

19. D. A. Fisher, Ontogenesis of hypothalamic-pituitary-thyroid function in the human fetus, in: "Pediatric Endocrinology," F. Delange, D. A. Fisher, and P. Malvaux, eds., Karger, Basel (1985).

20. B. Nataf, and M. Sfez, Debut du fonctionnement de la thyroide foetale du rat, C R Soc Biol (París) 156:1235 (1961).

21. J. Bernal, and F. Pekonen, Ontogenesis of the nuclear 3,5,3'-triiodothyronine receptor in the human fetal brain, Endocrinology 114:677 (1984).

22. F. Escobar del Rey, R. M. Pastor, J. Mallol, G. Morreale de Escobar, Effects of maternal iodine deficiency on the L-thyroxine and 3,5,3'-triiodo-L-thyronine contents of rat embryonic tissues before and after onset of fetal thyroid function, Endocrinology 118:1259 (1986).

23. F. Escobar del Rey, J. Mallol, M. R. Pastor, and G. Morreale de Escobar, Effects of maternal iodine deficiency on thyroid hormone economy of lactating dams and pups: Maintenance of

normal cerebral 3, 5,3'-triiodo-L-thyronine concentrations in pups during major phases of brain development, Endocrinology 121:803 (1987).

24. C. Ruiz de Oña, M. J. Obregón, F. Escobar del Rey, and G. Morreale de Escobar, Developmental changes in rat brain 5'-Deiodinase and thyroid hormones during the fetal period: The effects of fetal hypothyroidism and maternal thyroid hormones, Pediatr Res 24:588 (1988).

25. M. J. Obregón, C. Ruiz de Oña, A. Hernández, R. Calvo, F. Escobar del Rey, and G. Morreale de Escobar, Thyroid hormones and 5'-deiodinase in rat brown adipose tissue during fetal life, Am J Physiol 257:E625 (1989).

26. G. Morreale de Escobar, R. Calvo, M. J. Obregón, and F. Escobar del Rey, Contribution of maternal thyroxine to fetal thyroxine pools in normal rats near term, Endocrinology 126:2765 (1990).

27. R. Calvo, M. J. Obregón, C. Ruiz de Oña, B. Ferreiro, F. Escobar del Rey, and G. Morreale de Escobar, Thyroid hormone economy in pregnant rats near term: a "physiological" animal model of nonthyroidal illness?, Endocrinology 127:10 (1990).

28. R. Calvo, M. J. Obregón, C. Ruiz de Oña, F. Escobar del Rey, and G. Morreale de Escobar, Congenital hypothyroidism as studied in rats: Crucial role of maternal thyroxine but not of 3',3, 5-triiodothyronine in the protection of the fetal brain, J Clin Invest 86:889 (1990).

29. C. Ruiz de Oña, G. Morreale de Escobar, R. Calvo, F. Escobar del Rey, and M. J. Obregón, Thyroid hormone and 5'-deiodinase in the rat fetus late in gestation. Effects of maternal hypothyroidism, Endocrinology (In press, January 1991).

30. M. J. Obregón, G. Morreale de Escobar, and F. Escobar del Rey, Concentrations of triiodo-L-thyronine in the plasma and tissues of normal rats as determined by radioimmunoassay: Comparison with results obtained by and isotopic equilibrium technique, Endocrinology 103:2145 (1978).

31. M. J. Obregón, J. Mallol, F. Escobar del Rey, and G. Morreale de Escobar, Presence of thyroxine and triiodothyronine in tissues from thyroidectomized rats, Endocrinology 109:908 (1981).

32. J. van Doorn, F. Roelfsema, and D. van der Heide, Concentrations of thyroxine and 3,5,3'-triiodothyronine at 34 different sites in euthyroid rats as determined by the isotopic equilibrium technique, Endocrinology 117:1201 (1985).

33. F. Escobar del Rey, C. Ruiz de Oña, J. Bernal, M. J. Obregón, and G. Morreale de Escobar, Generalized deficiency of 3,5,3'-triiodo-L-thyronine (T3) in tissues from rats on a low iodine intake, despite normal circulating T3 levels, Acta Endocrinol (Copenh.) 120:490 (1989).

34. J. H. Dussault, and P. Coulombe, Minimal placental transfer of L-thyroxine in the rat, Pediatr Res 14:228 (1980).

35. J. D. Dubois, A. Cloutier, P. Walker, and J. H. Dussault, Absence of placental transfer of L-triiodothyronine in the rat, Pediatr Res 11:116 (1977).

36. J. J. DiStefano, Modeling approaches and models of the distribution and disposal of thyroid hormones, in: "Thyroid Hormone Metabolism," G. Hennemann, ed., New York & Basel, Marcel Dekker, Inc (1986).

37. B. Ferreiro, R. M. Pastor, and J. Bernal, T3 receptor occupancy and T3 levels in plasma and cytosol during rat brain development, Acta Endocrinol (Copenh.) 123:95 (1990).

38. L. R. Sweney, and B. L. Shapiro, Thyroxine and palatal development in the rat, Dev Biol 42:19 (1975).

39. R. J. Woods, A. K. Sinha, and R. Ekins, Uptake and metabolism of thyroid hormones by the rat fetus in early pregnancy, Clin Sci 67:359 (1984).

40. B. Bonet, and E. Herrera, Different responses to maternal hypothyroidism during the first and second half of gestation in the rat, Endocrinology 122:450 (1988).

41. A. Sechman, and S. Bobek, Presence of iodothyronines in the yolk of the hen's egg, Gen Comp Endocrinol 69:99 (1988).

42. M. Tagawa, and T. Hirano, Presence of thyroxine in eggs and changes in its content during early development of chum salmon, Gen Comp Endocrinol 68:129 (1987).

43. A. Pérez-Castillo, J. Bernal, B. Ferreiro, and Pans T., The early ontogenesis of thyroid hormone receptor in the rat fetus, Endocrinology 117:2457 (1985).

44. M. Luo, and J. H. Dussault, Immunocytochemical mapping of nuclear T3 receptor using a monoclonal antibody in the developing and adult rat brain, in: "Iodine and the Brain," G. R. DeLong, J. Robbins, and P. G. Condliffe, eds., Plenum Pres, New York (1989).

45. M. A. Haidar, and P. K. Sarkar, Ontogeny, regional distribution and properties of thyroid-hormone receptors in the developing chick brain, Biochem J 220:547 (1984).

46. D. Forrest, M. Sjöberg, and B. Vennström, Contrasting developmental and tissues specific expression of α and β thyroid hormone receptor genes, EMBO J 9:1519 (1990).

47. B. S. Baker, and J. R Tata, Accumulation of proto-oncogene C-erb-A related transcripts during Xenopus development: association with early acquisition of response to thyroid hormone and estrogen, EMBO J 9:879 (1990).

48. J. P. Geloso, Fonctionnement de la thyroide et correlations thyreohypophysaires chez le foetus de rat, Ann d'Endocrinologie 28:1 (1967).

49. J. P. Geloso, and G. Bernard, Effects de l'ablation de la thyroide maternelle ou foetale sur la taux des hormones circulantes chez le foetus de rat, Acta Endocrinol (Copenh) 56:561 (1967).

50. R. Calvo, M. J. Obregón, F. Escobar del Rey, and G. Morreale de Escobar, In vivo assessment of the role of the rat placenta near term as a barrier for the transfer of thyroxine and 3,5, 3'-triiodothyronine to the fetus, Advances in Perinatal Thyroidology, Longboat Key, Florida (Abstract) (1990).

51. J. H. E. Hendrich, S. P. Porterfield, and V. A. Galton, Pituitary-thyroid function of fetuses of hypothyroid and growth hormone treated hypothyroid rats, Horm Metab Res 11:362 (1979).

52. P. Santisteban, M. J. Obregón, A. Rodríguez-Peña, L. Lamas, F. Escobar del Rey, and G. Morreale de Escobar, Are iodine-deficiency rats euthyroid?, Endocrinology 110:1780 (1982).

53. M. J. Obregón, P. Santisteban, A. Rodríguez-Peña, A. Pascual, P. Cartagena, A. Ruiz-Marcos, L. Lamas, F. Escobar del Rey, and G. Morreale de Escobar, Cerebral hypothyroidism in rats with adult-onset iodine deficiency, Endocrinology 115:614 (1984).

54. C. Ruiz de Oña, M. J. Obregón, R. Calvo, G. Morreale de Escobar, and F. Escobar del Rey, Chronic maternal iodine deficiency in rats: Role of iodothyronine 5' deiodinases (5'D) in thyroid hormone economy during the fetal and suckling periods, XVIII Meeting European Thyroid Association, Copenhagen, Annales d'Endocrinologie (Abstract 125) 50:147 (1989).

55. B. Ferreiro, J. Bernal, C. G. Goodyer, and C. L. Branchard, Estimulation of nuclear thyroid hormone receptor saturation in human fetal brain and lung during early gestation, J Clin Endocrinol Metab 67:853 (1988).

56. R. Arem, M. M. Kaplan, G. Wiener, S. G. Kaplan, and S. Reichlin, Tissue thyroxine and triiodothyronine concentrations in human non-thyroidal illness, 61st Meeting of the American Thyroid Association, Phoenix, Arizona p T-4 (Abstract) (1986).

57. M. G. Karmarkar, D. Prabakaran, M. M. Godbole, and M. M. Ahuja, Thyroid hormone contents an 5 & 5' monodeiodinase activities in developing human cerebral cortex, in: "Iodine and the Brain," G. R. DeLong, J. Robbins, and P. G. Condliffe, eds., Plenum Press, New York (1989).

58. H. L. Su, P. Ling, R. K. Yang, and H. C. Chao, Ontogenesis of nuclear T3 receptor in human fetal brain, in: "Iodine and the Brain," G. R. DeLong, J. Robbins, and P. G. Condliffe, eds., Plenum Press, New York (1989).

59. L. W. González, and P. L. Ballard, Identification and characterization of nuclear 3,5,3'-triiodothyronine binding sites in fetal human lung, J Clin Endocrinol Metab 53:21 (1981).

60. N. B. Myant, Passage of thyroxine and triiodothyronine from mother to foetus in pregnant women, Clin Sci 17:75 (1958).

61. C. Osorio, and N. B. Myant, The passage of thyroxine hormone from mother to foetus and its relation to foetal development, Br Med Bull 16:159 (1961).

62. M. M. Grümbach, and S. H. Werner, Transfer of thyroid hormone across the human placenta at term, J Clin Endocrinol Metab 16:1392 (1956).

63. A. Costa, and J. P. Ravera, Studi sulla funzionalitá della tiroide nelle prime etá della vita, Minerva Nipiol 11:69 (1961).

64. J. E. Kearns, and W. Hutson, Tagged isomers and analogues of thyroxine (Their transmission across the human placenta and other studies), J Nucl Med 4:453 (1963).

65. D. A. Fisher, H. Lehman, and D. Lackey, Placental transfer of thyroxine, J Clin Endocrinol Metab 24:393 (1964).

66. D. A. Fisher, Panel discussion on hyperthyroidism in the pregnant woman and neonate, J Clin Endocrinol Metab 27:1639 (1967).

67. S. Raiti, G. B. Holzman, R. L. Scott, and R. M. Blizzard, Evidence for the placental transfer of tri-iodothyronine in human feings, N Engl J Med 277:456 (1967).

68. J. H. Dussault, V. V. Row, G. Lickrish, and R. Volpé, Studies of serum triiodothyronine concentration in maternal and cord blood: transfer of triiodothyronine across human placenta, J Clin Endocrinol Metab 29:595 (1969).

69. T. Vulsma, M. H. Gons, and J. de Viljder, Materna-fetal transfer of thyroxine in congenital hypothyroidism due to a total organification defect of thyroid agenesis, N Engl J Med 321:13 (1989).

70. A. J. McMichael, J. B. Potter, and B. S. Hetzel, Iodine deficiency, thyroid function and reproductive failure, in: "Endemic Goiter and Endemic Cretinism," J. B. Stanbury, and B. S. Hetzel, eds., Wiley and Sons, New York (1980).

71. G. W. Greenman, M. O. Gabrielson, J. Howard-Flanders, and M. A. Wessel, Thyroid function in pregnancy. Fetal loss and follow-up evaluation of surviving infants, N Engl J Med 267:426 (1962).

72. R. Wölter, P. Nöel, P. De Cock, Craen M., C. H. Eernould, P. Malvaux, F. Verstraeten, J. Simons, S. Mertens, N. van Broek, and M. Vanderschveren-Lodewyckk, Neuropsychological study in treated thyroid dysgenesis, Acta Pediatr (Scand.) 277:41 (1980).

73. J. Glorieux, Mental development of patients with congenital hypothyroidism defected by screening. Quebec experience, in: "Research in Congenital Hypothyroidism," F. Delange, D. A. Fisher, and D. Glinoer, eds., NATO ASI Series. Plenum Press, New York (1989).

74. P. Rochiccioli, f. Alexandre, and B. Roge, Neurological development in congenital hypothyroidism, in: "Research in Congenital Hypothyroidism," F. Delange, D. A. Fisher, and D. Glinoer, eds., NATO ASI Series. Plenum Press, New York (1989).

75. J. H. Dussault, Action of thyroid hormones on brain development, in: "Research in Congenital Hypothyroidism," F. Delange, D. A. Fisher, and D. Glinoer, eds., NATO ASI Series. Plenum Press, Newy York (1989).

76. P. O. D. Pharoah, K. Connolly, B. S. Hetzel, and R. P. Ekins, Maternal thyroid function and motor competence in the child, Dev Med Child Neurol 23:76 (1981).

77. K. J. Connolly, and P. O. D. Pharoah, Iodine deficiency, maternal thyroid levels in pregnancy and developmental disorders in children, in: "Iodine and the Brain," G. R. DeLong, J. Robbins, and P. G. Condliffe, eds., Plenum Press, New York (1989).

78. M. Hamburgh, L. A. Mendoza, I. Bennet, P. Krupa, Y. S. Kim, R. Kahn, K. Hogreff, and H. Francfort, Some unresolved questions of brain-thyroid relationships, in: "Thyroid Hormone and Brain Development, " G. D. Grave, ed., Raven Press, New York (1977).

79. G. Morreale de Escobar, A. Ruiz-Marcos, and F. Escobar del Rey, Thyroid Hormone and the Developing Brain, in: "Congenital Hypothyroidism," J. H. Dussault and P. Walker, eds., Marcel Dekker, New York (1983).

80. S. A. Stein, D. R. Shanklin, P. M. Adams, G. M. Mihailoff, M. B. Palnitkar, and B. Anderson, Thyroid hormone regulation of specific mRNAs in the developing brain, in: "Iodine and the Brain," G. R. DeLong, J. Robbins, and P. G. Condliffe, eds., Plenum Press, New York (1989).

81. A. Muñoz, A. Rodríguez-Peña, A. Pérez-Castillo, B. Ferreiro, J. G. Sutcliffe, and J. Bernal, Effects of neonatal hypothyroidism on rat brain gene expression, Endocrinology (in press)

THYROID FUNCTION AND HYPERFUNCTION IN THE

PREGNANT WOMAN

Gerard N. Burrow

University of California, San Diego
School of Medicine
9500 Gilman Drive (0602)
La Jolla, California 92093-0602

Thyroid Function During Pregnancy

Complex changes in thyroid function occur in the pregnant woman as a result of hormonal fluctuation and varying metabolic demands.[1] In counterpoint, thyroid dysfunction may affect pregnancy outcome. Increases in thyroxine-binding globulin are mainly responsible for the changes that occur in the parameters of thyroid function.

All forms of thyroid disease are four to five times more common in women than in men and thus are not rare during gestation. The clinical diagnosis of thyroid disease may be obscured by pregnancy while the laboratory diagnosis may be obscured by the changes in thyroid function tests. The presence of the fetus complicates the therapeutic decision once the diagnostic problem has been solved.

Many of the changes in the laboratory tests of thyroid function are secondary to an increase in thyroxine-binding globulin which is induced by the elevated estrogens found in pregnancy. Prior to the development of chemical determinations of thyroid function, the goiter and elevated BMR found in pregnant women suggested that the thyroid gland in pregnancy was hyperactive. The histologic picture of follicular hypertrophy and hyperplasia in the thyroid during pregnancy further reinforced this impression. Goiter in pregnancy is now thought to be due to compensatory

Advances in Perinatal Thyroidology, Edited by B.B. Bercu and
D.I. Shulman, Plenum Press, New York, 1991

enlargement in iodine-poor areas because of the increased iodine excretion secondary to the increased glomerular filtration role during pregnancy. The BMR increases during the first month of gestation and continues to rise slowly until the eighth month with a total 15 - 20% increase under rigorous basal metabolic conditions. About four-fifths of this increase can be accounted for by the increased oxygen consumption of the fetoplacental unit and the rest by the increase in maternal cardiac output.

As a result of these observations, there has been general acceptance that normal pregnancy is a euthyroid state. Net thyroxine turnover and, inferentially, thyroid hormone production are unchanged in pregnancy.[2] The net thyroxine turnover was 90 ug/day in nonpregnant women and 97 ug/day in pregnant women. When expressed as the daily turnover per square meter of body surface, there was no difference. Nevertheless, possible thyroid hyperfunction has been suggested by the lower TSH concentration in early pregnancy with elevated HCG concentrations and thyroid stimulatory activity in the sera of normal pregnant women.[3]

The fundamental status of the thyroid gland during pregnancy has important clinical implications. Hypothyroid animals have an increased incidence of fetal wastage. The stillbirth rate in 244 pregnant hypothyroid women was double that of the control group. Nevertheless, myxedematous women have been reported to carry their pregnancies to term successfully. In addition to maternal thyroid status, there is also the question of the placental transfer of thyroid hormone to the fetus. The available evidence suggests that T3 and T4 do cross the placental barrier but in inadequate amounts. The stage of gestation and aging of the placenta may play an important role in placental transfer.

However, the issue of thyroid status during pregnancy is still not resolved. In a recent study of 606 pregnant women, the serum T3 and serum T4 concentrations did not increase as much as would be expected by the increase in TBG.[4] Relative hypothyroxinemia was present in approximately one-third of the pregnant women with an apparent compensatory increase in serum TSH concentrations. In a study of 12 pregnant women receiving thyroid hormone for primary hypothyroidism, the serum TSH concentration increased in all patients with a significant decrease in the mean serum-free thyroxine index.[5] The dose of thyroxine was increased in nine of the 12 patients during gestation because of the increase in serum TSH concentration. The authors questioned whether ordinarily there is an increase in thyroxine production during the first trimester in normal pregnancy. Studies in pregnant rats have demonstrated a decrease in serum T3 and serum T4 concentrations during the last stages of pregnancy.[6] They have interpreted the changes as being similar to non-thyroidal illness in decreasing the catabolic processes. At this time the question of thyroid status during normal pregnancy remains unresolved.

Thyrotoxicosis During Pregnancy

Hyperthyroidism tends to be more easily controlled in the pregnant woman with relapses occurring in the postpartum period. There is a significant increase in low birth weight neonates and a suggestion of increased congenital anomalies in the offspring of mothers with untreated Graves' disease.[7,8]

There is no convincing evidence that fertility is a problem in mild or moderate hyperthyroidism. Once pregnancy ensues, there is disagreement whether fetal wastage is increased in thyrotoxic patients. The balance of available evidence suggests that mild to moderate thyrotoxicosis does not increase fetal mortality. Both toxemia of pregnancy in the mother and Down syndrome in the offspring have been reported to be increased in thyrotoxic pregnancies, but the studies were not well-controlled and the correlation appears doubtful.[9,10,11]

Etiology and Course

Graves' disease is the most common form of thyrotoxicosis occurring in the pregnant woman. Other causes of hyperthyroidism, including trophoblastic tumors and hydatiform mole, occur but are much less common. Hyperthyroidism in the pregnant woman may be associated with severe vomiting and raises the question of the relation between hyperemesis gravidarum and thyrotoxicosis.[12] Interestingly, about half the patients with hyperemesis gravidarum have elevated thyroid hormone concentrations.[13,14,15] In most patients the hyperthyroxinemia is transient and values return to normal pregnancy levels. Both HCG and TSH share a common alpha subunit, and the elevated serum HCG concentration has been proposed as the agent in hyperemesis gravidarum. Hyperthyroxinemia does not occur in the ordinary "morning sickness" of pregnancy.[16] The hyperemesis always resolves after delivery, but it is not clear that treating the hyperthyroxinemia relieves the vomiting.[17,18] Treatment should be considered in women who continue to be hyperthyroxinemic and hyperemetic after twenty weeks gestation. Recently erythrocyte zinc concentrations have been reported to differentiate the hyperthyroxinemia in hyperemesis gravidarum from Graves' disease.[19]

Course of Graves' Disease

Studies on the immune response in the pregnant woman have provided an explanation for the observation that women with Graves' disease tend to undergo a remission during pregnancy with an exacerbation of the thyrotoxicosis in the postpartum period. The fetus carries a full complement of paternal antigens. To protect the fetal allograft, humoral and cell-mediated immunity are depressed during normal pregnancy.[20,21,22] These changes are thought to result in an amelioration of Graves' thyrotoxicosis during pregnancy.[23] The removal of the increased fetal suppressor T cell function at the time of delivery might result in the postpartum exacerbation of the autoimmune disease.[24] Soluble factors from activated fetal suppressor cells cross the placenta but disappear at delivery with presumably a transient enhancement of the maternal immune response.[25,26]

Diagnosis

Pregnancy, like thyrotoxicosis, is a hyperdynamic state which makes the clinical diagnosis of thyrotoxicosis in the pregnant woman difficult. Heat intolerance, warm skin, tachycardia and systolic flow murmur may occur in the euthyroid pregnant woman. The difficulty in clinical diagnosis is compounded by difficulties in laboratory diagnosis which make it difficult to confirm the diagnosis. The presence of the eye changes of Graves' or pretibial myxedema are helpful but are not diagnostic of thyrotoxicosis. A resting pulse above 100 that fails to slow during a Valsalva maneuver should also rouse suspicion.

Laboratory Diagnosis

The serum thyroxine concentration is elevated in pregnancy, but values above 15 ug/dl are suggestive of hyperthyroidism. The recent development of the sensitive TSH immunometric assay is extremely helpful in the laboratory diagnosis of hyperthyroidism in the pregnant woman. It is reliable below 0.1 Uu/ml and not affected by the increase in TBG concentration. Other tests like a resin T3 uptake can be helpful and may be combined mathematically with the serum thyroxine value to obtain a free T4 index. The nonequilibrium dialysis methods for the estimation of free thyroxine may be helpful, but both tests can be affected by the TBG increase.[27,28] In a pregnant woman who is clearly thyrotoxic clinically with normal serum thyroxine values, the possibility of T3 toxicosis should be considered.

Therapy

Although both the mother and fetus tolerate mild to moderate degrees of thyrotoxicosis relatively well, there is an increased risk of thyroid storm, and the pregnant woman with thyrotoxicosis should be treated. Since radioactive iodine ablation of the thyroid is contraindicated, the choice lies between surgery and antithyroid drug therapy.

Surgery

There are several reasons why subtotal thyroidectomy is not the therapy of choice for the pregnant woman with hyperthyroidism. First, there is a surgical and anesthetic risk which is probably higher than the fatal risk associated with drug therapy.[29,30] Secondly, although complications of hypoparathyroidism and recurrent laryngeal paralysis are uncommon, they do occur and are difficult to correct. Unfortunately, this becomes a self-fulfilling prophecy. As less surgery is done, the complication rate increases.

If the decision is made to perform a subtotal thyroidectomy on a pregnant woman, surgery is often delayed until after the first trimester. The spontaneous abortion is highest during the first trimester, and there is concern that surgery during this period might increase the risk. However, surgery need not be avoided during the first trimester if indicated.[29] After appropriate preparation, subtotal thyroidectomy can be performed without incident.[31] If surgery is required for other reasons in a pregnant patient with thyrotoxicosis, regional epidural anesthesia is an option.[32]

If the pregnant woman with thyrotoxicosis is to have surgery, she should be observed carefully for signs of hypothyroidism postoperatively with serum TSH and T4 determinations. At the first laboratory indication of hypothyroidism, she should be started on 0.125 mg thyroxine and monitored at 4-6 week intervals. Subtotal thyroidectomy should probably be reserved for pregnant patients with antithyroid drug reactions or poor compliance.[33]

Anti-thyroid Drug Therapy

Even if the decision is made that subtotal thyroidectomy is indicated, the hyperthyroidism must first be controlled with antithyroid drugs.

Beta-blocking Agents

Beta-blocking agents, such as propranolol, have been used in the treatment of thyrotoxicosis.[34] Large numbers of pregnant women with hypertension have been treated with beta-blocking agents without incident.[34,35] However, there have been reports of intrauterine growth retardation with a small placenta, impaired response to anoxic stress and postnatal bradycardia and hypoglycemia.[36,37,38] For these reasons, long-term treatment has not been recommended in pregnant women but there are enough data to indicate that they may be used safely.[35] Cardioselective blockers, such as atenolol, may be used but no clear advantage has been demonstrated over propranolol.[39,40]

Thioamide Therapy

Thioamides are the mainstay of antithyroid drug therapy. They inhibit the synthesis of thyroid hormone by blocking iodination of the tyrosine moiety, but not the release. Therefore, clinical response to thioamides does not occur until the thyroid hormone stored in the colloid is utilized. As a consequence, the time required to achieve control depends upon the amount of iodinated thyroglobulin stored in the gland. Some clinical improvement may be noted after the first week of therapy and euthyroidism may be attained after 4-6 weeks.

Propylthiouracil (PTU) and methimazole both have relatively short durations of action. Although most women may be maintained on a single daily dose, some patients require the thioamides every eight hours or less for adequate control of the hyperthyroidism.[41] Propylthiouracil has the advantage of partially blocking the deiodination of T4 to T3 in addition to the inhibition of thyroid hormone synthesis. In addition, there is at least the suggestion that methimazole may be associated with aplasia cutis in the offspring.[42,43]

Complications of Thioamide Therapy

Skin rash, pruritus, drug fever and nausea are the most common drug reactions to thioamide administration occurring in about two percent of patients. These minor side effects usually occur during the first month of therapy. The development of one of these reactions does not require cessation of thioamide therapy. However, the patient should be switched from PTU to methimazole. If agranulocytosis occurs, the antithyroid drug should be stopped immediately. The drug-related leukopenia tends to occur between four and eight weeks of treatment. Although agranulocytosis occurs in about 3/1000 treated patients, it is fatal in less than 1/10,000. About ten percent of patients with Graves' disease will have leukopenia, and a polymorphonuclear leukocyte count should be obtained before antithyroid drug therapy is started.

Placental Transfer of Thioamides

Propylthiouracil crosses the placenta without difficulty and blocks the fetal thyroid.[44] In a few instances the thioamides may cause transient hypothyroidism in the fetus with goiter formation. Extrapolation suggests that about one to five percent of children exposed to thioamides develop transient hypothyroidism. Under ordinary circumstances sufficient maternal thyroid hormone appears to cross the placenta to prevent fetal goiter.[45] The attempt should be made to treat the pregnant woman with hyperthyroidism with as low a dose of thioamide as possible. It is better to err on the side of undertreatment of thyrotoxicosis during pregnancy. Finally, early diagnosis and treatment of neonatal hypothyroidism appear to prevent the sequelae of hypothyroidism in utero.

The question is raised whether children exposed to thioamides in utero attain optimal intellectual development. To answer this question we administered intelligence tests to 29 children who had been exposed to PTU in utero and compared them to 32 non-exposed siblings. No significant differences were noted between the two groups. With careful attention thioamides can be given to pregnant women without adversely affecting subsequent intellectual development in the offspring.

REFERENCES

1. G. N. Burrow, Thyroid status in normal pregnancy. An editorial, J. Clin. Endocrinol. Metab. 71:274 (1990).
2. J. T. Dowling, W. G. Appleton, and J. T. Nicoloff, Thyroxine turnover during human pregnancy, J. Clin. Endocrinol. Metab. 27:1749 (1967).
3. N. Yoshikawa, M. Nishikawa, M. Horimoto, M. Yoshimura, S. Sawaragi, Y. Horikoshi, I. Sawaragi, and M. Inada, Thyroid-stimulating activity in sera of normal pregnant women, J. Clin. Endocrinol. Metab. 69:891 (1989).
4. D. Glinoer, P. De Nayer, P. Bourdoux, M. Lemone, C. Robyn, A. Van Steirteghem, J. Kinthaert, and B. Lejeune, Regulation of maternal thyroid during pregnancy, J. Clin. Endocrinol. Metab. 71:276 (1990).
5. S. J. Mandel. P. R. Larsen, E. W. Seely, and G. A. Brent, Increased

need for thyroxine during pregnancy in women with primary hypothyroidism, N. Engl. J. Med. 323:91 (1990).

6.   R. Calvo, M. J. Obregon, C. Ruiz de Ona, B. Ferreiro, F. Escobar del Rey, and G. Morreale de Escobar, Thyroid hormone economy in pregnant rats near term: a "physiological" animal model of nonthyroidal illness?, Endocrinol. 127:10 (1990).

7.   K. R. Niswander, M. Gordon, and H. W. Berendes, "The Women and Their Pregnancies," W. B. Saunders Company, Philadelphia (1972).

8.   N. Momotani, K. Ito, N. Hamada, Y. Ban, Y. Nishikawa, and T. Mimura, Maternal hyperthyroidism and congenital malformation in the offspring, Clin. Endocrinol. 20:695 (1984).

9.   C. W. McLaughlin, Jr. and L. S. McGoogan, Hyperthyroidism complicating pregnancy, Am. J. Obstet. Gynecol. 45:591 (1943).

10. Y. Ueda, K. Hayaski, Y. Kishimoto, and S. Mizusawa, Sexual function in women with diseases of the thyroid gland, J. Jpn. Obstet. Gynecol. Soc. 11:48 (1964).

11. C. R. Myers, An application of the control group method to the problem of the etiology of mongolism, Proc. Am. Assoc. Mental Def. 62:142 (1938).

12. B. H. Valentine, C. Jones, and A. J. Tyack, Hyperemesis gravidarum due to thyrotoxicosis, Postgrad. Med. J. 56:746 (1980).

13. R. Bouillon, M. Naesens, F. A. Van Assche, L. De Keyser, P. De Moor, M. Renaer, P. De Vos, and M. De Roo, Thyroid function in patients with hyperemesis gravidarum, Am. J. Obstet. Gynecol. 143:992 (1982).

14. R. K. H. Chin and T. T. H. Lao, Thyroxine concentrations and outcome of hyperemetic pregnancies, Br. J. Obstet. Gynecol. 95:507 (1988).

15. T. T. Lao, R. K. H. Chin, N. S. Panesar, and R. Swaminathan, Observations on thyroid hormones in hyperemesis gravidarum, Asia-Oceania J. Obstet. Gynecol. 14:449 (1988).

16. A. J. Evans, T. C. Li, C. Selby, and W. J. Jeffcoate, Morning sickness and thyroid function, Br. J. Obstet. Gynecol. 93:520 (1986).

17. R. Dozeman, F. E. Kaiser, and O. Cass, Hyperthyroidism appearing as hyperemesis gravidarum, Arch. Int. Med. 143:2202 (1983).

18. B. Kirshon, W. Lee, and D. B. Cotton, Prompt resolution of hyperthyroidism and hyperemesis gravidarum after delivery, Obstet. Gynecol. 71:1032 (1988).

19. T. T. H. Lao, R. K. H. Chin, R. Swaminathan, N. S. Panesar, and C. S. Cockram, Erythrocyte zinc in differential diagnosis of hyperthyroidism in pregnancy, Br. Med. J. 294:1064 (1987).

20. A. E. Beer and R. E. Billingham, Immunobiology of mammalian

reproduction, Adv. Immunol. 14:1 (1971).

21. N. Amino, R. Kuro, O. Tanizawa, F. Tanaka, C. Hayashi, K. Kotani, M. Kawashima, K. Miyai, and Y. Kumahara, Changes of serum anti-thyroid antibodies during and after pregnancy in autoimmune thyroid diseases, Clin. Exp. Immunol. 31:30 (1978).

22. N. Amino, O. Tanizawa, K. Miyai, F. Tanaka, C. Hayashi, M. Kawashima, and K. Ichihara, Changes of serum immunoglobulins IgG, IgA, IgM, and IgE during pregnancy, Obstet. Gynecol. 52:415 (1978).

23. T. F. Davis and L. Weiss, Autoimmune thyroid disease and pregnancy, Am. J. Reprod. Immunol. 1:187 (1981).

24. V. Sridama, F. Pacini, S. Yang, A. Moawad, M. Reiley, and L. J. de Groot, Decreased levels of helper T cells: A possible cause of immunodeficiency in pregnancy, N. Engl. J. Med. 307:352 (1982).

25. E. J. Froelich, J. S. Goodwin, A. D. Bankhurst, and R. C. Williams, Pregnancy, a temporary fetal graft of suppressor levels in autoimmune disease, Am. J. Med. 69:329 (1980).

26. N. Amino, K. Miyai, T. Yamamoto, R. Kuro, F. Tanaka, O. Tanizawa, and Y. Kumahara, Transient recurrence of hyperthyroidism after delivery in Graves' disease, J. Clin. Endocrinol. Metab. 44:130 (1977).

27. J. A. Souma, D. C. Niejadlik, S. Cottrell, and S. Rankle, Comparison of thyroid function in each trimester of pregnancy with the use of triiodothyronine uptake, thyroxine iodine, free thyroxine and free thyroxine index, Am. J. Obstet. Gynecol. 116:905 (1973).

28. L. R. Witherspoon, S. E. Shuler, M. M. Garcia, and L. A. Zollinger, An assessment of methods for the estimation of free thyroxine, J. Nucl. Med. 21:529 (1980).

29. J. B. Brodsky, E. N. Cohen, B. W. Brown, Jr., M. L. Wu, and C. Whitcher, Surgery during pregnancy and fetal outcome, Am. J. Obstet. Gynecol. 138:1165 (1980).

30. A. B. Weingold, Surgical diseases in pregnancy, Clin. Obstet. Gynecol. 26:793 (1983).

31. R. F. Emslander, R. E. Weeks, and G. D. Malkasian, Jr., Hyperthyroidism and pregnancy, Med. Clin. North Am. 58:835 (1974).

32. S. H. Halpern, Anaesthesia for caesarean section in patients with uncontrolled hyperthyroidism, Can. J. Anesth. 36:454 (1989).

33. R. Innerfield and C. S. Hollander, Thyroidal complications of pregnancy, Med. Clin. North Am. 61:67 (1977).

34. J. L. Bullock, R. E. Harris, and R. Young, Treatment of thyrotoxicosis during pregnancy with propranolol, Am. J. Obstet. Gynecol. 121:242 (1975).

35.   P. C. Rubin, Beta-blockers in pregnancy, N. Engl. J. Med. 18:73
        (1983).
36.   G. R. Gladstone, A. Hordof, and W. M. Gersony, Propranolol
        administration during pregnancy: Effects on the fetus, J. Pediatr.
        86:962 (1975).
37.   A. Habib, and J. S. McCarthy, Effects on the neonate of propranolol
        administered during pregnancy, J. Pediatr. 91:808 (1977).
38.   S. C. Pruyn, J. P. Phelan, and G. C. Buchanan, Long-term
        propranolol therapy in pregnancy: Maternal and fetal outcome, Am.
        J. Obstet. Gynecol. 135:485, (1979).
39.   P. C. Rubin, L. Butters, D. M. Clark, D. J. Sumner, R. A. Low, B.
        Reynolds, D. Steedman, J. L. Reid, Placebo-controlled trial of
        atenolol in treatment of pregnancy-associated hypertension, Lancet
        1:431 (1983).
40.   B. Sandstrom, Antihypertensive treatment with the adrenergic beta-
        receptor blocker metoprolol during pregnancy, Gynecol. Invest.
        9:195 (1978).
41.   M. A. Greer, W. C. Meihoff, and H. Studer, Treatment of
        hyperthyroidism with a single daily dose of propylthiouracil, N. Engl.
        J. Med. 272:888 (1965).
42.   Q. Mujtaba and G. N. Burrow, Treatment of hyperthyroidism in
        pregnancy with propylthiouracil and methimazole, Obstet. Gynecol.
        46:282 (1975).
43.   M. J. Stephan, D. W. Smith, J. W. Ponzi, and E. R. Alden, Origin of
        scalp vertex aplasia cutis, J. Pediatr. 101:850 (1982).
44.   B. Marchant, B. E. W. Brownlie, D. M. Hart, P. W. Horton, and W.
        D. Alexander, The placental transfer of propylthiouracil,
        methimazole and carbimazole, J. Clin. Endocrinol. Metab. 45:1187
        (1977).
45.   T. Vulsma, M. H. Gons, and J. J. M. de Vijlder, Maternal-fetal
        transfer of thyroxine in congenital hypothyroidism due to a total
        organification defect or thyroid agenesis, N. Engl. J. Med. 321:13
        (1989).

POSTPARTUM THYROID DISEASE

Nobuyuki Amino, M.D.

Department of Laboratory Medicine
Osaka University Medical School
Osaka, 553, Japan

INTRODUCTION

Graves' disease was first described by Caleb H.
Parry in 1825. The first case among his 6 patients seen
in 1786 had experienced palpitation, neck swelling and
protrusion of eyes after delivery(1). Thus the first
patients with Graves' (Parry's) disease described in the
medical literature was of postpartum onset. In 1840,
Karl A. von Basedow reported 4 cases of Graves'
(Basedow's) disease(2). Interestingly, one of his 3
female patients was also a case of postpartum onset.
Reviewing these old reports, it is likely that postpartum
onset of Graves' disease may be frequently observed in
the general population. However, little is known about
the mechanism of postpartum onset of Graves' disease.

Postpartum hypothyroidism has long been known to
occur as a result of hypopituitarism since the report by
Sheehan in 1937(3). In 1948, however, Roberton reported
the frequent occurrence of hypothyroidism after delivery
in association with endemic goiter and suggested that
iodine deficiency in the postpartum period might be one
factor inducing these changes(4). In 1955 Fraser and
Garrod(5) reported four cases of primary myxedema dating
from "postpartum shock or hemorrhage." They did not
discuss its possible relation to autoimmune abnormalities
since the concept of autoimmune thyroiditis was unknown
at that time.

Recently we found pregnancy markedly influenced the
clinical course of autoimmune thyroid diseases and, in
1968, observed the first case of transient postpartum
hypothyroidism(6). Since then many cases have been

Advances in Perinatal Thyroidology, Edited by B.B. Bercu and
D.I. Shulman, Plenum Press, New York, 1991

reported as well as other types of postpartum thyroid dysfunction. Now it is well recognized in many countries in the world. This entity has been called postpartum autoimmune thyroid syndrome, or more simply postpartum syndrome or postpartum thyroiditis.

This chapter deals with recent progress in the effect of pregnancy on autoimmune thyroid disease with special emphasis on the various types of postpartum thyroid dysfunction.

POSTPARTUM CHANGES IN THYROID FUNCTION IN HASHIMOTO'S DISEASE

Figure 1 shows the spontaneous changes of free T_4 index (FT_4I), goiter size and titer of anti-thyroid microsomal antibody (McAb) in a pregnant patient with Hashimoto's disease. Goiter size and McAb titer decreased in late pregnancy when compared with values in early pregnancy. After delivery, however, destructive thyrotoxicosis developed at 1-3 months postpartum and then transient hypothyroidism was observed at 4-6 months postpartum. In association with postpartum aggravation of thyroid disease McAb titer increased markedly and enlargement of thyroid gland was observed after delivery.

Figure 2 shows various changes in the free T_4 index after delivery in 15 patients with Hashimoto's disease(8). Patients were divided into the following three groups according to the severity of hypothyroidism 2 to 6 months postpartum: in group (a), the maximum level of serum TSH was more than 50 $\mu U/ml$; in group (b), the serum TSH was 10 to 50 $\mu U/ml$; and in group (c), the serum TSH was less than 10 $\mu U/ml$. In group (a), four of six patients showed marked thyrotoxicosis 1 to 3 months postpartum. After thyrotoxicosis, these patients all developed hypothyroidism 3 to 4 months postpartum and then their TSH levels returned to the normal range 6 to 8 months postpartum. Two other patients developed overt hypothyroidism without preceding thyrotoxicosis. In group (b), four of five patients showed mild thyrotoxicosis at 2 to 4 months postpartum and then developed mild hypothyroidism with a serum TSH of less than 50 $\mu U/ml$ at 3 to 5 months postpartum. None of the four patients in group (c) developed hypothyroidism, although two showed very mild transient thyrotoxicosis 3 to 4 months postpartum. All cases of thyrotoxicosis resulted from thyroid destruction. These data indicate that postpartum hypothyroidism is the consequence of postpartum transient thyrotoxicosis and marked thyroid destruction in the early postpartum period leading to more obvious hypothyroidism.

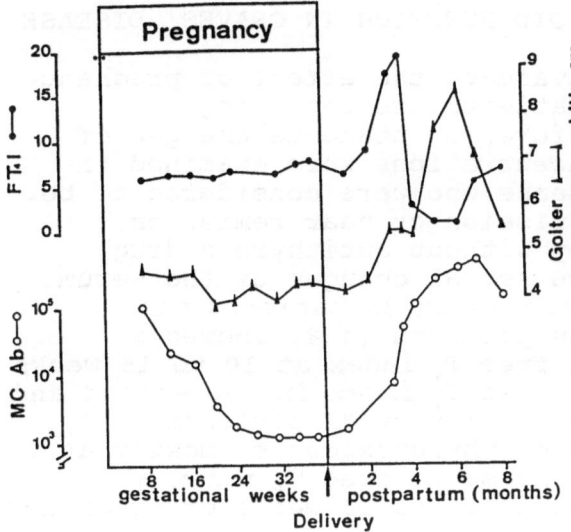

Figure 1.
Changes of free T_4 index (FT_4I), goiter size and anti-thyroid microsomal antibody during and after pregnancy in a patient with Hashimoto's disease.

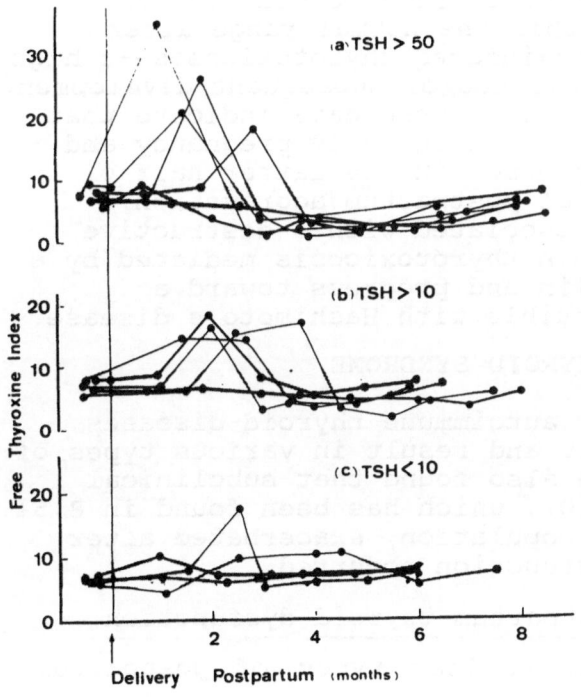

Figure 2.
Change of free T_4 index after delivery in 15 patients with Hashimoto's disease. (See text)

POSTPARTUM CHANGE IN THYROID FUNCTION IN GRAVES' DISEASE

It is difficult to evaluate the effect of pregnancy on Graves' disease when patients are receiving antithyroid drugs. Therefore, spontaneous changes of serum thyroid hormone concentrations were examined in patients with Graves' disease who were considered to be in a state of clinical remission or near remission, during and after pregnancy without antithyroid drug treatment. Figure 3 shows serial changes of the serum free T_4 index in 41 pregnancies in 35 patients with Graves' disease. Eighteen patients (44%) showed a transient increase of the free T_4 index at 10 to 15 weeks of pregnancy but a normal free T_4 index in the second and third trimesters(9). After delivery 32 patients (78%) developed varying degrees of thyrotoxicosis, mostly at 2 to 4 months postpartum. The serum free T_3 index also showed a similar elevation at 10 to 15 weeks of pregnancy and after delivery.

These patients with Graves' disease were divided into the following four groups according to postpartum changes of thyroid function: (1) persistent thyrotoxicosis--free T_4 index more than 20 with high radioactive iodine uptake and thyrotoxicosis persisting for more than 2 months; (2) transient thyrotoxicosis--a high free T_4 index with normal or high radioactive iodine uptake (RAIU) and transient thyrotoxicosis; (3) no change--free T_4 index within the normal range after delivery; (4) destruction-induced thyrotoxicosis--a high free T_4 index with low RAIU and/or subsequent development of transient hypothyroidism. These data indicate that Graves' disease was aggravated in early pregnancy and after delivery, and suppressed in the latter half of pregnancy. At the time of postpartum aggravation, Graves' disease may be associated with a destructive thyrotoxicosis rather than thyrotoxicosis mediated by a stimulating immunoglobulin and progress toward a pathological state compatible with Hashimoto's disease.

POSTPARTUM AUTOIMMUNE THYROID SYNDROME

As described above, autoimmune thyroid diseases exacerbate after delivery and result in various types of thyroid dysfunction. We also found that subclinical autoimmune thyroiditis(10), which has been found in 8.5% of women in the general population, exacerbated after delivery and thyroid dysfunction occurred.

1. Various types of postpartum thyroid dysfunction

Figure 4 illustrates various types of postpartum

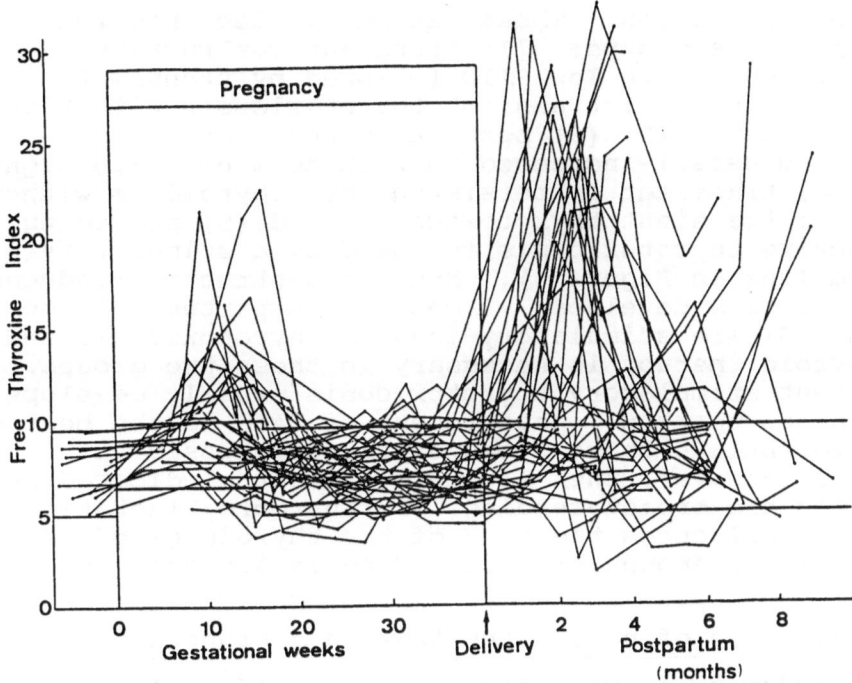

Figure 3. Spontaneous sequential changes of serum free T_4
           index during and after pregnancy in patients
           with Graves' disease who were considered to be
           in a state of remission or near remission.

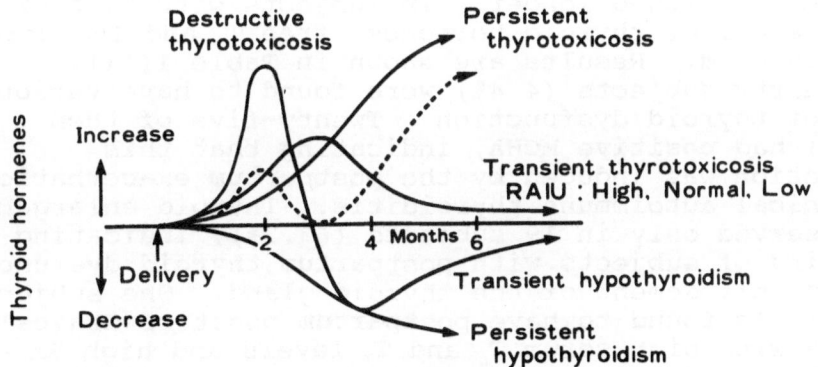

Figure 4. Various types of postpartum thyroid dysfunction.
          (Modified from Reference 8)

thyroid dysfunction. These can be divided into the
following five groups: (1) transient destructive
thyrotoxicosis with low RAIU followed by transient
hypothyroidism; (2) transient thyrotoxicosis with high,
normal or low RAIU; (3) persistent thyrotoxicosis
(Graves' disease)-increased thyroid hormones with high
RAIU; (4) transient or persistent hypothyroidism without
preceding transient thyrotoxicosis; and (5) preceding
destructive thyrotoxicosis followed by a stimulatory one
(dotted line in Figure 4). Most of destruction-induced
thyrotoxicosis develops 1-3 months postpartum. Groups
(3) and (5) are stimulatory-induced thyrotoxicosis and
antithyroid therapy is necessary in these two groups.
Persistent stimulatory thyrotoxicosis usually develops 3-
4 months postpartum. Measurement of RAIU is the best way
to differentiate these types of thyrotoxicosis at
present. However, the RAIU test is contraindicated in
women who are breast-feeding. Furthermore, lactation
markedly influences the RAIU of the thyroid gland, since
a significant amount of radioiodine is secreted into
milk.

2. Prevalence of postpartum thyroid dysfunction

In order to examine the prevalence of postpartum
thyroid dysfunction in the general population, thyroid
abnormalities were examined in 680 consecutive mothers at
3 months postpartum. The neck was palpated and serum T_4,
T_3, TSH, anti-thyroglobulin hemagglutination antibody
(TGHA), and anti-thyroid microsomal hemagglutination
antibody (MCHA) were measured in all subjects. Any
subjects in whom the transverse width of the thyroid
gland was more than 4.0 cm was considered to have a
goiter. When the subjects had abnormal thyroid function,
enlarged thyroid gland and/or positive titers of MCHA,
they were followed longer. In subjects with high or low
serum levels of thyroid hormones, free T_4 and TBG were
also measured. Results are shown in Table 1(11).

Thirty subjects (4.4%) were found to have various
types of thyroid dysfunction. Twenty-five of them
(83.3%) had positive MCHA, indicating that this
dysfunction was induced by the postpartum exacerbation of
subclinical autoimmune thyroiditis. Thyroid enlargement
was observed only in 19 subjects (63.3%), indicating that
one-third of subjects with postpartum thyroid dysfunction
have no enlargement of the thyroid gland. One subject
(0.15%) was found to have postpartum onset of Graves'
disease with high serum T_4 and T_3 levels and high RAIU.
One subject (0.15%) was found to have irreversible
persistent hypothyroidism which was confirmed by the
follow-up study. The remaining 28 (93.3%) subjects had

Table 1. Prevalence of postpartum thyroid dysfunction amoung
          680 consecutive postpartum women

Type of thyroid Abnormality	Patient No. (%)	Number of cases		
		Goiter	TGHA	MCHA
Transient thyrotoxicosis	13 (1.91)	6	5	11
Transient thyrotoxicosis followed by transient hypothyroidism	7 (1.03)	4	3	6
Persistent thyrotoxicosis	1 (0.15)	1	0	0
Transient hypothyroidism	8 (1.18)	8	3	7
Persistent hypothyroidism	1 (0.15)	1	1	1
Total	30 (4.41)	19	12	25

(Modified from Reference 11)

transient thyroid dysfunction. Destructive-type
thyrotoxicosis was found in 20 subjects (2.9%) and
postpartum hypothyroidism was found in 16 subjects
(2.6%). The incidence of postpartum hypopituitarism (so
called Sheehan's syndrome) has been estimated to be
0.003% (4 of 124,752 deliveries)(12). Thus, postpartum
primary hypothyroidism is 1000 times more frequent than
postpartum secondary hypothyroidism. Even if we include
only persistent postpartum hypothyroidism, this figure is
still far higher than that for postpartum pituitary
hypothyroidism.
        The high prevalence of postpartum thyroid
dysfunction was not only observed in Japan(11) but also
in other countries. In Sweden(13), the United States(14)
and Canada(15) the incidence was reported to be 5-7%. A
much higher prevalence of 17% was reported in Wales(16),
but this was in a preselected population including women
with positive anti-thyroid antibodies.

3. Treatment for postpartum thyrotoxicosis and
hypothyroidism

        Most cases of postpartum thyrotoxicosis are self
limited, and thyrotoxic symptoms are transient. Normal
postpartum women have various symptoms which may obscure
the presentation of postpartum thyroid dysfunction. The
frequency of fatigue and palpitation are increased in the
thyrotoxic patients compared to controls; the severity of

symptoms is also greater in some patients. When patients
report moderate or severe symptoms of thyrotoxicosis, a
mild tranquilizer or ß-blocker may be used temporarily.
When patients are revealed to have persistent stimulative
thyrotoxicosis (postpartum Graves' disease with high RAIU
and positive stimulating antibody), an antithyroid drug
should be prescribed; the early initiation of drug
therapy is associated with early remission.
Propylthiouracil (PTU) is preferred over methimazole,
since less PTU is transferred in breast milk and it has
a negligible effect on the neonate in lactating women.

     As for postpartum hypothyroidism, hypometabolic
symptoms are less severe compared to those of long-
standing hypothyroidism. Replacement therapy with a
submaximal dose of T_3 may be useful for postpartum
hypothyroidism when patients report severe hypometabolic
symptoms, since spontaneous recovery of thyroid function
can be predicted by the significant increase of serum T_4
during treatment(17). One example of T_3 therapy is shown
in Figure 5. In this patient, hypothyroidism developed
two months postpartum and T_3 therapy was started at three
months postpartum. During this therapy, serum TSH was

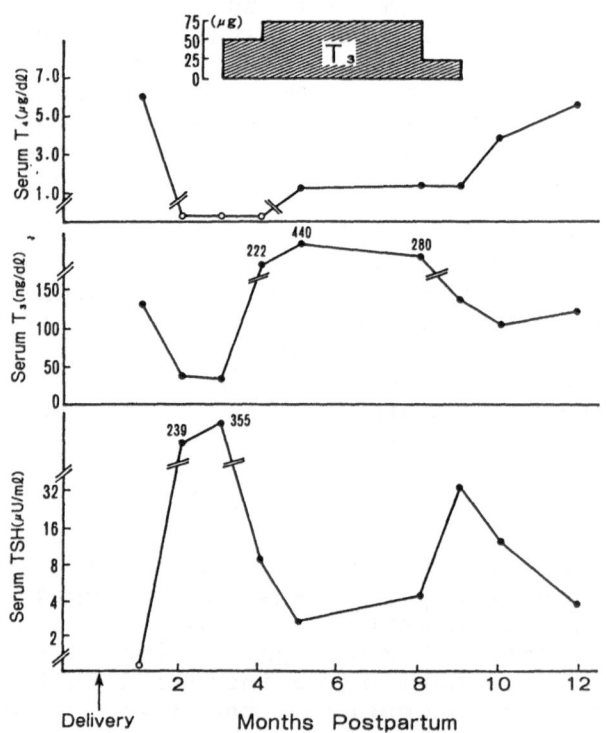

Figure 5.
Treatment with
submaximal dose
of T_3 in a patient
with postpartum
hypothyroidism.

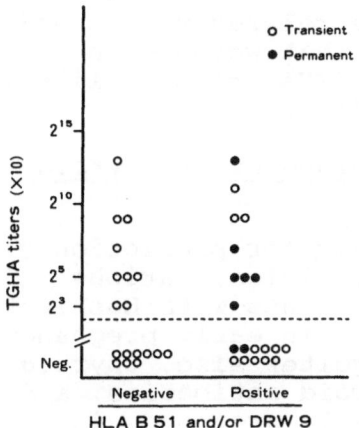

Figure 6.
The relationships
between HLA-B51
and/or HLA-DRW9
and anti-
thyroglobulin
(TGHA) titers in
women with transient
and permanent post-
partum hypothyroidism.

always detectable. A significant increase of serum T_4 was observed at five months postpartum. This is a good sign of spontaneous recovery of thyroid function. Finally T_3 therapy was discontinued at nine months postpartum and the patient recovered completely to the euthyroid state.

4.   Long term follow up study

Most postpartum hypothyroidism is transient but little is known about its long term clinical course. In order to clarify this point we followed 44 patients with postpartum hypothyroidism (59 postpartum episodes; mean age of mothers at delivery, 28.2 years) for five or more years (mean interval after delivery, 8.7 years; range, 5-16 years)(18). Forty-nine episodes (83%) in 34 women were followed by recovery within 1 year postpartum, and those women remained euthyroid thereafter (group A); 10 women [10 episodes (17%)] developed permanent hypothyroidism during the follow-up period (group B). Five women in group B recovered during the first year, but became hypothyroid again later; the other 5 women in Group B remained persistently hypothyroid. HLA typing revealed significantly higher frequencies of HLA-DR3, -DRW8, -DRW9, -A26, -BW46, and -BW67, and significantly lower frequencies of HLA-DR2, -BW52, -BW62, and -CW7 in women with postpartum hypothyroidism than in normal women. Of 9 women with postpartum hypothyroidism who had HLA-DRW9 and/or -B51 associated with TGHA titers of 2^3 x 10 or higher, 6 developed permanent hypothyroidism in Figure 6. Thus long term follow-up is essential for women with postpartum hypothyroidism because of the risk of permanent hypothyroidism. The results suggest that

some immunogenetic factors may be related to the etiology
of postpartum hypothyroidism and that women with HLA-DRW9
and/or -B51 and higher titers of TGHA are more likely to
develop permanent hypothyroidism(18).

5.   Prediction of postpartum development of thyroid dysfunction

        In order to establish a method for prediction of the
postpartum development of hypothyroidism, various
parameters were measured in 40 patients with Hashimoto's
disease; these women were observed in early pregnancy and
the postpartum period(8). Age, goiter size, thyroid
function tests, titer of antithyroid antibodies, and
peripheral lymphocyte count in early pregnancy in
patients who developed postpartum hypothyroidism (group
A) were compared with those who did not develop
hypothyroidism (group B). Only the titer of MCHA was
found to be significantly different between the two
groups. Individual values of the MCHA titer are shown in
Figure 7. An interesting observation was that 13 of 26
patients (50%) in group A but only 2 of 14 patients (14%)
in group B delivered a female baby. Thirteen of 15
patients (87%) with female babies developed postpartum
hypothyroidism, whereas 13 of 25 patients (52%) with male
babies developed postpartum hypothyroidism. Considering
these data, it is suggested that postpartum
hypothyroidism may occur more frequently in patients with
a female baby and a high titer of MCHA (more than 2^7 x
10) in early pregnancy. Furthermore, postpartum
hypothyroidism can also be predicted in patients who have
had previous postpartum hypothyroidism.

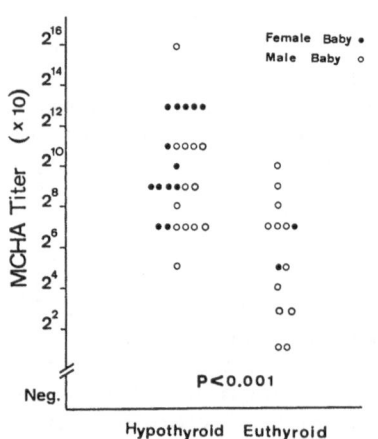

Figure 7.
Comparison of titers
of antithyroid
microsomal antibody (MCHA)
in early pregnancy
in Hashimoto's
patients who developed
postpartum hypo-
thyroidism and those
who did not develop
hypothyroidism.

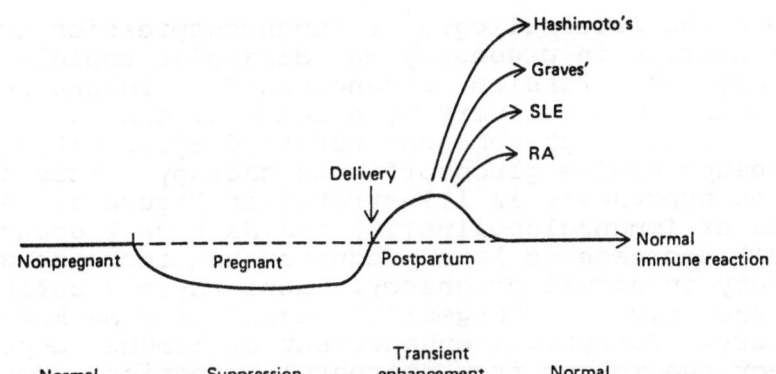

Figure 8. Immune rebound hypothesis regarding the postpartum onset of autoimmune diseases. Possible immunosuppression in pregnancy may disappear at delivery. "Transient enhancement" of the immune reactions may occur after delivery.

It is very important to predict the postpartum onset of Graves' disease, which is also thought to develop from sub-clinical autoimmune thyroiditis. Therefore, the patients with subclinical autoimmune thyroiditis were screened by measurement of MCHA in early pregnancy. Three hundred and eighty-three (13.9%) of 2760 consecutive pregnant women seen in our maternity clinic had positive titers of MCHA. Of these, 125 had previously diagnosed Graves' disease, 36 had Hashimoto's disease and one was newly diagnosed as being hypothyroid. The remaining 221 subjects (8.0%) were found to have subclinical autoimmune thyroiditis. Sixty-five of these patients were followed monthly during pregnancy and after delivery until 6 months postpartum.

Forty-two (64.6%) patients developed various types of thyroid dysfunction after delivery. Postpartum onset of Graves' disease was found in 4 (6.2%) of 65 patients with subclinical autoimmune thyroiditis. One had positive anti-TSH receptor antibody (TRAb). One other patient had suppressed serum TSH levels in the first and second trimesters. Considering these results, screening tests for MCHA first and then for TRAb and TSH in pregnancy may be clinically useful for prediction of the postpartum onset of Graves' disease.

6. Mechanism of postpartum aggravation

Many factors related to gestation have been reported to have immunosuppressive activities and may work to

protect the fetal allograft. Immunosuppression induced by
these factors in pregnancy may disappear rapidly at
delivery and "transient enhancement" of immune reaction
may occur after delivery by a somewhat similar mechanism
to the "rebound phenomenon" observed after withdrawal of
immunosuppressive glucocorticoid therapy. This immune
rebound hypothesis is illustrated in Figure 8. Serum
levels of immunoglobulins(19) and NK/K cell counts
(20,21) decrease in late pregnancy but increase after
delivery in normal pregnancy. Anti-thyroid antibodies
also show similar changes(22), increasing markedly after
delivery. Postpartum enhancement of immune response may
protect the mother from puerperal infection. However,
overt disease may develop if she has a subclinical form
of autoimmune disease. Diseases occurring after delivery
should, therefore, be examined from the standpoint of an
autoimmune abnormality.

REFERENCES

1.   C.H. Parry, Enlargement of the thyroid gland in
     connection with enlargement or palpitation of the
     heart. In Collection from the Unpublished Papers of
     the late Caleb Hilliel Parry 2:111 London (1825).
2.   K.A. von. Basedow, Exophthalmos durch Hypertrophie
     des Zellgewebes in der Augenhohle. Wochenschr.
     Heilk. 6:197, 220 (1840).
3.   H.L. Sheehan, Simmonds's disease due to post-partum
     necrosis of the anterior pituitary. Q. J. Med.
     8:277 (1939).
4.   H.E.W. Roberton, Lassitude, coldness, and hair
     changes following pregnancy, and their response to
     treatment with thyroid extract. Brit. Med. J.
     2:Suppl:2275 (1948).
5.   R. Fraser and O. Garrod, Primary myxoedema
     apparently dating from postpartum shock. Brit. Med.
     J. 2:1484 (1955).
6.   N. Amino, K. Miyai, M. Fukuchi and Y. Kumahara,
     Transient hypothyroidism associated with increased
     antimicrosomal antibodies. Endocrinol. Jap. 22:141
     (1975).
7.   N. Amino, K. Miyai, R. Kuro et al., Transient
     postpartum hypothyroidism: fourteen cases with
     autoimmune thyroiditis. Ann. Intern. Med. 87:155
     (1977).
8.   N. Amino and K. Miyai, Postpartum autoimmune
     endocrine syndromes. In Davies TF (ed), Autoimmune
     Endocrine Disease, John Wiley & Sons, New York, pp.
     247-272 (1983).
9.   N. Amino, O. Tanizawa, H. Mori et al., Aggravation

of thyrotoxicosis in early pregnancy and after delivery in Graves' disease. J. Clin. Endocrinol. Metab. 55:108 (1982).

10. H. Yoshida, N. Amino, K. Yagawa et al., Association of serum antithyroid antibodies with lymphocytic infiltration of the thyroid gland: studies of seventy autopsied cases. J. Clin. Endocrinol. Metab. 46:859 (1978).

11. N. Amino, H. Mori, Y. Iwantani et al., High prevalence of transient postpartum thyrotoxicosis and hypothyroidism. N. Eng. J. Med. 306:849 (1982).

12. H.L. Sheehan, The frequency of post-partum hypopituitarism. J. Obstet. Gynaecol. (Br. Commonw.) 72:103 (1965).

13. R. Jansson, S. Bernander, A. Karlsson, K. Zevin, G. Nilsson, Autoimmune thyroid dysfunction in the postpartum period. J. Clin. Endocrinol. Metab. 58:681 (1984).

14. T.F. Nikolai, J. Brosseau, M.A. Kettrick, R. Roberts, E. Beltaos, Lymphocytic thyroiditis with spontaneously resolving hyperthyroidism (silent thyroiditis). Arch. Intern. Med. 140:478 (1980).

15. P.G. Walfish and J.Y.C. Chan, Post-partum hyperthyroidism. Clin. Endocrinol. Metab. 14:417 (1985).

16. H.Y. Fung, M. Kologlu, K. Collison et al., Postpartum thyroid dysfunction in Mid Glamorgan. Brit. Med. J. 296:241 (1988).

17. N. Amino, Are silent thyroiditis and postpartum silent thyroiditis forms of chronic thyroiditis or different (new) forms of viral thyroiditis?: a commentary. In: Hamburger J.I., Miller (eds), Controversies in Clinical Thyroidology J.M., Springer-Verlag, New York.31:6 (1981).

18. J. Tachi, N. Amino, H. Tamaki et al., Long term follow-up and HLA association in patients with postpartum hypothyroidism. J. Clin. Endocrinol. Metab. 66:48 (1988).

19. N. Amino, O. Tanizawa, K. Miyai et al., Changes of serum immunoglobulins IgG, IgA, IgM and IgE during pregnancy. Obstet. Gynecol. 52:415 (1978).

20. S. Asari, Y. Iwatani, N. Amino, O. Tanizawa and K. Miyai, Peripheral K cells in normal human pregnancy: decrease during pregnancy and increase after delivery. J. Reprod. Immunol. 15:31 (1989).

21. Y. Iwatani, N. Amino, J. Tachi et al., Changes of lymphocyte subsets in normal pregnant and postpartum women: Postpartum increase of NK/K (Leu 7) cells. Am. J. Reprod. Immunol. 18:52 (1988).

22.   N. Amino, R. Kuro, O. Tanizawa et al.   Changes of
      serum anti-thyroid antibodies during and after
      pregnancy in autoimmune thyroid diseases.   Clin.
      Exp. Immunol.   31:30 (1978).

TRANSFER AND METABOLISM OF THYROID-RELATED SUBSTANCES IN

THE PLACENTA

Charles H. Emerson and Lewis E. Braverman

University of Massachusetts Medical School
Worcester, Massachusetts, USA 01655

INTRODUCTION

The primary role of the placenta is to transfer substances between the mother and fetus. Another function is to metabolize fetal substrates, perhaps including thyroid hormones. This paper will outline general aspects of placental transfer (PT), discuss methods that have been used to evaluate "thyroid-related" substances (TRSs), and review the current information regarding their PT.

PLACENTAL TRANSFER - GENERAL PRINCIPALS

According to a model based on the principal of mass-action, with similarities to the formula of Moya and Thorndyke (Reynolds, 1979), the unidirectional rate of transfer of a substance across the placenta is a function of the "effective concentration" (EC) of the substance on the proximate side of the membrane, the overall rate constant (K) for the flux of that substance, and the area of the placental membrane. As noted by Reynolds (Reynolds, 1979) this formula of needs to be interpreted with care. Both K and the EC dependent on many factors (Faber and Thornburg, 1983). The EC is a construct for the concentration of the substance which is available for transfer. It is dependent on the concentration of the substance, either the free or readily disassociated form. This in turn may be influenced by the rate and characteristics of placental blood flow. This varies with the type of placentation. In hemochorial placentas with villus chorionic projections (human and non-human primates), maternal blood may form subchorial lakes as it circulates through the placenta (Ramsey and Donner, 1980). These lakes are thought to bathe the chorionic villus tree, the membrane across which maternal-fetal transfer occurs (Ramsey and Donner, 1980). Alternatively, maternal blood flow in the human placenta has been described as a large network of channels or an extensive vascular bed (Pearson, 1979). In some species with either

Advances in Perinatal Thyroidology, Edited by B.B. Bercu and
D.I. Shulman, Plenum Press, New York, 1991

hemochorial or epitheliochorial placentation, but labyrinthine
internal structure, a countercurrent system of blood flow has
been proposed (Ramsey, 1982). Conceivably these perfusion
patterns could favor the release of substances from plasma
binding proteins by increasing the transit time and
concentration gradient (See Ekins, 1990).

K, the overall rate constant for the flux of a substance,
is inversely proportional to the thickness of the placental
membranes which separate maternal and fetal blood. The
membranes of hemochorial placentas are entirely fetal in
origin. In contrast, epitheliochorial placental membranes
have more tissue layers and are derived from both the mother
and the fetus (Ramsey and Donner, 1980). However,
epitheliochorial placental membranes are not invariably
thicker than hemochorial placental membranes (Faber and
Thornburg, 1983). Humans have hemochorial placentas as do
most species used in thyroid research. This includes non-
human primates, rats, and guinea pigs. An exception is the
sheep, a species with epitheliochorial placentation.

K is also directly proportional to the "permeability" of
placental membranes. "Permeability" is used here in a broad
sense, not to indicate that diffusion is the only mechanism of
transfer. Simple diffusion and carrier-mediated transport,
either active or facilitative, are the most important means of
transcellular placental transfer. For mechanisms that involve
transcellular diffusion K is, after correction for the area of
the placental membrane, equal to the permeability constant
(Rowland and Tozer, 1980). According to Faber, "hemochorial
placentas have transfer properties that are very different
from those of the more complex epitheliochorial placentas
(Faber and Thornburg, 1983). However, since they are not
necessarily thicker than hemochorial placentas, it is likely
that the maternal layers present in epitheliochorial placentas
have a different permeability than the fetal layers which are
the exclusive component of hemochorial placentas. Water
soluble substances diffuse poorly if their molecular weight is
greater than 100 or they are ionized (Reynolds, 1979). For
some substances, transfer is influenced by the nature of their
charge (anionic vs cationic) (Berhe et al.1987). Lipid
soluble molecules whose molecular weight is less than 600 -
1000 diffuse freely (Reynolds, 1979). The chemical nature of
T4 does not favor its rapid PT. Its molecular weight is more
than 700 and, although it has a hydrophobic ring core, it also
has a ionizable phenolic hydroxyl and an alpha amino acid
group. Its solubility in water is low, but so is its
solubility in many organic solvents (The Merck Index).

It is unclear if PT also occurs through paracellular
pathways. Stulc (Stulc, 1988) states that paracellular
channels exist and are wider in hemochorial than in
epitheliochorial placentas. Steven and Samuel (Steven and
Samuel, 1979) cite anatomical evidence for narrow paracellular
channels in the maternal layer of the sheep placenta but,
because of their syncytial nature, doubt that these are
present in the fetal portion of placenta membranes. Wynn
(Wynn, 1975) indicates that extracellular spaces (? channels)

exist in the fetal as well as the maternal placental
membranes. He notes that they are not continuous but rather
separated by tight intracellular junctions.

Net transfer across the placenta is equal to the sum of
the bidirectional transfer rates. Substances whose transfer
is due to diffusion will be transferred in the direction of
the concentration gradient. For substances of either fetal or
maternal origin, this is maximized if a countercurrent system
of blood flow is present. In addition, and even in the
absence of a concentration gradient, substances that are
transferred by carrier-mediated transport may be
asymmetrically transferred. Asymmetrical transfer has been
demonstrated for certain amino acids (Schneider et al.1987;
Yudilevich and Sweiry, 1985). It may be due to the fact that
the area of microvillous surface (maternal-facing) is greater
than the basal (fetal-facing) membrane rather than to
differences between the two surfaces in carrier concentration
(Schneider et al.1987). These concepts may be relevant when
considering the PT of iodide since it has been proposed that
this ion is concentrated on the fetal side of the placenta
(Roti, Gnudi and Braverman, 1983; Logothetopoulos and Scott,
1956; Nathanielsz et al.1973).

It is evident that PT is a complex process. K, the
overall rate constant for transplacental flux, should be an
ideal way to quantitate placental "permeability". However,
its measurement is influenced by many factors including the
age of gestation and technical artifacts. Therefore, it is
more useful to express placental flux in relation to that of
standard such as water or antipyrine, a highly lipid soluable
compound with a molecular weight of 188.

DESIGN OF STUDIES DEALING WITH PLACENTAL TRANSFER OF THYROID-
RELATED SUBSTANCES

The TRSs include thyroid hormones and other
iodothyronines, iodide, thyrotropin (TSH), thyrotropin-
releasing hormone (TRH), thyroid stimulating immunoglobulins
(TSI), and thyrotropin binding inhibitory immunoglobulins
(TBII). Many methods have been used to evaluate the PT of
these substances.

In Vitro and In Situ Models The most direct methods are
those which utilize in vitro perfusion or diffusion systems.
Intact placentas, isolated placental cotyledons, and tissue
fragments from humans, sheep, rats, rabbits, and guinea pigs
have all be used to study PT of amino acids, glucose, fatty
acids, and drugs (Yudilevich and Sweiry, 1985). However, they
have rarely been employed for TRSs. In contrast, the PT and
metabolism of thyroid hormones and TRH has been studied using
in situ models (Nogimori et al.1985; Cooper et al.1983)
(Castro et al.1985). When the guinea pig or rabbit is used
for in situ studies, the fetal side of the placenta is
perfused through the umbilical vein. The effluent is
collected through one umbilical artery and either returned to
the perfusion reservoir (recirculating system) or collected in
aliquots (one pass system). In situ studies do not require
oxygenization of the perfusion buffer and are more

physiological than in vitro models because the maternal and
fetal vascular connections within the placenta are not
disrupted. Substances can be added either to the perfusion
reservoir or infused into the maternal circulation and
maternal-fetal or fetal-maternal transfer measured. When a
recirculating system is employed, the perfusion reservoir is
representative of the fetal compartment and no provision needs
to be made for production or clearance of the substance by
fetal tissues.

   Fetal bioassay   An early method of studying the PT of
TRS was to measure bioassay responses in the fetus. For
example, the concept that TSH does not cross the placenta is
based almost entirely on the observation that maternal TSH
administration, or perturbations that alter TSH concentrations
in maternal serum, do not alter fetal thyroid weight. Fetal
bioassay studies have several limitations. The most important
are their lack of precision and the fact that results can be
misinterpreted if differences between maternal and fetal
tissue sensitivity are not appreciated. Thus, if fetal
tissues are less sensitive than maternal tissues to the TRS
being tested, PT will be underestimated. Conversely, PT will
be overestimated if fetal tissues are more sensitive than
maternal tissues to the test substance. As noted below, this
factor has not been fully appreciated in studies of TRH PT.

   Plasma Concentrations of TRSs in the Mother and Fetus.
Other methods to study the PT of TRSs require measuring their
concentrations in maternal and fetal plasma, usually after
injection of the labeled TRS into the fetus and/or mother.
These are difficult studies because it is also necessary to
measure a large number of pharmacokinetic parameters in the
mother and fetus (DiStefano et al.1973). This is not always
possible because of technical problems or ethical
considerations. Even if all of the necessary parameters can
be measured, their utility may be compromised by high error
estimates (DiStefano et al.1973).

   Sometimes the maternal-fetal plasma concentration
gradient (MFPG) is used as the major endpoint of the study.
Intuitively it would seem that the smaller the MFPG, the
greater the permeability of the TRS in the placenta. However,
this is not always the case. This can be demonstrated with
compartmental model in which, for purposes of simplification,
only the unbound fraction of the substance is considered. The
variables in this model are as follows.

   Cm   - The concentration of the TRS in maternal
          plasma

   Ko   - The overall rate constant for bidirectional
          transfer of the TRS across the placenta

   Pf   - The fetal tissue production rate of the TRS

   Cf   - The concentration of the TRS in fetal plasma

Clf - The clearance of the TRS due to elimination by
fetal tissues

In this model, the amount of TRS entering the fetal plasma is
equal to the sum of Pf and the unidirectional flux of the TRS
from mother to fetus. This is equal to Ko times Cm.
Similarly, the amount of substance leaving the fetal plasma is
equal to the sum of the elimination rate by fetal tissues and
the unidirectional flux of the substance from fetus to mother.
The elimination rate by fetal tissue is equal to Clf times Cf
and the flux from fetus to mother is equal to Ko times Cf.
Under steady state conditions the following equation can be
written since the amount of the TRS entering and leaving fetal
plasma are equal.

$$Pf + Ko(Cm) = Ko(Cf) + (Clf \times Cf)$$

Solving this equation for each of the unknowns and
substituting values demonstrates that the MFPG (i.e. Cm/Cf) is
a function not only of placental permeability, as represented
by K, but also of the Clf and Pf. Moreover, depending on the
relationship between different parameters, the effect of
changes in placental permeability on the MFPG may be less
than, equal to, or greater than the effect of similar changes
in the clearance by fetal tissues. In one situation placental
permeability does not affect the MFPG. This is when fetal
tissues do not clear or produce the substance in question. In
this state the MFPG will be unity no matter how large or small
K is provided it is greater than 0.
    The idealized model should not be viewed as a faithful
representation of the in vivo state. Steady state conditions
cannot be reached in vivo. This is because fetal growth, and
maturation of systems that produce and degrade TRSs, is a
dynamic process that continues throughout gestation. In
addition, the equation is difficult to apply to experimental
animal models. For example, in trying to determine the fetal
elimination rate there are problems in distinguishing between
loss of a substance due to degradation by fetal tissues, and
loss related to fetal-maternal transfer. Nevertheless, the
model is a useful way to illustrate that placental
permeability is only one factor that affects the MFPG and that
even poorly transferred substances will accumulate in the
fetus provided they are not degraded by fetal tissues.
    As discussed in the next section, not only the fetus but
also the placenta can generate TRSs de novo. Moreover, the
placenta can also degrade TRSs. The preceding equation could
be modified to show these processes. From a conceptual
standpoint it may be clearer, however, to define Ko in a way
that reflects transfer as limited not only by "permeability"
but also by degradation within placental membranes.

PLACENTAL METABOLISM OF THYROID-RELATED SUBSTANCES

In 1979 Beard wrote the following (Beard, 1979).

> "The placenta tends to be regarded as a semipermeable
> membrane through which substances are exchanged between
> the mother and the fetus. Little thought has been given
> to other factors, which might influence this metabolic
> activity and, hence, the functional capacity of the
> placenta."

Although this was written in reference to the another matter,
it could apply as well to more recent information regarding
thyroid hormone metabolism in the placenta.

Louros et al (London, Money and Rawson, 1963), using
placental slices, were one of the first to show that the
placenta deiodinated T4. However, they were unable to
identify the T4 deiodination products. Later Bonavac et al
(Banovac et al.1980) found that placental homogenates
generated both T3 and rT3 when incubated with
supraphysiological amounts of T4 but tissue specific activity
was very low. Shortly thereafter Roti et al (Roti et
al.1982b; Roti et al.1981a) working in our laboratory
incubated rat and human placental homogenates with labeled T4,
T3, and rT3. During the incubation the homogenates were
enriched with a sulfhydral reducing agent, a critical
modification from previous work. Under these conditions, T4
and T3 were rapidly deiodinated in the tyrosyl (inner) ring
position but rT4 was stable. Thus T4 was converted to rT3,
and T3 was converted first to 3,3'-diiodothyronine (3,3'-T2)
and then to 3'-monoiodothyronine (3'-T1). These findings have
been confirmed by other workers (Yoshida et al.1984; Suzuki et
al.1986). In addition, using a sensitive iodide release assay
under conditions in which inner ring deiodinase enzymes are
saturated by unlabeled T3, it has been shown that the placenta
contains some outer ring deiodinase activity (Kaplan and Shaw,
1984). Interestingly, this activity is higher on the maternal
than the fetal side of the placenta.

The in-situ perfused guinea pig placenta model has been
used to study placental deiodinase activity. In the earliest
paper, little evidence for deiodination was obtained (London,
Money and Rawson, 1963). The reason for this is not clear;
perhaps the methodology did not distinguish between thyroid
hormone and its metabolites. More recently, Cooper et al
(Cooper et al.1983) showed that T4 is deiodinated to rT3 when
perfused through the fetal side of the placenta. The
fractional deiodination rate for a single pass was 0.4
percent. When dual-labeled T4 and rT3 was administered to the
mother, rT3 appeared in the placental effluent. More than 90
percent of this was derived, not from rT3 that had been
infused into the maternal circulation, but from T4 that had
been infused into the maternal circulation. This indicates
that rT3 in the "fetal compartment" was generated from
maternal T4 as it crossed the placenta. Cooper et al (Cooper
et al.1983) identified considerable label that was tentatively

identified as T4; it is not clear if efforts were made to
precisely characterize this. We have demonstrated that T3 is
much more rapidly deiodinated than T4 when perfused into the
fetal side of the guinea pig placenta (Castro et al.1985).
When T4 is perfused at a concentration of 44 nM, less than 1
percent is deiodinated. In contrast, 27 percent of T3 is
deiodinated when perfused at a concentration of 0.14 nM and 15
percent of T3 is deiodinated when perfused at a concentration
of 140 nM (Castro et al.1985). Similarly T3 is deiodinated
more rapidly than T4 in placental homogenates (Roti et
al.1982b; Roti et al.1981a). However, the difference is not
as great as in the perfused placenta.

The demonstration that the placenta contains an active
inner ring deiodinase activity raises several questions. One
is whether this activity has an important influence on the
peripheral metabolism of thyroid hormones in the fetus.
Supporting this is the fact that the placenta receives 40 to
50 percent of the fetal cardiac output (Rudolph and Heymann,
1970). In addition, the nature of this activity is compatible
with the unique profile of iodothyronines in fetal serum. The
ability to convert T4 to rT3, the marked activity against T3,
and the lack of activity against rT3 are all consistent with
the observation that fetal plasma T3 concentrations are high
and T3 concentrations are low. Just prior to term fetal
plasma T3 concentrations rise modestly (Nathanielsz et
al.1973). This may be related, in part, to an increase in
fetal T3 production. In addition, there are two factors that
decrease the ability of the placenta to clear T3 from fetal
plasma. One is a fall in the specific activity of the inner
ring deiodinase (Roti et al.1982a; Yoshida et al.1985). The
other is a decline in the fractional distribution of the fetal
cardiac output to the placenta (Rudolph and Heymann, 1970).

Another question regarding the role of placental
deiodinase activity is whether it influences PT of thyroid
hormones. Inner ring deiodinase destroys thyroid hormone
bioactivity and along with permeability, could limit PT. This
has been referred to as deiodination-limited transfer (DLT).
DLT is defined as the rate of transfer of an iodothyronine
under conditions in which deiodination is completely inhibited
minus the rate of transfer of a substance under conditions in
which deiodination is not inhibited (Emerson, 1989). Both
permeability and DLT probably affect PT of thyroid hormones in
species with hemochorial placentation. Thus, in the rat there
is a similar decrease in fetal plasma rT3 concentrations
following maternal thyroidectomy as there is in a model of
fetal hypothyroidism (El-Zaheri et al.1981). This decrease is
probably not related to changes in maternal-fetal rT3 flux
because rT3 crosses the placenta poorly and the gradient
favors fetal-maternal rather than maternal-fetal flux. Cooper
et al (Cooper et al.1983) found evidence for DLT of T4 in the
guinea pig but suggested that only a small fraction of the T4
that passed through the placental membranes was deiodinated.
However, although the nondeiodinated fraction was thought to
be T4 it was not characterized. DLT is less likely to be a

factor in the sheep because its epitheliochorial placentation
may limit access of maternal thyroid hormones to fetal
membranes, the putative site of inner ring deiodinase
activity.

PLACENTAL TRANSFER OF THYROID-RELATED SUBSTANCES: CURRENT
CONCEPTS

The quantitative index of placental permeability is K.
However, absolute K values are of little use in discussing the
PT of TRS. This in not only because few K values for TRSs
have been published but also because absolute K values do not
provide a good sense of the effect of K on the maternal
contribution to the fetal pool. Even for substances with
moderately low permeability, this may be substantial, both in
absolute and relative terms, provided the production and
clearance of the substance by fetal tissues is low.
Conversely, once K reaches a critical threshold, further
increases do not greatly affect the percentage of the fetal
pool that is derived from maternal sources. Therefore, when
discussing specific TRSs, placental permeability will be
described in subjective terms.

Iodide Iodide rapidly appears in fetal serum after
injection into the mother, regardless of species (Roti, Gnudi
and Braverman, 1983). Moreover, studies in rabbits, guinea
pigs, and sheep indicate that a fetal-maternal gradient is
generated after bolus injection of labeled iodide into the
fetus and/or mother. In one of these studies, thiocyanate was
found to abolish this gradient (Logothetopoulos and Scott,
1956). This has been interpreted as an indication that iodide
is actively transported to the fetus by the placenta
(Logothetopoulos and Scott, 1956) and that this transport
mechanism is thiocyanate sensitive. However, it is not clear
if these observations were made under steady state conditions.
An alternative hypothesis, currently being tested in our
laboratory, is that the fetal-maternal gradient results from
differences between the maternal and fetal elimination rate
constants for iodide, and that more active thyroidal uptake of
iodide by the mother is responsible for this difference.

The high permeability of iodide in the placenta, and the
fact that the fetal thyroid can trap iodide as early as 10 to
15 weeks of gestation (Roti, Gnudi and Braverman, 1983), is
the major reason why inadvertent administration of radioactive
iodide to pregnant women is a cause of congenital
hypothyroidism (Green et al.1971).

Thyroxine, Triidothyronine, and Related Iodothyronines
The PT of iodothyronines is an integral aspect of perinatal
and fetal thyroidology. Accordingly it is featured in many
chapters in this monograph, including those dealing with
maternal hypothyroidism, fetal thyroid agenesis, and the use
of antithyroid drugs in pregnancy. The questions relating to
this cannot be reduced to whether or not the placenta is
permeable to thyroid hormones. Rather, the type and amount of
iodothyronine, the species, and the time of gestation must be
specified. Moreover, placental permeability itself is of less

interest than the size and relative contribution of the mother
to the fetal pool. This in turn leads to the question of
whether this contribution, even if small, has an important
influence on the development of the fetus or ameliorates the
effect of fetal thyroid failure.

Whereas T3 is the bioactive form of thyroid hormone at
the nuclear level, T4 is probably a more bioactive form in
fetal plasma. T4/T3 ratios are much higher in fetal than in
adult plasma, and T4 is the major source of T3 in fetal
tissues. In sheep, a species with epitheliochorial
placentation, little or no PT of T4 occurs during the second
half of gestation (Roti, Gnudi and Braverman, 1983). The most
extensive studies of PT in species with hemochorial
placentation have been performed in the rat. Obregon et al
(Obregon et al.1984) reported that rat embryos accumulate T4
of maternal origin and, elsewhere in this monograph, Morreale
de Escobar presents evidence that fetal tissues continue to
derive T4 and T3 from maternal T4 throughout gestation.
However, the percentage of fetal T4 that is derived from
maternal sources declines to about 17 % at term. This is
probably due not only to changes in the PT of T4, but also to
increases in fetal T4 production. Dussault and Coulombe
(Dussault and Coulombe, 1980) have estimated that less than 1
percent of fetal plasma T4 is derived from maternal sources at
term. The reason for the differences between these estimates
is not clear. Other studies in the rat suggest that the
amount of T4 in fetal plasma that is derived from maternal
sources is small at term. However, transfer is easily
detected when the mother is treated with supraphysiological
amounts of T4 (Roti, Gnudi and Braverman, 1983). For example
we have noted, in attempting to administer replacement doses
of T4 to pregnant rats with maternal and fetal hypothyroidism,
that the elevated fetal serum TSH concentrations were not
affected unless the maternal replacement dose was
supraphysiological as shown by the fact that serum TSH
concentrations in the mother were also suppressed (El-Zaheri
et al.1981). In guinea pigs, as in rats, PT can easily be
demonstrated when T4 is administered in supraphysiological
doses (Peterson and Young, 1952). However, in the last week
of gestation, the guinea pig fetus maintains a nearly fourfold
fetal-maternal free T4 concentration gradient (Castro et
al.1986), suggesting that the placenta is relatively
impermeable to T4 at this time.

Studies of T4 transfer in humans are limited and
indirect. Recently Vulsma et al (Vulsma, Gons and de Vijlder,
1989) found that neonates whose thyroids were completely
nonfunctional had T4 concentrations in their plasma that were
about 40% of those in normal neonates. Their data indicates
that the fetus derives some T4 from the mother in late
gestation. This is not necessarily as high as 40 percent of
the blood production rate of the normal fetus, an estimate
based on plasma T4 concentration data. This is because there
is a fetal-maternal gradient in plasma free T4 concentrations
in late gestation (Fisher and Klein, 1981; Castro et al.1986).

This, in proportion to placental T4 permeability, causes fetal T4 losses. These losses ensure that estimates of fetal T4 production that are based on plasma T4 concentrations will be falsely depressed.

The data of Vulsma et al (Vulsma, Gons and de Vijlder, 1989) complement early work showing some PT of labeled T4 just prior to delivery (Grumbach and Werner, 1956). A question of great interest is whether this degree of T4 transfer is limited to the preterm period or is characteristic of most of gestation. Early studies by Carr (Carr et al.1959) indicate that amelioration of fetal hypothyroidism could be achieved if very large amounts of dessicated thyroid, sufficient to induce maternal hyperthyroidism, were administered. More recently, however, Ramsay et al (Ramsay, Kaur and Krassas, 1983) reported that doses of T4 which were moderately supraphysiological and sufficient to cause maternal hyperthyroxinemia did not prevent antithyroid drug induced fetal hypothyroxinemia. In summary, there are conflicting data regarding PT of T4 in humans. If the studies in rats can be extrapolated to humans it seems likely that, in early gestation, the fetus derive plasma T4 and tissue T3 from maternal T4. The function of T4 in early gestation is not clear. During the second and third trimesters, T4 transfer seems to be so limited that fetal hypothyroidism can only be corrected if clinically significant maternal hyperthyroidism is induced. However, some T4 transfer may occur during the latter part of gestation and this could ameloriate the effects of fetal thyroid agenesis on brain development. One way to explain the various studies is to postulate that there are two transfer mechanisms for T4. One of these is saturated at levels of transfer which are sufficient to make a partial contribution to the fetal T4 pool. The second mechanism is passive diffusion and its rate is such that, in cases of thyroid agenesis, the fetal T4 pool can be restored to normal only by inducing maternal hyperthyroxinemia. The hypothesis that a limited capacity transport mechanism for T4 exists has not been tested but, in view of the fact that transport mechanisms exist for certain amino acids (Yudilevich and Sweiry, 1985), it is not completely implausible.

T3 is less effective than T4 in preventing PTU-induced fetal goiter in rats (Knobil and Josimovich, 1959) and guinea pigs (Peterson and Young, 1952; Postel, 1957). Furthermore, although an early study suggested that administration of T3 to humans lowered cord blood T4 (Raiti et al.1967), a more recent report indicates that T3 administration does not prevent high cord blood TSH concentrations in infants of PTU treated mothers (Ibbertson, Seddon and Croxson, 1975). These studies suggest that PT of T3 is less than that of T4. However, similar results would be obtained if T3 was relatively ineffective compared to T4 in suppressing TSH in the fetus. More direct studies using the in situ perfused guinea pig placenta show that, when T3 is infused into the mother to achieve concentrations that are more than tenfold higher than normal, T3 transfer to the fetal side of the placenta can

still not be detected (Emerson, 1989). PT of T3 in sheep is
also severely limited as demonstrated by studies which show
that almost all of the fetal T3 pool is derived from fetal
sources (Fisher et al.1972). The fact that PT of T3 is very
low is only one of several reasons why maternal administration
of T3 is not a rational way to treat fetal hypothyroidism.
Plasma T3 concentrations are quite low even in the normal
fetus, and fetal tissues seem to derive most of their T3 from
plasma T4.

Although rT3 is not considered to be an active form of
thyroid hormone, it is of interest because fetal life is the
only time that this iodothyronine circulates in high
concentrations. The physiological significance of this is not
clear. There are several observations which indicate that the
permeability of rT3 in the placenta is low. Infusion of large
amounts of rT3 into the maternal circulation has little effect
on fetal serum rT3 concentrations (El-Zaheri et al.1981).
Moreover, the MCR of rT3 in the fetus is also low (Chopra,
Sack and Fisher, 1975), despite the fact that there is a large
gradient favoring fetal-maternal transfer. Finally, rT3 is
poorly eliminated from the perfusion reservoir when its
contents are recirculated through the fetal side of the guinea
pig placenta (Emerson, 1989).

TSH   Many studies indicate that alterations in maternal
TSH concentration have no effect on fetal thyroid weight or
morphology (Roti, Gnudi and Braverman, 1983). Others show
that labeled TSH is not transferred either in the maternal-
fetal or fetal-maternal direction (Roti, Gnudi and Braverman,
1983). Thus, placental permeability to TSH is very low. This
is not surprising in view of its high molecular weight.

TRH   In rats, monkeys, and humans maternal TRH
administration increases fetal serum TSH concentrations
(Azukizawa et al.1976) (Roti et al.1981b) (Roti, Gnudi and
Braverman, 1983). In contrast, no increase in fetal serum TSH
is seen when TRH is administered to sheep (Thomas et al.1975).
These "fetal bioassay" studies indicate that placental
permeability is sufficient to permit TRH transfer in species
with hemochorial placentation but not those with
epitheliochorial placentation. However, placental
permeability of TRH is probably more limited than these
studies would suggest. In the perfused guinea pig placenta,
only about 11 percent of the TRH is extracted in a single
pass. In contrast the extraction of water is 57 percent.
This is consistent with a similar diffusion process for TRH
and water since their transfer ratio, and the ratio of the
square root of their molecular weights, is similar. About
half of the TRH which diffuses into the placenta is degraded
to deamido-TRH (Nogimori et al.1985), a metabolite that has
little TSH releasing activity.

It is remarkable considering that TRH is partially
degraded by the placenta and is not freely permeable (i.e.
transfer rate < bulk flow rate) that the fetal TSH response to
maternal TRH administration is several times greater than the
maternal TSH response (Azukizawa et al.1976; Roti et

al.1981b). Although this may be partially due to differences
in TRH and TSH pharmacokinetics in the mother and fetus, it
supports the concept that the fetal pituitary is
hyperresponsive to TRH (Roti et al.1981b).
    Thyroactive Immunoglobulins   In many species, but not
the sheep, the fetus acquires maternal antibodies, conferring
passive immunity on the newborn. Placental transfer of
immunoglobulins is an apparent violation of the rule that high
molecular weight substances are poorly transferred to the
fetus. In rabbits and guinea pigs ligation of the vitelline
blood vessels reduces antibody transfer indicating that a
functional yolk sac placenta is important for this process
(Faber and Thornburg, 1983). Its selectivity is attested to
by the fact that epitopes within the Fc portion of IgG are
important for transfer. The process in humans is less clear.
The human lacks a functional yolk sac (Faber and Thornburg,
1983) but there is no question that maternal IgG, including
the thyroid stimulating immunoglobulins of Grave's Disease and
TBII is transferred to the fetus (Dirmikis and Munro, 1975).

REFERENCES

Azukizawa, M., Murata, Y., Ikenoue, T., Martin, C.B.Jr. and
Hershman, J.M., 1976, Effect of thyrotropin-releasing hormone
on secretion of thyrotropin, prolactin, thyroxine, and
triiodothyronine in pregnant and fetal rhesus monkeys, J.
Clin. Endocrinol. Metab., 43:1020.

Banovac, K., Bzik, L., Tislaric, D. and Mekso, M., 1980,
Conversion of thyroxine to triiodothyronine and reverse
triiodothyronine in human placenta and fetal membranes,
Hormone Research, 12:253.

Beard, R.W. The effects of diabetes on placental transfer. In:
Placental Transfer, edited by Chamberlain, G.V.P. and
Wilkinson, A.W. Tunbridge Wells: Pitman Medical Publishing ,
1979, p. 205.

Berhe, A., Bardsley, W.G., Harkes, A. and Sibley, C.P., 1987,
Molecular charge effects on the protein permeability of the
guinea-pig placenta, Placenta, 8:365.

Carr, E.A.Jr., Deierwaltes, W.H., Raman, G., Dodson, V.N.,
Tanton, J., Betts, J.S., Stambaugh, R.A., Spafford, N.R. and
Tanner, K., 1959, The Effect of Maternal Thyroid Function on
Fetal Thyroid Function and Development, J. Clin. Endocrinol.
Metab., 19:1.

Castro, M.I., Braverman, L.E., Alex, S., Wu, C.F. and Emerson,
C.H., 1985, Inner-ring deiodination of 3,5,3'-triiodothyronine
in the in situ perfused guinea pig placenta, J. Clin. Invest.,
76:1921.

Castro, M.I., Alex, S., Young, R.A., Braverman, L.E. and
Emerson, C.H., 1986, Total and free serum thyroid hormone

concentrations in fetal and adult pregnant and nonpregnant
guinea pigs, Endocrinology, 118:533.

Chopra, I.J., Sack, J. and Fisher, D.A., 1975, 3,3',5'-
Triiodothyronine (reverse T3) and 3,3',5-triiodothyronine (T3)
in fetal and adult sheep: studies of metabolic clearance
rates, production rates, serum binding, and thyroidal content
relative to thyroxine, Endocrinology, 97:1080.

Cooper, E., Gibbens, M., Thomas, C.R., Lowy, C. and Burke,
C.W., 1983, Conversion of Thyroxine to 3,3',5'-
Triiodothyronine in the Guinea Pig Placenta: in Vivo Studies,
Endocrinology, 112:1808.

Dirmikis, S.M. and Munro, D.S., 1975, Placental Transmission
of Thyroid-stimulating Immunoglobulins, Br. Med. J., 2:665.

DiStefano, J.J.III, Durando, A.R., Jang, M., Jenkins, D.,
Johnson, D.J., Mak, P., Marshall, T., Mons, B., Warsavsky, A.
and Fisher, D.A., 1973, Estimates and estimation errors of
hormone secretion, transport and disposal rates in the
maternal-fetal system, Endocrinology, 93:324.

Dussault, J.H. and Coulombe, P., 1980, Minimal Placental
Transfer of L-Thyroxine (T4) in the Rat, Pediatr. Res.,
14:228.

Ekins, R., 1990, Measurement of free hormones in blood,
Endocr. Rev., 11:5.

El-Zaheri, M.M., Vagenakis, A.G., Hinerfeld, L., Emerson, C.H.
and Braverman, L.E., 1981, Maternal thyroid function is the
major determinant of amniotic fluid 3,3',5'-triiodothyronine
in the rat, J. Clin. Invest., 67:1126.

Emerson, C.H. Role of the placenta in fetal thyroid
homeostasis. In: Research in Congenital Hypothyroidism, edited
by Delange, F., Fisher, D.A. and Glinoer, D. New York: Plenum
Press, 1989, p. 31-43.

Faber, J.J. and Thornburg, K.L. Placental Physiology Structure
and Function of Fetomaternal Exchange, New York:Raven Press,
1983. pp. 115-137.

Fisher, D.A., Dussault, J.H., Erenberg, A. and Lam, R.W.,
1972, Thyroxine and Triiodothyronine Metabolism in Maternal
and Fetal Sheep, Pediatr. Res., 6:891.

Fisher, D.A. and Klein, A.H., 1981, Thyroid Development and
Disorders of Thyroid Function in the Newborn, N. Engl. J.
Med., 304:702.

Green, H.G., Gareis, F.J., Shepard, T.H. and Kelley, V.C.,
1971, Cretinism associated with maternal sodium iodide I 131
therapy during pregnancy, Am. J. Dis. Child., 122:247.

Grumbach, M.M. and Werner, S.C., 1956, Transfer of thyroid hormone across the human placenta at term, J. Clin. Endocrinol. Metab., 16:1392.

Ibbertson, H.K., Seddon, R.J. and Croxson, M.S., 1975, Fetal hypothyroidism complicating medical treatment of thyrotoxicosis in pregnancy, Clin. Endocrinol., 4:521.

Kaplan, M.M. and Shaw, E., 1984, Type II iodothyronine 5'-deiodination by human and rat placenta in vitro, J. Clin. Endocrinol. Metab., 59:1808.

Knobil, E. and Josimovich, J.B., 1959, Placental transfer of thyrotropic hormone, thyroxine, triiodothyronine, and insulin in the rat, Ann. NY. Acd. Sci., 75:895.

Logothetopoulos, J. and Scott, R.F., 1956, Active iodide transport across the placenta of the guinea-pig, rabbit and rat, J. Physiol., 132:365.

London, W.T., Money, W.L. and Rawson, R.W., 1963, Placental transport of I131-labeled thyroxine and triiodothyronine in the guinea pig, J. Clin. Endocrinol. Metab., 73:205.

Nathanielsz, P.W., Comline, R.S., Silver, M. and Thomas, A.L., 1973, Thyroid function in the foetal lamb during the last third of gestation, J. Endocrinol., 58:535.

Nogimori, T., Alex, S., Baker, S. and Emerson, C.H., 1985, Thyrotropin-releasing hormone metabolism and extraction by the perfused guinea pig placenta, Endocrinology, 117:565.

Obregon, M.J., Mallol, J., Pastor, R., Morreale de Escobar, G. and Escobar del Rey, F., 1984, L-thyroxine and 3,5,3'-triiodo-L-thyronine in rat embryos before onset of fetal thyroid function, Endocrinology, 114:305.

Pearson, J.F. Gas exchange. In: Placental Transfer, edited by Chamberlain, G.V.P. and Wilkinson, A.W. Tunbridge Wells: Pitman Medical Publishing, 1979, p. 108-117.

Peterson, R.R. and Young, W.C., 1952, The problem of placental permeability for thyrotropin, propylthiouracil and thyroxine in the guinea pig, Endocrinology, 50:218.

Postel, S., 1957, Placental transfer of perchlorate and triiodothyronine in the guinea pig, Endocrinology, 60:53.

Raiti, S., Holzman, G.B., Scott, R.I. and Blizzard, R.M., 1967, Evidence for the Placental Transfer of Tri-iodothyronine in Human Beings, N. Engl. J. Med., 277:456.

Ramsay, I., Kaur, S. and Krassas, G., 1983, Thyrotoxicosis in pregnancy: results of treatment by antithyroid drugs combined with T4, Clin. Endocrinol., 18:73.

Ramsey, E.M. and Donner, M.W. Placental Vasculature and Circulation, Philadelphia:W.B. Saunders, 1980.

Ramsey, E.M. The Placenta Human and Animal, New York:Praeger Publishers, 1982. Ed. 2 pp. 31-32.

Reynolds, F. Transfer of drugs. In: Placental Transfer, edited by Chamberlain, G.V.P. and Wilkinson, A.W. Tunbridge Wells: Pitman Medical Publishing, 1979, p. 168-170.

Roti, E., Fang, S.L., Green, K., Emerson, C.H. and Braverman, L.E., 1981a, Human placenta is an active site of thyroxine and 3,3',5-triiodothyronine tyrosyl ring deiodination, J. Clin. Endocrinol. Metab., 53:498.

Roti, E., Gnudi, A., Braverman, L.E., Robuschi, G., Emanuele, R., Bandini, P., Benassi, L., Pagliani, A. and Emerson, C.H., 1981b, Human cord blood concentrations of thyrotropin, thyroglobulin, and iodothyronines after maternal administration of thyrotropin-releasing hormone, J. Clin. Endocrinol. Metab., 53:813.

Roti, E., Braverman, L.E., Fang, S.L., Alex, S. and Emerson, C.H., 1982a, Ontogenesis of placental inner ring thyroxine deiodinase and amniotic fluid 3,3',5'-triiodothyronine concentration in the rat, Endocrinology, 111:959.

Roti, E., Fang, S.L., Braverman, L.E. and Emerson, C.H., 1982b, Rat placenta is an active site of inner ring deiodination of thyroxine and 3,3',5-triiodothyronine, Endocrinology, 110:34.

Roti, E., Gnudi, A. and Braverman, L.E., 1983, The placental transport, synthesis and metabolism of hormones and drugs which affect thyroid function. [Review], Endocr. Rev., 4:131.

Rowland, M. and Tozer, T.N. Clinical pharmacokinetics, Philadelphia:Lea and Febiger, 1980. pp. 17.

Rudolph, A.M. and Heymann, M.A., 1970, Circulatory changes during growth in the fetal lamb, Circulation Research, 26:289.

Schneider, H., Proegler, M., Sodha, R. and Dancis, J., 1987, Asymmetrical transfer of a-aminoiosbutyric acid (AIB), leucine and lysine across the in vitro perfused human placenta, Placenta, 8:141.

Steven, D.H. and Samuel, C.A. The anatomy of placental transfer. In: Placental Transfer, edited by Chamberlain, G.V.P. and Wilkinson, A.W. Tunbridge Wells: Pitmann Medical Publishing, 1979, p. 2-6.

Stulc, J., 1988, Is there control of solute transport at placental level, Placenta, 9:19.

Suzuki, M., Yoshida, K., Sakurada, T., Kaise, N., Kaise, K., Fukazawa, H., Nomura, T., Itagaki, Y., Yonemitsu, K., Yamamoto, M. and et, a.l., 1986, Effect of changes in thyroid state on metabolism of thyroxine by rat placenta, Endocrinol. Jpn., 33:37.

Thomas, A.L., Jack, P.M.B., Mannus, J.G. and Nathanielsz, P.W., 1975, Effect of synthetic thyrotropin releasing hormone on thyrotropin and prolactin concentrations in the peripheral plasma of the pregnant ewe, lamb fetus and neonatal lamb, Biol. Neonat., 26:109.

Vulsma, T., Gons, M.H. and de Vijlder, J.J., 1989, Maternal-fetal transfer of thyroxine in congenital hypothyroidism due to total organification defect or thyroid agenesis, N. Engl. J. Med., 321:13.

Wynn, R.M. Fine structure of the placenta. In: The Placenta and its Maternal Supply Line, edited by Gruenwald, P. Baltimore: University Park Press, 1975, p. 56-79.

Yoshida, K., Suzuki, M., Sakurada, T., Kitaoka, H., Kaise, N., Kaise, K., Fukazawa, H., Nomura, T., Yamamoto, M., Saito, S. and Yoshinaga, K., 1984, Changes in thyroxine monodeiodination in rat liver, kidney and placenta during pregnancy, Acta Endocrinologica (Copenhagen), 107:495.

Yoshida, K., Suzuki, M., Sakurada, T., Shinkawa, O., Takahashi, T., Furuhashi, N., Kaise, N., Kaise, K., Kitaoka, H., Fukazawa, H., Nomura, T., Itagaki, Y., Yamamoto, M., Saito, S. and Yoshinaga, K., 1985, Human placental thyroxine inner ring monodeiodinase in complicated pregnancy, Metabolism, 34:535.

Yudilevich, D.L. and Sweiry, J.H., 1985, Transport of amino acids in the placenta, Biochim. Biophys. Acta, 822:169.

REGULATION OF THYROID FUNCTION IN PREGNANCY:

MATERNAL AND NEONATAL REPERCUSSIONS

Daniel Glinoer

Laboratory of Radioisotopes
Thyroid Investigation Clinic
Hospital Saint-Pierre
Free University of Brussels
Brussels, Belgium

Pregnancy constitutes a unique experimental model in humans, wherein a normal thyroid is faced with a triple challenge, as a result of 3 separate factors acting in concert to stimulate the gland : 1) the increased thyroid hormone-binding capacity of serum, 2) the effects of increased levels of hCG on TSH and on the thyroid, and 3) the reduced availability of iodine for the maternal thyroid, at least in conditions with a limited dietary iodine supply.

THE THYROIDAL ECONOMY DURING PREGNANCY

Adjustment to the changes in TBG levels

Important modifications occur in thyroidal economy due to the marked increase in circulating levels of the major T4 transport protein (TBG) in response to high estrogen levels (1). As a consequence, the TBG content of the extra-cellular distribution space increases from ~2700 to ~7400 nmol. To maintain a stable free hormone level, the increase in the binding capacity of the system leads to an adjustment of the extrathyroidal thyroxine pool, from ~1000 to ~2700 nmol. In pregnancy, changes in TBG levels take place over a period of 3 months, and it can be calculated that the thyroidal adjustment represents an overall enhancement in T4 output above baseline values of ~40 % during the first, ~60 % during the second and ~75 % during the third month, that is 1-3 %/day during the first 3 months. These figures theoretically represent small "extraloads" on the glandular adaptation mechanisms, and probably explain why T4 kinetic studies in pregnancy have failed to unequivocally demonstrate higher T4 production rates (2). There are, however, many examples of hypothyroid patients - adequately substituted with l-T4 before pregnancy - in whom hypothyroidism developed during gestation, if substitution doses were not rapidly raised in order to comply with the increased hormone requirements .

Advances in Perinatal Thyroidology, Edited by B.B. Bercu and
D.I. Shulman, Plenum Press, New York, 1991

In a recent prospective study of a large cohort of healthy pregnant subjects performed in an area with a moderately low iodine intake (average : 50 μg/day), a crucial finding was that after 10 weeks gestation, serum T4 (and, to a lesser extent, also T3) did not catch up, and total hormone levels trailed behind TBG changes. During the second half of gestation, a plateau was reached, with total T4 and T3 remaining below the expected theoretical values in most women, and free hormone concentrations near the lower reference limit of non pregnant subjects. Furthermore, the study provided clear information that variable patterns of thyroidal adjustment take place in normal pregnancy. In brief, at least one third of the subjects were characterized as having relative hypothyroxinemia, preferential T3 secretion (with increasing T3/T4 ratios), and a higher setting of the pituitary thyrostat, i.e. a pattern of increased glandular stimulation. The data also confirmed earlier reports indicating that pregnancy was associated with a decrease in free T4 and T3 levels by about 30 % in late pregnancy compared to values in early pregnancy and to those in non pregnant women. Finally, there was a clear tendency, at the level of the individual subject, to maintain its pattern of glandular adaptation throughout gestation : for example, a woman who was in the lower tertile for free T4 concentrations during early gestation, had a greater than 80 % probability to remain in the lower (or middle) tertile during late gestation; conversely a woman in the upper tertile for free T4 concentrations during late gestation, had a greater than 90 % probability to have been in the upper (or middle) tertile during early gestation, i.e. when the adaptation to the changes in serum TBG were most prominent (3).

Thyrotropic effects of hCG

On the basis of structural similarities between hCG and TSH, it has been suggested that hCG possesses intrinsic TSH-like activity (4) and several recent studies have emphasized the TSH-like effect of high hCG levels in normal pregnancy (5,6). In our recent work (3), the data were consistent with an intrinsic TSH-like activity of hCG and suggested that hCG acts directly on the thyroid in normal pregnancy. We reported clear evidence of a decrease in serum TSH corresponding to peak hCG levels, with a mirror image between TSH and hCG changes in individual samples, from 8-14 weeks gestation. The results also indicated a linear relationship between hCG and free T4 concentrations. The stimulatory effect of hCG was, however, relatively weak : a 10,000 IU/L hCG increment corresponded to a mean free T4 increment of 0.6 pmol/L and, in turn, to a lowering of TSH of 0.1 mU/L. Hence, during the first trimester of gestation, undetectable serum TSH levels were observed in 13 % of healthy subjects (in comparison to only 1 % during the last trimester), but free T4 levels in the thyrotoxic range were exceedingly rare. The effect of hCG to stimulate the gland directly was probably confined to the first half of gestation. However, further studies are required to assess whether hCG activity is not more pronounced and/or more prolonged in clinical circumstances such as gemellar pregnancies or pregnancy in women with thyroidal abnormalities (goiter, nodules, etc.).

Metabolism of iodine

Pregnancy is accompanied by a decrease in the availability of iodide for the maternal thyroid, due to increased renal clearance and losses to the feto-placental complex during late gestation, resulting in a relative iodine deficiency state (7). Recommended daily iodine intakes during gestation should probably equal at least 150 µg/day. In countries such as the U.S., where the average iodine intake is high, the effects of relative iodine deficiency will therefore not be observed. On the contrary, in many European countries with a daily iodine intake of only 50-100 µg/day, it can be anticipated that a marginally low iodine supply might become relatively insufficient in a prolonged physiological condition in which the maternal thyroid requirements as well as iodide losses are increased. Even though direct proof of relative iodine deficiency, at the individual level, could not be demonstrated in our recent work, the data suggested that the marginal iodine intake in Brussels had an overall permissive role in allowing for the regulatory changes in the thyroidal economy to be enhanced.

GOITROGENESIS AND PREGNANCY

Changes in thyroid volume during gestation

Data on alterations in thyroid volume (TV) during pregnancy are scanty. Goiter is rarely observed in the U.S. In Europe, many clinicians give the impression that goiter is perhaps more frequent among pregnant women; systematic studies, however, were lacking. In our recent work (3), ultrasonographies of the thyroid were carried out in several hundred healthy pregnancies at initial presentation and TV was reevaluated at delivery. The data indicated that the size of each lobe, and, hence, total volume did increase significantly by an average of 20 %. True goiter, defined as TV greater than 22 ml, was found in 9 % of the cohort at delivery. Changes in size were common, occurring in 70 % of women and were correlated to factors such as initial size (negatively), hCG stimulation (positively) and biochemical indices of thyroid stimulation, in particular high T3/T4 ratio and TG levels (positively). The best indicator of goitrogenesis was the increment in TG levels. Our work confirmed that TG levels were increased during the first, but mostly during the last trimester of pregnancy. At delivery, two thirds of the cohort had TG levels above the upper limit of normality (30 µg/L) and in 9 % TG levels were in excess of 100 µg/L.

Recovery during postpartum

Our studies indicated that, at least in conditions of marginally low iodine intake, pregnancy constitutes a goitrogenic stimulus and provided additional arguments to suggest that iodine supply be increased if the daily intake was below 100-150 µg of iodine. Are the alterations reversible during the postpartum period ? This crucial question remains open but we have proposed the hypothesis that if the changes occurring during gestation were not entirely reversible and were to be repeated during later pregnancies, they may represent a key clue to the understanding of the higher prevalence of thyroid disorders in women compared to men, at least in predisposed subjects.

To date 3 sets of information can be provided to support this hypothesis. First, serum TG was reevaluated 6 months postpartum in 100 unselected healthy women form the cohort investigated : in one third of women with elevated TG at delivery, serum TG was still abnormally increased. Second, women with a greater than 35 % increase in TV during gestation were recalled 1 year postpartum and TV reevaluated : in 8/10 cases, the glandular volume had not reverted to normal (and in some cases had even further increased!) (8). Third, we investigated during pregnancy 120 women presenting mild thyroid abnormalities and were able to show that these women were in average 2 years older and had had more frequent gestations, in comparison to healthy controls. Even though not definitely conclusive at present time, the data strongly suggest that the thyroidal "stress" of pregnancy may indeed not be entirely reversible during postpartum and that the number of pregnancies may play a role in the development of thyroid abnormalities (unpublished observations).

NEONATAL REPERCUSSIONS OF MATERNAL THYROID ALTERATIONS

We investigated thyroid function parameters in 250 mothers 3 days after delivery and compared them to those of their newborns in cord serum. To exclude alteration directly related to prematurity or disease, only healthy full-term newborns were considered. In addition to the characteristic low T3 syndrome at birth, neonates exhibited higher TSH (7.4 vs 2.0 mU/L; $p < 0.001$) and TG levels (72 vs 51 µg/L; $p < 0.001$), in spite of the fact that their total and free T4 levels were not only within the reference ranges, but significantly higher than in the mothers (T4 : 11.9 vs 11.0 µg/dl and free T4 : 1.5 vs 1.1 ng/dl; $p < 0.001$). We were also able to show that in newborns, TSH and TG concentrations increased in parallel with maternal values. Thus, in newborns from mothers with TSH in the upper tertile of the population, mean TSH was 25 % higher compared to TSH in those born to mothers in the lower tertile. Similarly, in newborns from mothers with TG greater than 60 µg/L, mean TG was 40 % higher compared to TG in those born to mothers with normal TG levels. The results indicated for the first time that the TSH surges at birth in full-term healthy newborns are not independently regulated but are associated with similar alterations in maternal thyroid function (9). The common stimulatory factor is most probably relative iodine deficiency. Hence, even in conditions of only marginally low iodine intake, with all the parameters exploring maternal thyroid function remaining within reference limits, pregnancy constitutes a stimulus for both the maternal and neonatal thyroids.

In conclusion, the alterations in maternal thyroid function during gestation are intricate and far from fully understood. They include adjustment of thyroidal economy due to changes in serum TBG, as well as direct stimulation of the gland due to elevated hCG levels. Furthermore, in areas of marginally low iodine intake, gestation is associated in a significant number of healthy women with relative hypothyroxinemia, increased TG and an enlarged thyroid. The glandular "stress" does not resolve "automatically" during postpartum and may therefore constitute one of the environmental factors leading to thyroid pathology in predisposed subjects. Finally, the gestational stimuli on maternal thyroids have functional consequences on neonatal thyroids, providing a link in the regulatory mechanisms involved.

REFERENCES

1. J. Robbins, S.-Y. Cheng, M.C. Gerschengorn, D. Glinoer, H.J. Cahnmann, and H. Edelnoch, Thyroxine transport proteins of plasma. Molecular properties and biosynthesis, Recent Progr Horm Research. 34 : 477 (1978).

2. J.T. Dowling, W.G. Appleton, and J.T. Nicoloff, Thyroxine turnover during human pregnancy, J Clin Endocrinol Metab. 27 : 1749 (1967).

3. D. Glinoer, P. De Nayer, P. Bourdoux, M. Lemone, C. Robyn, A. Van Steirteghem, J. Kinthaert, and B. Lejeune, Regulation of maternal thyroid during pregnancy, J Clin Endocrinol Metab. 71 : 276 (1990).

4. J.G. Kenimer, J.M. Hershman, and H.P. Higgins, The thyrotropin in hydatidiform moles is human chorionic gonadotropin, J Clin Endocrinol Metab. 40 : 482 (1975).

5. N. Yoshikawa, M. Nishikawa, M. Horimoto, M. Yoshimura, S. Sawaragi, Y. Horikoshi, I. Sawaragi, and M. Inada, Thyroid-stimulating activity in sera of normal pregnant women, J Clin Endocrinol Metab. 69 : 891 (1989).

6. M. Kimura, N. Amino, H. Tamaki, N. Mitsuda, K. Miyai, and O. Tanizawa, Physiologic thyroid activation in normal early pregnancy is induced by circulating hCG, Obstet Gynecol. 75 : 775 (1990).

7. J.D. Potter, Hypothyroidism and reproductive failure, Surg Gynecol Obstet. 150 : 251 (1980).

8. D. Glinoer, P. De Nayer, P. Bourdoux, J. Kinthaert, J.P. Grun, and B. Lejeune, Regulation of maternal thyroid during and after pregnancy, J Endocrinol Invest. 13 (suppl. 2-5) : 174 (1990).

9. D. Glinoer, I. Laboureur, P. De Nayer, F. Delange, P. Bourdoux, J. Kinthaert, J.P. Grun, and B. Lejeune, Maternal and neonatal thyroid functions in conditions of marginally low iodine intake, Proc 10th Intern Thyroid Congress. (Den Haege) (in press, 1991).

CONGENITAL HYPOTHYROIDISM:

NEW INSIGHT REGARDING ETIOLOGY

Jean H. Dussault

Director Screening Program for Congenital
Hypothyroidism. The Quebec Network for Genetic
   Medicine
Director, Ontogenesis and Molecular Genetic Unit
CHUL Laval University

GENETIC

The etiology of congenital hypothyroidism is unknown in most
cases although there is a greater incidence of thyroid disease
in families of infants with sporadic cretinism (1). The
occurrence of thyroid dysgenesis in more than one member of a
family has not helped to resolve the genetic factors involved.
In Japan an association with HLA type AW24 has been reported
whereas in North America and Europe such an association has not
been confirmed (2,3). On the other hand, Shepard reported an
increased incidence of non-tasters of phenylthiocarbamide in
families of a member with sporadic hypothyroidism (4). Since
the ability to taste Phenylthiocarbamide is a recessive
inherited trait and since there is a lower incidence of non-
tasters among blacks, a lower incidence of the disease was
expected in the black population, a fact that was confirmed in
North America (5).

In conclusion, improved methodology in chromosomal typing is
probably needed before a final statement call be made
concerning any genetic predisposition for congenital
hypothyroidism.

AUTOIMMUNITY

The possible role of an autoimmune phenomenon to explain
congenital hypothyroidism dates from a clinical description in
the early 1960's of a mother with thyroid autoantibodies who
had given birth to several newborns with congenital
hypothyroidism (6,7); it appears that the disease was transient
since during childhood the therapy was discontinued and thyroid
function tests were normal for age (8). These studies were
later expanded to include 121 mothers of newborns with
congenital hypothyroidism (9). Twenty-five percent were

positive for at least one type of thyroid antibody, 14% with
thyrocytoxic antibodies that may have caused intrauterine
autoimmune thyroiditis. On the other hand a subsequent study
done in 104 hypothyroid newborns showed that only one had a
detectable titer of anti-microsomal antibodies (10). Since the
early 1980's numerous studies have demonstrated the presence of
either thyrotropin binding inhibiting immunoglobulin (TBII) or
thyrotropin stimulating blocking immnunoglobulin (TSI Block) in
the sera of newborns with transient hypothyroidism (11). With
complete recovery the antibodies disappeared from the newborn
circulation indicating their maternal origin. Finally, Van Der
Gaag recently reported a Turkish family with evidence of
autoimmune hypothyroidism where two affected infants with
congenital hypothyroidism (as well as the mother) had an
antibody that blocked the growth promoting effect of TSH
(TGI-Block) (12). There reports have raised the interest of many
groups in order to confirm the role of autoimmunity in the
etiology of congenital hypothyroidism.

TBII ACTIVITY

We have evaluated TBII activity in the sera of mothers of
newborns with congenital hypothyroidism (MCH) (N=108) of which
22 were transient and 86 permanent and compared them to normal
pregnant women (NPW). The sera of 4/22 mothers of infants with
transient hypothyroidism were demonstrated to have more than 10%
inhibition of the binding of labelled TSH to its receptor. On
the other hand only 7/86 mothers of the infants with permanent
hypothyroidism tested positive by these criteria of which 4 were
borderline. None of the 19 normal pregnant women (NPW) were
positive.

TGI-BLOCK

Serum samples from infants with congenital hypothyroidism and
their mothers from the Quebec program were tested for TGI-Block
activity by the Feulgen cytophotometry method. In summary, one
tests the blocking effect of immunoglobulins on the growth
promoting effect of TSH on guinea pig thyroid follicular cells
in vitro by measuring DNA content. Fifteen out of 33 mothers and
8 of 17 infants were positive (13). Of 16 paired sera 75% were
concordant (14 permanent, 2 transient). Expanding this study to
the Dutch program (14) 11/13 MCH were positive as were 8/11
newborns using a more sensitive assay (dose response). On follow
up 3-5 years after diagnosis, 2/15 of the mothers were still
positive whereas in 4/14 infants TGI-Block was still present.

[H^3] THYMIDINE UPTAKE

Another way to study the effect of immunoglobulins on TSH
induced growth on thyroid cells is to use a rat cell line (FTRL$_5$)
and to study the incorporation of [H^3]-thymidine into those
cells. The results are expressed as DPM of [H^3] thymidine/ug of
DNA. The mean value obtained in 46 MCH was 1.06 versus 0.98 in

normal women (N=21) whereas in the NPW group (N=27) immunoglobulins seem to have a very significant effect on [3H] thymide incorporation with a mean of 0.60. We have shown that hCG was not responsible for this phenomena. If this "factor" is an immunoglobulin it must be of the class IgG_3 type which does not bind to protein A but crosses the placenta.

I^{131} UPTAKE

This(ese) factor (s) might explain the results we obtained using the I^{131} uptake TSH stimulation as a model to study the effects of immunoglogulins of MCH. In that study we found that MCH IgG stimulated the I^{131} uptake compared to NPW: 170% vs 100% (15). Indeed our results using this model could not be duplicated by two different groups of investigators, Cluovato et al (16) and Brown et al (17), in a much smaller number of samples.

OTHER THYROID ANTIBODIES

Anti-microsomal antibodies

Numerous studies have shown the presence of anti-microsomal antibodies in the MCH group (10-15%) but this incidence is not different than in normal pregnant women (10).

CA_2 antibodies

On the other hand Boyage and colleagues (14) have looked at the incidence of CA_2 (antibodies against the second colloid antigen). In the Quebec population 50% of the MCH sera were positive versus 36% of the Dutch population of which 33% of the newborns were positive. Three to five years after diagnosis 47% of the maternal sera were still positive compared to 33 % for the children. In a normal population, pregnant or not pregnant, the prevalence of those antibodies is between 8 and 10%.

Cytotoxic antibodies

Finally Bogner et al (18) determined antibody-dependent cell mediated cytotoxicity in 61 patients with congenital hypothyroidism and 46 of their mothers. In the newborns 33% were positive versus 24% of their mothers. This activity was still present in 25% of older children (mean age 15). In 96% of the mother-infant pairs concordance existed between the cytotoxic activity in mothers and their children.

SUMMARY AND CONCLUSION (Table I)

Some of the results presented here are still controversial especially those obtained using the $FRTL_5$ cell lines. One must understand that none of the studies were done under the same conditions (concentration of TSH, time of TSH deprivation, time of incubation, etc). The only statement that can be made is that in some specific conditions one can distinguish MCH and NPW

TABLE I

Thyroid antibodies in congenital hypothyroidism

Antibody	Sera (% positive)	
	Maternal	Newborn
Anti-microsomial	10 - 15%	1 - 5%
TBII	10%	5%
TGI Block	50 - 80%	50%
CA_2	30 - 50%	30 - 50%
Cytotoxic	25%	33%
I^{131} uptake	100%	---
[H^3] thymidine incorporation	100%	---

groups and that the causal factor is present in the NPW group!
This approach is probably still valid to distinguish mothers at

risk but does not imply a role of autoimmunity. On the other
hand, if the methodology employed for the TGI-Block assay is
valid, then the results are more significant (more than 50% of
MCH). Since this assay is still controversial one must look at
the data obtained for the CA_2 antibodies and the antibody-
dependent cell-mediated cytotoxicity where the percentage of
positivity varies between 25 and 50%. From these data one could
speculate that in a significant number of cases of congenital
hypothyroidism transplacental passage of self-reactive T
lymphocytes directed against thyroid tissue occurs during the
first 12 weeks of fetal life prior to the development of the HLA
system. The survival of these lymphocytes and an inherited
defect of immunoregulation in the fetus inducing a cell-mediated
immune response has to be accepted as well (19).

On the other hand, the presence of CA_2 antibodies and cytotoxic
antibodies at a later age implies that the disease is not only
maternally induced. Thus other etiological factors such as
genetic predisposition, environment, viral (CA2 antibodies are
mostly present after an episode of sub-acute thyroiditis) (14).

In summary, only detection of thyrocytotoxic antibodies in all
the mothers of congenital hypothyroid infants or the detection
of a defect in immunoregulation of the hypothyroid newborn will
establish autoimmunity as the sole cause of congenital
hypothyroidism. For now we can only state that autoimmunity is
one of the numerous factors in the etiology of congenital
hypothyroidism which is heterogeneous.

REFERENCES

1. B.Childs and L.I. Gartner, Etiologic factors in sporadic cretinism: An analysis of ninety cases, Ann Hum Genet 19:90 (1954, 1955).
2. P. Cimino, R. Banks, N. MacLaren et al, HLA and congenital hypothyroidism. N Engl J Med 303: 1177 (1980).
3. K. Miyai, H. Mizuta, O. Nose et al, Increased frequency of HLAOAw24 in congenital hypothyroidism in Japan. N Engl J Med 303: 226 (1980).
4. T.H. Shepard, S.M. Gartler, Increased incidence of non-tasters of phenylthiocarbamide among congenital cretins. Science 131: 929 (1960).
5. A.L. Brown, P.M. Fernhoff, J. Milner et al, Racial differences in the incidence of congenital hypothyroidism. J Pediatr 99: 934 (1981).
6. R.M. Blizzard, R.W. Chandler, G.H. Landing, et al, Maternal autoinununization to thyroid as a probable cause of athyreotic cretinism. N Engl J Med 263: 327 (1960).
7. J.M. Sutherland, V.M. Esselborn, R.L. Burket, B.T. Skillman, J.T. Benson, Familial nongoitrous cretinism due to maternal antithyroid antibody: report of a family. N Engl J Med 263: 336 (1960).
8. R.E. Goldsmith, A.J. McAdams, P.R. Larsen, M. MacKenzie, E.V. Hess, Familial autoimmune thyroiditis: Maternal-fetal relationship and the role of generalized autoirnmunity. J Clin Endocrinol Metab 37: 265 (1973).
9. R.W. Chandler, R.M. Blizzard, W. Hung, M. Kyle, Incidence of thyrotoxic factors and other antithyroid antibodies in the mothers of cretins. New Engl J Med 267: 376 (19@2).
10. J.H. Dussault, J. Letartre, H. Guyda, C. Laberge, Lack of influence of thyroid antibodies on thyroid function in the newborn infant and on a mass screening program for congenital hypothyroidism, J Pediatr 96: 385 (1980).
11. N. Matsuura, Y. Yamada, Y. Nohara, et al, Familial neonatal transient hypothyroidism due to maternal TSH-binding inhibitor immunoglobulins. N Engl J Med 303: 738 (1980).
12. R.D. Van Der Gaag, H.J. Frish, M. Weissel, H.A. Drexhage, Congenital hypothyroidism in a Turkish family: The role of immmunoglobulins blocking the trophic effects of TSH and maternal-foetal relationship. Acta Endocrinol 111: 44- (986).
13. R.D. Van Der Gaag, H.A. Drexhage, J.H. Dussault, Role of maternal immunoglobulin blocking TSH -induced thyroid growth in sporadic forms of congenital hypothyroidism. Lancet 1:246 (1985).
14. S.C. Boyages, J.W. Lens, R.D. Van Der Gaag, G.F. Maberly, C.J. Eastman and H.A. Drexhage, Sporadic and endemic congenital hypothyroidism: evidence for autosensitization, in: Research in Congenital Hypothyroidism F. Delange, D.. Fisher and D. Glinoer, eds., Plenum Press, New York : 123 (1989).
15. J.H. Dussault and D, Bernier, [125]I Uptake by FRITL5 cells: a screening test to detect pregnant women at risk of giving birth to hypothyroid infants, Lancet 2: 1029 (1985).

16. L.Chiovato, P. Vitti, C. Marcocci, G. Fenzi, L. Giusti, F.
    Santini, P Bassi, M. Ciampi, M. Tonacchera, and A.
    Pinchera: TSH-blocking antibodies and congenital
    hypothyroidism. in Congenital Hypothyroidism F. Delange, D.
    Fisher and D. Glinoer, eds., Plenum Press, New York: 141
    (1989).

MATERNALLY TRANSFERRED THYROID DISEASE IN THE INFANT: RECOGNITION AND TREATMENT

Thomas P. Foley, Jr.

Department of Pediatrics
University of Pittsburgh
3705 Fifth Avenue at DeSoto Street
Pittsburgh, Pennsylvania 15213-2583, USA

INTRODUCTION

Throughout pregnancy the fetus is dependent upon the placental transport of nutrients from the mother to support growth and differentiation. By this same mechanism the fetus is exposed to a variety of maternal factors that may adversely affect thyroid embryogenesis and function during pregnancy and postnatal life. The maternal factors include altered concentrations of iodide, immunoglobulins primarily in the G class (IgG), drugs, environmental goitrogens and other substances that are poorly understood or defined.

The most common neonatal thyroid disease world-wide that is acquired from the mother is the oldest thyroid disease known to mankind - endemic goiter and cretinism from iodine deficiency.[1,2] In regions where iodine deficiency no longer exists, a group of diseases mediated by placental transport of IgG are the most common cause of a spectrum of transient and permanent thyroid disorders presenting as hypothyroidism or hyperthyroidism.[3,4] Most of these diseases are poorly understood both as to why the IgG impairs thyroid gland growth and embryogenesis, and the reason for the presence of IgG in the mother who rarely has clinical thyroid disease. Neonatal disease may result from the maternal use of iodide and antithyroid drugs that cause transient, usually mild hypothyroidism in the fetus and neonate.[5-7] In this review these diseases and a new disease that we recently reported as Congenital Transient Dyshormonogenesis (CTD)[8] will be discussed.

Both the proven and hypothesized maternally derived, transplacentally acquired thyroid diseases of the infant may be classified according to the

Table 1.   MATERNALLY   DERIVED   TRANSPLACENTALLY
ACQUIRED NEONATAL THYROID DISEASES

I.   Disorders of Iodine Metabolism
A. Iodine Deficiency Disorders
B. Iodine Intoxication

II.   Immunoglobulin G Mediated Disorders
A. Hyperthyroidism: Neonatal Graves Disease
1. Monoclonal Thyrotropin Receptor Antibodies
2. Polyclonal Thyrotropin Receptor Antibodies
B. Hypothyroidism
1. Thyrotropin Receptor Blocking Antibodies
2. Antibody Dependent Cell-Mediated Cytotoxicity
3. Thyroid Growth Inhibitory Immunoglobulins
4. Antibodies to Second Colloid Antigen

III.   Drugs
A. Antithyroid drugs
B. Therapeutic Doses of Radioiodine

IV.   Environmental Goitrogens

V.   Idiopathic
A. Congenital Transient Dyshormonogenesis
B. Miscellaneous

pathogenic factors or mechanisms that induce thyroid disease in the infant.
(Table 1)

DISORDERS OF IODINE METABOLISM

Endemic goiter and cretinism has been known since antiquity.
Although iodine supplementation programs in much of the world have
eradicated this disease, where these programs have not been instituted it is a
very prevalent problem, particularly in mountainous regions and areas of
glaciation where the water from melting glaciers washed iodine from the soil.
Although the optimal daily iodine intake is estimated to range between 150 and
300 μg, euthyroidism is maintained when the daily intake of iodine is between
50 and 1,000 μg. Iodine deficiency disorders (IDD) are associated with a
urinary iodine excretion that is less than 50 μg/gm of creatinine.[1,9] Neonatal
and maternal thyromegaly among patients living in endemic goiter regions in
the absence of maternal Graves disease should be considered as IDD. The

diagnosis is confirmed by testing the iodine content of urine. If the infant is euthyroid, treatment with iodine alone is appropriate; if the infant has primary hypothyroidism, as indicated by an elevated levels of thyrotropin (TSH) and low free thyroxine (FT4), the addition of low dose thyroxine (T4), 25 μg/day, for two to four weeks with iodine therapy would be indicated.

There are several reports in the literature that therapeutic doses of iodine by systemic or vaginal administration to the mother causes neonatal goiter and primary hypothyroidism.[6,10,11] With this general awareness, maternal sources of iodine infrequently cause neonatal goiter or hypothyroidism today. However, high circulating levels of iodide in the neonate may develop after the use of iodide-containing radiocontrast dyes or antiseptics applied daily to the skin or mucous membranes of the infant and cause primary hypothyroidism. This problem has been appreciated only recently.[10] With this general awareness, maternal sources of iodine infrequently cause neonatal goiter or hypothyroidism today.

The thyroidal response to excessive iodide intake in infants is very similar to the experimental data reported in normal rats.[5] (Figure 1) Iodide was added to drinking water of pregnant and nursing rats in concentrations that delivered an estimated 1 to 2 mg of iodide daily. After weaning until age 60 days iodide was added to the drinking water of the pups. A significant elevation of serum TSH in the pups was found before birth and during the first 18 days of postnatal life compared to controls. T4 values were lower than controls from day 1 through day 10 in the iodide-treated pups, but comparable to controls thereafter. These data in normal rats and the clinical experience in newborn infants suggest that the immature thyroid is very susceptible to the chronic inhibitory effects of iodide on hormonogenesis. This effect may be caused by immaturity of the iodide oxidation and protein iodination system in the thyroid, for the young rat develops resistance to the inhibitory effects of iodide, and like the normal child and adult is not susceptible to iodide-induced hypothyroidism.

These patients are best managed by the avoidance of any iodide-containing medications during pregnancy and the early postpartum period. However, if iodide therapy after birth is essential, thyroid function of the infant must be monitored very closely and T4 therapy initiated as long as iodide is used and there is an elevated serum TSH value.

IMMUNOGLOBULIN G MEDIATED DISORDERS: HYPERTHYROIDISM

The observation that neonatal hyperthyroidism occurred when the mother had either active Graves disease, or a history of previously treated

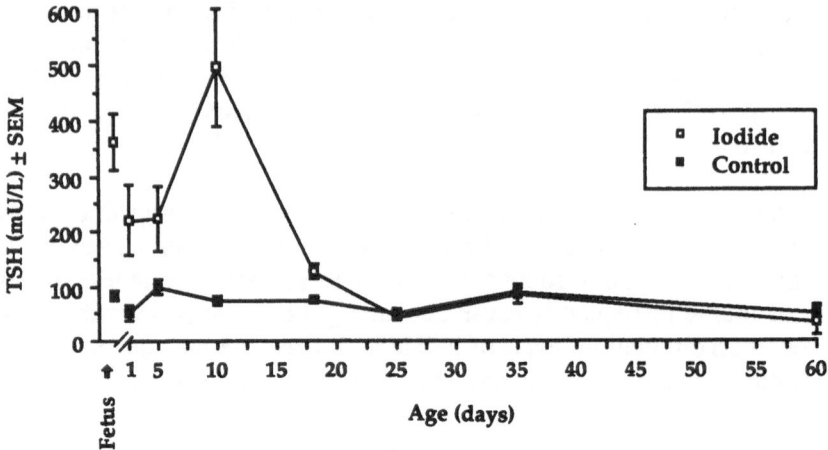

Diagram of data from Theodoropoulos et al Science 1979, 205:502

Figure 1. Effect of iodide on TSH in rats is reproduced from data previously published.[5] The TSH response to pharmacologic amounts of iodide given to pregnant and postpartum rats on fetal and neonatal TSH are compared to controls.

Graves disease, strongly supported the hypothesis that a humoral factor, passively transferred from mother to fetus, was the cause of Graves disease.[12,13] The disease may occur in the neonate even when the mother has Hashimoto's disease and is euthyroid on therapy or hypothyroid. Thyroid-stimulating antibodies (TSAb) of maternal origin are known to cross the placenta, bind to and stimulate the TSH receptor, and cause the hyperthyroidism of Graves disease in the adult and neonate. Recent studies reported that TSAb activity in serum from adults with Graves disease was restricted to the single IgG_1 subclass.[14] During pregnancy the maternal-to-fetal transport of IgG gradually increases until the fetal and maternal levels are comparable around thirty weeks of gestation.[15] Fetal TSAb concentrations will be maximal during the last trimester. The higher the TSAb values, the more likely the neonate is clinical thyrotoxic, the more severe the disease, and the greater the duration of the neonatal disease. Usually the neonatal disease is transient, and the pattern of the clinical course reflects the 1 to 2 week circulating half-life of IgG. In our experience using a thyroid binding inhibitory immunoglobulin (TBII) method to measure TSH receptor antibodies (TRAb),[16,17] predominately TSAb in Graves disease, we found that the values in mother and her newborn infant were similar at diagnosis, but decreased steadily in the infant during the first two months of age. (Figure 2)

The disease should be recognized before clinical signs are apparent if a careful history for maternal Graves disease is obtained and the newborn thyroid is carefully examined. Maternal TSAb stimulates the fetal thyroid to secrete excessive amounts of T4; however, fetal T4 is converted to reverse triiodothyronine (rT3) instead of T3,[18] so the presence of fetal hyperthyroidism must be caused by the direct thyroidal secretion of T3 from TSAb stimulation. TSAb causes thyromegaly which is detected in the fetus by ultrasound, and easily on clinical examination after birth. The onset of clinical thyrotoxicosis may not be seen until two to three days of age or later for two reasons: maternal antithyroid drugs easily cross the placenta and protect the fetus and neonate at birth, but are metabolized within hours so that thyrotoxicosis may occur by the second day of life;[4]   secondly, the maturation of 5'monodeiodinase activity occurs promptly after birth to cause an increase in the monodeiodination of T4 to T3 during the first day of life.[18]   The combination of direct thyroidal secretion of T3 by the thyroid that no longer is blocked by antithyroid drugs and increased T3 production by peripheral conversion from T4 shortly after birth results in thyrotoxic levels of T3 as early as the first 24 hours of age.

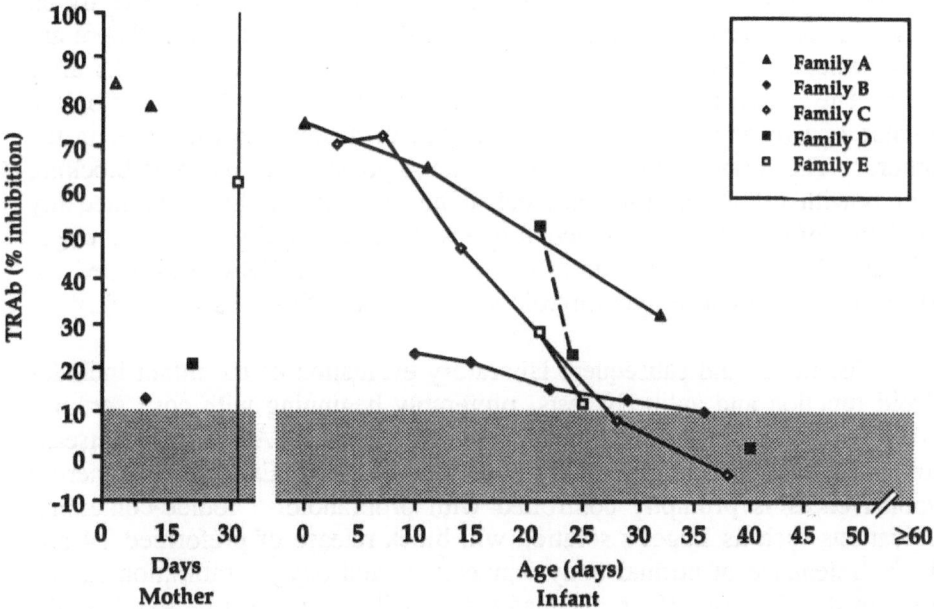

Figure 2.   TSH receptor antibody determinations in serum from mothers with a history of either active Graves disease or treatment for Graves disease and serially in their infants during the first two months of life are compared.   TRAB was measured in unextracted serum by a TBII method.[16,17]

Table 2. SYMPTOMS AND SIGNS OF NEONATAL GRAVES DISEASE

Irritability	Enlargement of reticuloendothelial
Tachypnea	system: liver, spleen, lymph nodes
Tachycardia	Small for gestational age
Vomiting	Poor weight gain or excessive weight
Diarrhea	loss with normal caloric intake
Flushing	Onset from birth or delayed by weeks
Voracious appetite	Thrombocytopenia,hypoprothrombinemia
Thyromegaly	Hypertension
Exophthalmos	Accelerated skeletal maturation
Low grade fever	Frontal bossing
Periorbital edema	Synostosis

The clinical symptoms and signs involve multiple systems and often are non-specific.[19] (Table 2) Neonatal Graves disease has been mistaken for congenital heart disease, withdrawal of a narcotic addicted infant and sepsis. In severe cases the disease is evident during fetal life as an intrauterine death, fetal hyperthyroidism with poor weight gain, tachycardia and thyromegaly on ultrasound, or premature delivery with clinical symptoms at birth. There are reports where clinical hyperthyroidism is delayed until several weeks after birth.[20] The reason for the delayed onset has been shown to be caused by the presence of two or more populations of IgG, or a polyclonal disease, in the mother.[21] The infant acquired from mother a potent TSH receptor blocking antibody with a short half life that blocked the TSH receptor and prevented any stimulation of the thyroid by a second IgG with TSAb activity that had a longer half life. Once the blocking antibody was metabolized, TSAb stimulated the TSH receptor and cause hyperthyroidism around the fifth to sixth week.[22]

The initial and subsequent laboratory evaluation of the infant includes thyroid function and antibody tests, preferably beginning with cord serum.[23] (Table 3) Therapy is designed to control the transient, TSAb-mediated disease without the development of hypothyroidism.[24] (Table 4) Excessive adrenergic responsiveness is promptly controlled with propranolol. Iodide-containing medications such as Lugol's solution will block release of preformed T4 and T3. Maintenance of normal body temperature and oxygen saturation should be controlled by use of an incubator that will automatically regulate the temperature, humidity and oxygen delivery inside the incubator in response to the temperature and oxygen saturation of the infant. In moderate and severe thyrotoxicosis antithyroid drugs are given to inhibit further hormonogenesis. Dexamethasone may be used in severe thyrotoxicosis to block thyroid hormone

Table 3.   THYROID EVALUATION IN NEONATAL GRAVES DISEASE

Cord serum or serum at diagnosis, and maternal serum
  T4, Free T4, T3 and TSH
  TRAb by TBII or TSAb methods
  Thyroid peroxidase and thyroglobulin antibodies
  CBC, differential and platelet counts

Subsequent evaluation of the affected infant
  Serial thyroid function every one to three days until stable, then once
    or twice a week
  TRAb by a microvolume method twice a week until normal

---

secretion and control the inflammatory response when exophthalmos is present although its efficacy has not been proven in this disease.   Of theoretical advantage is the known effect of propylthiouracil (PTU), propranolol, and dexamethasone to inhibit the conversion of T4 to T3, an effect that has not been documented in the infant.

Table 4.   TREATMENT FOR NEONATAL GRAVES DISEASE

Propranolol: 2 mg/kg/day in four divided doses

Iodine 5% and 10% KI: one drop (8 mg) every eight hours

Propylthiouracil: 5 to 10 mg/kg/day in three divided doses

Sedation, adequate oxygenization and appropriate hydration

Dexamethasone: 0.3 mg/kg/day in three divided doses in severe thyrotoxicosis
  and for prominent exophthalmos

Digitalization in the presence of high output heart failure

Increase PTU and iodide by 50 to 100% if there is no therapeutic response
  within 48 hours

Monitor thyroid function for control of hyperthyroidism and prevention of
  hypothyroidism during therapy

Follow the infant over several months for recurrent disease

The response to therapy usually is prompt. Tests of thyroid function are important to monitor so to avoid the development of hypothyroidism. The dose of iodide and PTU should be adjusted to maintain normal thyroid function. Usually they are gradually tapered every three to seven days as the severity of the disease declines with the decrease in TRAb activity. If there is negligible response to initial therapy, the dose of iodide and PTU should be increased, and the addition of dexamethasone considered if there are no contraindications to its use. The infant should be followed by pediatric cardiologists to identify the development of high output heart failure and the need for digitalization and decrease in the dose of propranolol. An occasional patient may have a prolonged course during infancy which has been postulated to result from an IgG that would enhance or prolong the TSAb activity in serum. Since a rare patient may have congenital, or permanent, Graves disease, the infant should be followed for several months with serial thyroid function. These infants usually have a very strong family history of Graves disease, and may inherit the disease with an early onset. There are no tests to predict which infant has permanent disease since the etiology of this entity is unknown. The prognosis for the transient disease is excellent. Poorly controlled or untreated chronic thyrotoxicosis during infancy is associated with growth retardation, craniosynostosis with mental retardation, and advanced skeletal maturation [4,19,24].

IMMUNOGLOBULIN G MEDIATED DISORDERS: HYPOTHYROIDISM

Thirty years ago there was a report of a mother with Hashimoto's thyroiditis who delivered children with congenital hypothyroidism (CH) with variable severity.[25,26] For the first time autoimmune thyroid disease in the mother was reported to cause CH. Several years later when this author evaluated her four surviving children who had CH when they were neonates (two had died in the first week of life), they were euthyroid on no therapy, although they had some intellectual impairment. The autoimmune nature of the disease in this family was further documented in a later study.[27] There have been several reports during the past decade in which TSH receptor blocking antibodies were detected in serum of mothers and infants with transient CH.[28] When the maternal IgG were recently characterized, either pure inhibitory or both inhibitory and stimulating activities were identified, and the bioactivity, both stimulating and inhibitory, resided in either the kappa or gamma light chains of the IgG.[29] In infants with CH, the inhibitory IgG had greater potency. There is no common terminology for these disorders that we have called congenital transient immunoglobulin-mediated hypothyroidism (CTIH).

Infants with CTIH are detected by newborn screening for CH, are found to have no thyroid tissue by technetium or iodide imaging, and on confirmatory serum testing are found to have primary hypothyroidism. Their presentation is very similar to infants with athyreotic CH except that thyroid tissue is palpable on examination and is detected by ultrasound if that test is performed, and TRAb by a TBII method are present in serum from infant and mother at diagnosis. Antibodies to thyroglobulin (TgAb) and thyroid peroxidase (TpAb) may be detected, usually in low titers. These infants should be treated with L-thyroxine either until after age 2 years, or until the TRAb in serum no longer is detected, when the dose of L-thyroxine may be decreased. If serum TSH remains normal on only 25 μg daily, then L-thyroxine may be discontinued and TSH values monitored weekly for 4 to 6 weeks.

Most well characterized, transplacentally acquired thyroid diseases of the neonate are transient, for the disorders are not associated with permanent damage the thyroid. Approximately 90% of infants with sporadic CH (SCH) have permanent disease. The etiology of these disorders, known collectively as thyroid dysgenesis, is poorly understood. Recent studies suggest that an immune-mediated mechanism, antibody dependent cell mediated cytotoxicity (ADCC) may be involved in the pathogenesis of SCH.[30] By this mechanism the infant acquires an IgG from the mother by placental transport. The maternal IgG in the presence of normal lymphocytes (effector cells) causes lysis of normal thyroid cells. The in vitro assay utilizes normal adult lymphocytes, adult thyroid cells that are labeled with ^{51}Cr and maintained in culture, and IgG preparations from mother and child. The release of ^{51}Cr into the media, indicative of thyrocyte lysis, is greater in the presence of IgG from the mothers and their infants or children with SCH compared to IgG from normal mothers and infants or children.[30] The assay is not as sensitive when rat thyroid cells, such as the Fischer Rat Thyroid Line (FRTL-5), are used, suggesting species specificity.

Among 61 patients with SCH and 46 mothers, ADCC was found in 31% of children and 24% of their mothers. There was concordance the cytotoxic activity in mothers and their children in 96%. In patients with athyreosis, an ectopic gland, and a normal gland, ADCC was found in 30%, 50%, and 7%, respectively. This transplacental immune-mediated hypothesis suggests that the mother has either subclinical autoimmune thyroid disease, or secretes an IgG that is active against the fetal but not the maternal thyroid gland.[30] This theory does not explain the observation that twice as many females have SCH compared to male infants.

During the past decade antibodies that stimulate thyroid follicular cell growth, or thyroid growth stimulating immunoglobulins (TGI) have been studied in several thyroid diseases associated with thyromegaly. In a manner analogous to studies of TRAb that inhibit the TSH receptor to cause hypothyroidism, investigators in The Netherlands hypothesize that TGI blocking antibodies in serum of the mother cross the placenta and inhibit fetal thyroid growth.[31] The assay determines the effect of IgG and TSH on cell division in the Feulgen cytochemical bioassay, and the TGI blocking antibodies decrease cell division that is stimulated by TSH.

In their initial report 34 mothers and 16 infants with SCH were tested for TGI blocking antibodies. There were 15 mothers (44%) and 8 (50%) infants who were positive for TGI that blocked TSH-induced thyroid growth. Up to three years later only three of seven mothers were still positive, suggesting that the TGI blocking antibodies do not persist, an observation that agrees with the rare occurrence of more than one affected infant with SCH in a family.[31] These data do not explain the predominance of affected female infants, and raise the question why the mother very infrequently has clinical thyroid disease if the antibody has a direct effect on the fetal thyroid gland.

Recently one study reports the presence of antibodies to the second antigen in colloid (CA_2) in serum from infants with SCH, their mothers and controls.[32] CA_2 antibodies were measured by immunofluorescence at the time of diagnosis of SCH or 3 to 5 years after diagnosis. Of 26 mothers and infants at diagnosis, CA_2 antibodies in serum were positive in 50% and 31%, respectively. After 3 to 5 years, 46% and 31% of mothers and children were positive. However, serum from only 3 of 26 maternal-infant pairs and 2 of 13 maternal-child pairs both tested positive, questioning the pathogenic role of this antibody in SCH.

Infants with SCH usually are identified by newborn screening programs. Once the diagnosis is confirmed with serum thyroid function tests and thyroid image, L-thyroxine therapy should be initiated promptly. We reported that an initial dose of L-thyroxine 50 μg daily (10-14 μg/kg/day) should be given to term infants with thyroid dysgenesis.[33] In many infants the dose needs to be decreased to 37.5 μg daily within one week of therapy, depending upon the results of day 3 and 7 thyroid function tests. However, doses of 5 to 6.5 μg/kg/day may be inadequate after age 2 months to maintain suppression of TSH below 5 mU/L. With the more sensitive TSH assays that will discriminate normal and low TSH values, we suggest that L-thyroxine doses be prescribed to maintain TSH values between 0.05 and 5 mU/L in the absence of clinical hyperthyroidism; usually the FT4 remains within the upper range of normal for infants, and the T3 value below 250 ng/dL (3.85 nM).[33] The ability for conversion of T4 to reverse T3 when there are excessive T4 concentrations protects the infant and child against clinical hyperthyroidism. When the patient is clinically euthyroid, linear growth is normal and body weight is appropriate for height, we recommend that only thyroid function tests (FT4, T3, TSH) need to be monitored. The prognosis is excellent for normal growth and development, and intelligence is not impaired compared to sibling and age-matched controls if severe intrauterine hypothyroidism has not occurred, therapy is started within one month of age, and the patient remains euthyroid with normal thyroid function throughout the first two years of age.[34]

MATERNAL DRUGS THAT AFFECT FETAL THYROID FUNCTION

We have known for years that antithyroid drugs administered during pregnancy to the mother may cause fetal and neonatal hypothyroidism and thyromegaly.[35] Current doses of PTU should be maintained at the lowest dose, preferably less than 200 mg daily, to avoid these side effects.[7,36] If there is evidence of fetal hyperthyroidism, the use of higher doses of PTU would be indicated since the drug readily crosses the placenta and would decrease fetal thyroid hormone secretion. However, further enlargement of the fetal thyroid gland may occur, and rarely has caused upper airway obstruction. Careful monitoring of maternal thyroid function to maintain mildly elevated T4 and T3 values and fetal heart rate to prevent tachycardia and promote normal weight gain are essential during pregnancy.[36]

Once the fetal thyroid can concentrate iodide around the 10 to 12th week of gestation, the administration of radioiodine in therapeutic doses can ablate the fetal thyroid gland to cause fetal hypothyroidism.[37] We reported the presence of persistent hyperthyrotropinemia in a boy whose mother received 8.5 mCi of ^{131}I-iodine between the nineteenth and twentieth week of gestation.[8] During hourly 24 hour sampling at age 7 years, TSH values ranged between 4.1 and 14.0 mU/L with the higher values during the night. This boy had no evidence of autoimmune thyroiditis, and had a low radioiodine uptake at age one year at an age when his TSH response to TRH was exaggerated, and basal TSH values ranged between 5.1 and 34.7 mU/L. This patient never had an abnormal total or free T4. Thyroid function has remained normal on 25 μg of L-thyroxine daily. We hypothesize that the maternal radioiodine damaged a sufficient amount of thyroid tissue that excessive stimulation by TSH of the remaining gland was required to maintain normal FT4 values and linear growth. The minimum dose of ^{131}I-iodine that is not associated with thyroid function abnormalities is not known. A dose of 14 mCi at the end of the first trimester induced hypothyroidism in an infant. This child developed mental retardation which may be caused in part by the late age of 18 months at diagnosis.[37] Although there is very limited experience, intra-amniotic therapy with 500 ug of L-thyroxine weekly caused partial suppression of TSH at birth in one patient, and seems to be a preferable method of L-thyroxine therapy for the fetus compared to fetal intramuscular injections which are more difficult and very limited in the amount of dose that can be administered.[38,39]

ENVIRONMENTAL GOITROGENS

There are specific goitrogenic factors in the diet and in the environment that are ingested by the mother and may cause endemic cretinism in the infant. These goitrogens are found in certain foods and in contaminated water supplies, particularly thiocyanates and sulfur-containing organic compounds.[1,2] Both clinically and experimentally the effect of goitrogens is an aggravation of existing iodine deficiency, and that these factors alone in an iodine-sufficient population do not cause neonatal goiter or hypothyroidism.[2]

CONGENITAL TRANSIENT DYSHORMONOGENESIS

The formation of T4, T3 and reverse T3 (rT3) from dietary iodide and tyrosyl residues of thyroglobulin (Tg) is dependent on specific energy-requiring, enzymatic reactions.[40] Several defects of thyroidal hormonogenesis are inherited in man as autosomal recessive traits, and are known collectively as familial thyroid dyshormonogenesis (FTD).[41] They are recognized by newborn screening programs as primary hypothyroidism, and infants are found to have normal or enlarged thyroid glands at diagnosis.[42] Except for TSH receptor and iodide transport defects, these infants have an elevated uptake of radioiodine. When a defect in the oxidation or organification of iodide to iodine is present, there is a prompt discharge of radioiodine from the thyroid after the administration of an iodide transport competing anion such as perchlorate or thiocyanate.[40,41] Hypothyroidism in these patients is permanent although the severity is quite variable. These diseases require therapy with thyroxine for life.

In a review of our twenty year experience in the diagnosis and management of CH, we identified five patients by newborn screening with a normal or enlarged thyroid gland, an elevated TSH and in three infants an elevated radioiodide uptake with discharge of radioiodide after oral perchlorate.[8] (Figure 3) Therapy with thyroxine suppressed TSH to normal, and these patients were diagnosed as FTD. The five patients were born either in 1975 or in 1980-1. After two years of age they were clinically euthyroid after medication was discontinued. Thyroid function tests and radioiodide uptake were normal and have remained normal.[8] In each mother and infant TRAb, TpAb and TgAb were not detected in serum.[43] In one mother on L-T4 for hypothyroidism and her son who has a partial oxidation defect and compensated hypothyroidism, we have preliminary evidence for an antibody that binds TSH.[43,44] We are uncertain as to the role of this antibody in the disease of this infant. These five infants with CTD constitute 3% of our population of 168 infants with primary CH.[8]

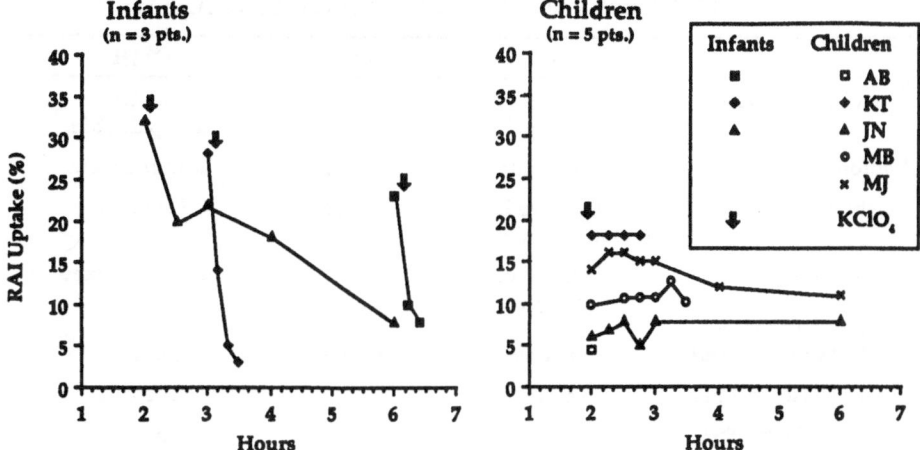

Figure 3. Radioiodide uptake and discharge of radioiodide after oral potassium perchlorate during infancy in patients with congenital transient dyshormonogenesis demonstrate the presence in three infants of an oxidation defect in thyroid hormonogenesis. The radioiodide uptake after therapy was discontinued during childhood is normal for each patient.

We hypothesize that this disorder is caused by a maternal factor, perhaps an IgG, that inhibits the sequential steps of iodide oxidation and binding to tyrosyl residues of Tg, and the oxidative coupling of iodotyrosines into iodothyronines. To cause a transient block of protein iodination, the maternal factor might interfere at several steps in hormonogenesis, such as the enzymes of the H_2O_2-generating system, thyroperoxidase (Tp), Tp activation by heme binding, or Tp binding to structures on the apical membrane.[40,41] An interesting analogy is the observation that antibodies against NADPH cytochrome c reductase inhibit iodination in vitro.[45] Of interest is the absence of thyroid disease in three of the five mothers, and no evidence of autoimmune thyroid disease in two mothers who were receiving thyroxine therapy for antibody-negative primary hypothyroidism and after thyroidectomy for papillary carcinoma of the thyroid.[8]

The differential diagnosis of congenital primary hypothyroidism in the neonate with thyroid tissue present in the normal anterior cervical location includes three specific diseases that are important to distinguish from transient thyroid dysfunction of the newborn and idiopathic hyperthyrotropinemia, or

Table 5. CONGENITAL PRIMARY HYPOTHYROIDISM

	FTD	CTD	CTIH
Thyroid gland size	Normal or enlarged	Normal or enlarged	Small or non-palpable
Radioiodine uptake	Elevated*	Elevated	Low or absent
Perchlorate discharge	Positive if oxidation defect	Positive	Negative
Thyroid image	Normal or enlarged	Normal or enlarged	Non-visualized
Thyroid antibodies:			
1. TpAb and TgAb	Negative	Negative	Positive
2. TRAb by TBII	Negative	Negative	Positive
3. TRAb by TSI	Negative	Negative	Inhibition
4. TRAb by TGI	Negative	Negative	Negative
Duration of Disease:	Permanent	Transient	Transient

FTD: Familial Thyroid Dyshormonogenesis
CTD: Congenital Transient Dyshormonogenesis
CTIH: Congenital Transient Immunoglobulin-Mediated Hypothyroidism

* Excluding TSH receptor and iodide trapping defects

compensated hypothyroidism, of infancy. The three diseases, FTD,[41] CTD,[8] and CTIH,[28] may be distinguished on maternal history, physical examination, laboratory tests at diagnosis and the duration of the disease. (Table 5) Thyroid function tests of infants with idiopathic hyperthyrotropinemia are normal except for an elevation for age of serum TSH.[46] In our experience the TSH values usually are below 20 mU/L, and gradually decrease during infancy. Thyroxine therapy is not required unless free T4 values decrease during serial two to four week tests of thyroid function. By definition, the thyroid image is normal. Transient disorders of thyroid function comprises a group of transient thyroid abnormalities that usually are identified by newborn screening programs that perform an initial T4 test; they are thought to be caused by immaturity of the hypothalamic-pituitary-thyroid feedback system.[18] Abnormal low T4 values with or without mildly elevated TSH values are seen in premature infants and very ill neonates.[18] The values usually return to normal within a few weeks, but low dose thyroxine therapy has been initiated for a few months in some infants with very low T4 values. These disorders are not thought to be maternally transmitted.

ACKNOWLEDGEMENTS

This work was supported in part by grants from the Swiss National Science Foundation number 32-9506.88, the Roche Research Foundation, the Renziehausen Trust, the Pediatric Endocrine Research Fund and USPHS grant number RR-84 for the General Clinical Research Center at Children's Hospital of Pittsburgh. We thank Ms Kathleen Willey and Ms Patricia Antonio for their assistance in the preparation of this manuscript, Ms Carlie White and Ms Vaishali Patel for technical assistance and the nurses and staff of the Clinical Research Center.

REFERENCES

1. B. S. Hetzel, "The prevention and control of iodine deficiency disorders," Elsevier, Amsterdam (1987).
2. F. M. Delange, Endemic cretinism, in: "Werner's The Thyroid," 5th ed., S. H. Ingbar and L. E. Braverman, eds., J. B. Lippincott Co., Philadelphia (1986).
3. D. A. Fisher, M. R. Pandian, and E. Carlton, Autoimmune thyroid disease: an expanding spectrum. Pediatr Clin North Am 34:907 (1987).
4. D. A. Fisher, Neonatal thyroid disease in the offspring of women with autoimmune thyroid disease. Thyroid Today 9(4):1 (1986).
5. T. Theodoropoulos, L. E. Braverman, and A. G. Vagenakis, Iodide-induced hypothyroidism: a potential hazard during perinatal life. Science 205:502 (1979).
6. A. Grüters, D. L'Allemand, P. H. Heidemann, P.Schürnbrand, Incidence of iodine contamination in neonatal transient hyperthyrotropinemia. Eur J Pediatr 140:299 (1983).
7. R. G. Cheron, M. M. Kaplan, P. R. Larsen, H. A. Selenkow, and J. F. Crigler Jr., Neonatal thyroid function after propylthiouracil therapy for maternal Graves' disease. N Eng J Med, 304:525 (1981).
8. T. P. Foley Jr., Etiology and pathogenesis of congenital hypothyroidism. Endokrynol Polska 41:343 (1990).
9. A. M. Ermans, Endemic goiter, in: "Werner's The Thyroid," S. H. Ingbar and L. E. Braverman, eds., J. B. Lippincott Co., Philadelphia (1986).
10. D. l'Allemand, A. Grüters, P. Beyer, and B. Weber, Iodine in contrast agents and skin disinfectants is the major cause for hypothyroidism in premature infants during intensive care. Horm Res 28:42 (1987).
11. J. P. Chanoine, M. Boulvain, P. Bourdoux, A. Pardou, H. V. Van Thi, A. M. Ermans, and F. Delange, Increased recall rate at

screening for congenital hypothyroidism in breast fed infants born to iodine overload mothers. Arch Dis Childh 63:1207 (1988).

12. J. M. McKenzie, Neonatal Graves' disease. J Clin Endocrinol Metab 24:660 (1964).

13. B. Rees Smith, S. M. McLachlan, and J. Furmaniak, Autoantibodies to the thyrotropin receptor. Endocr Rev 9:106-21 (1988).

14. A. P. Weetman, M. E. Yateman, P. A. Ealey, C. M. Black, C. B. Reimer, R. C. Williams, B. Shine, and N. J. Marshall, Thyroid-stimulating antibody activity between different immunoglobulin G subclasses. J Clin Invest 86:723 (1990).

15. R. Bernales and J. A. Bellanti, Fetal and Neonatal Immunology, in: "Fetal and Maternal Medicine," E. J. Quilligan and N. Kretchmer, eds., John Wiley and Sons, New York (1980).

16. K. Southgate, F. M. Creagh, M. Teece, C. Kingwood, and B. Rees-Smith, A receptor assay for the measurement of TSH receptor antibodies in unextracted serum. Clin Endocrinol (Oxf) 20:539 (1984).

17. T. P. Foley Jr, C. White, and A. New, Juvenile Graves' disease: Usefulness and limitations of thyrotropin receptor antibody determinations. J Pediatr 110:378 (1987).

18. D. A. Fisher and A. H. Klein, Thyroid development and disorders of thyroid function in the newborn. N Eng J Med 304:702 (1981).

19. J. S. Dallas and T. P. Foley, Hyperthyroidism, in: "Pediatric Endocrinology: A Clinical Guide," F. Lifshitz, ed., Marcel Dekker, New York (1990).

20. W. H. Hoffman, P. Sahasrananan, S. S. Ferandos, C. L. Burek, and N. R. Rose, Transient thyrotoxicosis in an infant delivered to a long-acting thyroid stimulator (LATS) and LATS-protector-negative, thyroid-stimulating antibody-positive patient with Hashimoto's thyroiditis. J Clin Endocrinol Metab 56:354 (1982).

21. M. Zakarija, J. M. McKenzie, and D. S. Munro, Immunoglobulin G Inhibitor of Thyroid-stimulating Antibody is a Cause of Delay in the Onset of Neonatal Graves' Disease. J Clin Invest 72:1352 (1983).

22. M. Zakarija, A. Garcia, and J. M. McKenzie, Studies on multiple thyroid cell membrane-directed antibodies in Graves's disease. J Clin Invest 76:1885 (1985).

23. H. Tamaki, N. Amino, K. Takeoka, Y. Iwatani, J. Tachi, M. Kimura, N. Mitsuda, K. Miki, O. Nose, O. Tanizawa, and K. Miyai, Prediction of later development of thyrotoxicosis or central hypothyroidism from the cord serum thyroid-stimulating hormone level in neonates born to mothers with Graves disease. J Pediatr 115:318 (1989).

24. T. P. Foley Jr, Thyroid Disease, in: "Current Pediatric Therapy," S.

S. Gellis and B. M. Kagan, eds., W. B. Saunders Co., Phila., 12th Edition (1985).

25.   R. M. Blizzard, R. W. Chandler, B. H. Landing, M. D. Pettit, and C. D. West, Maternal autoimmunization to thyroid as probable cause of athyrotic cretinism. N Eng J Med 263:327 (1960).

26.   J. M. Sutherland, V. M. Esselborn, R. L. Burket, T. B. Skillman, and J. T. Benson, Familial nongoitrous cretinism apparently due to maternal antithyroid antibody. N Eng J Med 263:336 (1960).

27.   R. E. Goldsmith, A. J. McAdams, P. R. Larsen, M. MacKenzie, and E. V. Hess. Familial autoimmune thyroiditis: Maternal fetal relationship and the role of generalized autoimmunity. J Clin Endocrinol Metab 37:265 (1973).

28.   N. Matsuura, Y. Yamanda, Y. Nohara, J. Konishi, K. Kasagi, K. Endo, H. Kojima, and K. Wattaya, Familial neonatal transient hypothyroidism due to maternal TSH-binding inhibitor immunoglobulins. New Eng J Med 303:738 (1980).

29.   M. Zakarija, J. M. McKenzie, and M. S. Eidson, Transient neonatal hypothyroidism: characterization of maternal antibodies to the thyrotropin receptor. J Clin Endocrinol Metab 70:1239 (1990).

30.   U. Bogner, A. Grüters, B. Sigle, H. Helge, and H. Schleusener, Cytotoxic antibodies in congenital hypothyroidism. J Clin Endocrinol Metab 68:671 (1989).

31.   R. D. Van der Gaag, H. A. Drexhage, and J. H. Dussault, Role of maternal immunoglobulins blocking TSH-induced thyroid growth in sporadic forms of congenital hypothyroidism. Lancet 1:246 (1985).

32.   P. van Trotsenburg, T. Vulsma, A. M. Bloot, R. D. Van der Gaag, J. W. Lens, H. A. Drexhage, and J. J. de Vijlder, Antibodies to 'second colloid antigen'. A study on the prevalence in sporadic forms of congenital hypothyroidism. Acta Endocrinol 121:659 (1989).

33.   Germak JA, Foley, TP Jr. Longitudinal assessment of L-thyroxine therapy for congenital hypothyroidism. J Pediatr 1990, 117:211-219.

34.   J. F. Rovet, Congenital hypothyroidism: Intellectual and neuropsychological functioning, in: "Psychoneuroendocrinology: Brain, Behavior, and Hormonal Interactions," C. S. Holmes, ed., Springer-Verlag Publishers, New York (1990).

35.   Roti, E., Gnudi, A., Braverman, L.E.: The placental transport, synthesis and metabolism of hormones and drugs which affect thyroid function. Endocr Rev 4:131 (1983).

36.   D. R. Hollingsworth, Hyperthyroidism in pregnancy, in: "Werner's The Thyroid," S. H. Ingbar and L. E. Braverman, eds., J. B. Lippincott Co., Philadelphia (1986).

37.   W. D. Fisher, M. L. Voorhess, and L. I. Gardner, Congenital hypothyroidism in infant following maternal I^{131} therapy, J Pediatr 62:132 (1963).

38.   A. J. Van Herle, R. T. Young, D. A. Fisher, R. P. Uller, and C. R. Brinkman, III, Intra-uterine treatment of a hypothyroid fetus. J Clin Endocrinol Metab 40:474 (1975).

39.   E. S. Lightner, D. A. Fisher, H. Giles, and J. Woolfenden, Intra-amniotic injection of thyroxine (T4) to a human fetus. Am J Obstet Gynecol 127:487 (1977).

40.   Dumont JE, Vassart G, Refetoff S. Thyroid disorders. In: The Metabolic Basis of Inherited Disease, Scriver CR, Beaudet AL, Sly WS, Valle D. (eds.) McGraw-Hill, New York, 1989, Chapter 73, pp. 1843-79.

41.   Foley TP Jr: "Familial Thyroid Dyshormonogenesis", In: Pediatric Thyroidology, Delange F, Fisher DA, Malvaux P, Eds., S. Karger AG Medical and Scientific Publishers, Basal, Switzerland, 1985, 174-88.

42.   D. A. Fisher, J. H. Dussault, T. P. Foley, Jr., A. H. Klein, S. LaFranchi, P. R. Larsen, M. L. Mitchell, W. H. Murphey, and P.G. Walfish, Screening for congenital hypothyroidism: Results of screening 1 million North American infants. J Pediatr, 94:700 (1979).

43.   T. P. Foley, Jr., J. Zeng, and T. Torresani, Congenital transient thyroid dyshormonogenesis: A post TSH receptor disorder, (in preparation).

44.   J. S. Dallas, T. P. Foley, Jr, and T. Torresani, The presence of thyrotropin-binding immunoglobulin G in sera from children and its effects on in vitro rat thyroid cell function. (submitted)

45.   D. Deme, A. Virion, N. A. Hammou, and J. Pommier, NADPH-dependent generation of H_2O_2 in a thyroid particulate fraction requires Ca^{2+}. FEBS Lett, 186:107 (1985).

46.   K. Miki, O. Nose, K. Miyai, H. Yabuuchi, and T. Harada, Transient infantile hyperthyrotropinaemia, Arch Dis Childh 64:1177, (1989).

THYROID FUNCTION IN THE PRETERM INFANT

Frank B. Diamond, Jr.[#] and Allen W. Root[*]

Departments of Pediatrics[#,*] and Biochemistry[*]
University of South Florida College of Medicine
and
All Children's Hospital
St. Petersburg, Florida

I.   INTRODUCTION

Postnatal thyroid function in the preterm infant is
dependent on the neonate's gestational maturity,
intrauterine growth and birth weight, and state of
health.   In general, in larger premature infants the
pattern of pituitary-thyroid function is qualitatively
similar but quantitatively less than that in the term
infant, whereas in the smaller (<1,000 g, <30 weeks)
preterm infant there is delay in activation of
pituitary-thyroid function (1, 2).

II.   THYROID HORMONES IN CORD BLOOD OF PRETERM INFANTS

In umbilical cord serum, levels of thyrotropin (TSH)
vary from undetectable to 4 uU/mL at 11 to 18 weeks
gestation, increase to 2.4 to 20 uU/mL by 22 to 34
weeks, and then decline moderately to term (range 0.5-
18.5 uU/mL) (2) (Table I).   In one study the mean
concentration of TSH in umbilical cord sera of 56
premature infants 20 to 30 weeks gestation was 8.7 ±
6.0 (SD) uU/mL (3).

In premature infants umbilical cord serum
concentrations of thyroxine (T4) are lower than those
found in term infants and less than in paired maternal
specimens (Table I).   There is a direct correlation
between serum T4 concentrations, gestational age and
birth weight in preterm infants between 30 and 45 weeks
gestational age (4); however, there is no correlation

TABLE I. THYROID HORMONE CONCENTRATIONS IN CORD AND MATERNAL SERUM

| Gestation Weeks | CORD SERUM | | | | | | MATERNAL SERUM | | |
	T4 ug/dl	Free T4 ng/dl	T3 ng/dl	Free T3 pg/dl	rT3 ng/dl	TSH uU/ml	T4 ug/dl	FT4 ng/dl	TSH uU/ml
11-18	2.6	1.9				2.4	12.9	3.0	4.2
13-24	2.6	1.9	<15	<103					
20-30	5.5					8.7			
30-31	6.5								
26-33					295	10			
17-33	6.2								
32-33	7.5	2.5					12.2	2.8	3.8
22-34	7.2					9.6			
25-34	9.9	3.6	30	109					
33	6.5		39		232	5.6			
28-37	8.1				264	13.5			
35-40	14.5	4.5	77	234					
38-40	11.2	2.9				8.9	11.5	2.3	4.3
38-40	8.2								
Term	10.2		50		224	11.6			

Data from references: 2, 3, 8, 11, 17, 27, 28, 32.

between T4 values and gestational age in premature
babies of 20-30 weeks gestation (3).   There is a
simultaneous rise in total and free T4 values during
gestation.   Serum concentrations of thyroxine binding
globulin (TBG) vary during the latter half of
gestation.   Maximal T4 binding capacity (MBC), a
reflection of TBG, increases between 8-12 weeks (2
ug/dl) and 20-24 weeks (8 ug/dl).   In term infants the
MBC is 24.7 ug/dL (5).   In infants with birth weights
<1,350 gms (30-32 weeks gestational age) TBG levels
correlate positively with birth weight and gestational
age (6).   In newborns with birth weights appropriate
for gestational age (AGA), levels of thyroxine binding
prealbumin (TBPA) and albumin increase with advancing
gestational age and birth weight (7).

After 30 weeks gestation cord serum concentrations of
triioodothyroxine (T3) increase slowly to a mean value
of 50 ng/dL (range 35-70 ng/dl) at term (8).   Free T3
values parallel total T3 levels.   There is a positive
correlation between cord serum T3 concentrations and
gestational age between 29 and 42 weeks (9).   Between
13 and 40 weeks the rate of increase in serum T3 levels
exceeds that of T4 so that T4/T3 and FT4/FT3 ratios in
fetal sera decrease progressively with increasing
gestational age.   Between 35-40 weeks the ratio of
fetal T4/T3 (188/1) and FT4/FT3 (192/1) exceed maternal
values of 102/1 and 87/1 respectively.   In the fetus,
T4 metabolism differs from that during extrauterine
life in that T4 is metabolized predominantly to reverse
T3 (rT3) (8).   Cord serum rT3 concentrations range from
65 to 340 ng/dL and decline progressively with
increasing gestational age between 29 and 42 weeks (9).
Reverse T3/T4 ratios are higher in premature than in
term infants.

III. POSTNATAL THYROID FUNCTION IN HEALTHY PRETERM
     INFANTS

In well preterm, AGA neonates, serum TSH levels
increase within 30 minutes after birth in response to
the postnatal decline in body temperature, but achieve
peak levels less than those in healthy term infants
(Table II).   In five preterm infants of gestational age
33 weeks and birth weight 1,950 grams, the mean cord
TSH level of 11 uU/mL exceeded that in matched full-
term infants (6.7 uU/mL) and increased significantly
two hours postnatally to a mean value of 31 uU/mL,
while full-term infants reached a TSH concentration of
52 uU/mL at the same point (10).   Uhrmann et al (11)

TABLE II. POSTNATAL CHANGES IN SERUM TSH CONCENTRATIONS (uU/mL) IN HEALTHY AGA PRETERM INFANTS

GESTATION	2h	24h	48h	72h	Week 1	Week 2	Week 3	Week 4	Week 5	Week 6
28.6		14		10	12		11	12	9	11
29.6			5.3		6.4	10.4				
<30		3.5	1.1	1.6	2.8	5.9		2.9		
30							8.8			
31.6			14.7		8.7	9.6				
32						7.1				
33.1		10.5	5.1	3.7	3.6	3.4	3.3			
33.2	31									
33.2					3.9					
33.7			16.5		3.5	2.5				
34						2.0				
Term	52									
Term			29.6		6.6	6.0				

Data from references 6, 10, 11, 13, 15, 30, 31.

reported that in 13 well AGA preterm infants (33 weeks gestational age) the mean TSH level increased from 5.6 in cord blood to 10.5 uU/mL at 24 hours, returning to cord values by 48 hours. TSH concentrations ranged from 3.3 uU/mL to 3.6 uU/mL between 1 and 3 weeks and were lower than levels in cord serum (11).

Kok et al (12) measured TSH concentrations in cord blood and 1, 3, 7, 14 and 21 days after birth in well AGA and small for gestational age (SGA) premature infants and in infants with respiratory distress syndrome (RDS). Thyrotropin levels increased on day 1 only in AGA and SGA premature infants over 30 weeks gestation. In AGA babies with gestational ages 26-29 weeks, TSH levels on day 1 were lower than values in AGA infants of 30-36 weeks gestation.

In response to the postnatal increase in TSH secretion, total T4 concentrations in healthy premature infants 30-34 weeks gestation increase in the first 24 hours, following a pattern similar to that in term infants but at lower values (Table III). Uhrmann et al (11) reported that in AGA premature infants T4 values increased from a mean cord level of 6.2 ug/dL to 9.3 ug/dL at 24 hours and were 7.8 ug/dL at 48 hours. Thereafter, T4 values varied from 7.1 ug/dL at one week to 7.8 ug/dL at 2 and to 8.8 ug/dL at 3 weeks of age. Jacobsen (7) reported that the relative rise in total T4 values in low birth weight infants was of the same magnitude as that in full term newborns in the first 24 hours. In premature infants with gestational age <30 weeks, T4 values decline to a nadir at 2-4 weeks, then return to cord levels by 8-12 weeks of age (14).

Despite reduced serum total T4 values in preterm as compared to term infants, free T4 levels are generally within or above the normal adult range in premature infants but lower than those in term infants (6, 14). Hirano et al (6) measured total and free T4 values in cord blood and at 2 days, 1 week and 2 weeks of age in 3 groups of low birth weight premature infants (Group 1: mean birth weight 848 grams, gestational age 29.6 weeks; Group 2: 1,217 grams, 31 weeks; Group 3: 1,763 grams, 33.7 weeks) and compared these data with those in term infants. Except in Group 3 at birth, the total T4 values in premature infants were significantly lower than those in term babies. There was significant rise in T4 concentrations two days after birth only in term infants. In group 2 and group 3 babies the postnatal rise in T4 was blunted, while in Group 1 infants T4

TABLE III. POSTNATAL CHANGES IN SERUM THYROXINE CONCENTRATIONS (ug/dL.) IN HEALTHY AGA PRETERM INFANTS

GESTATION	24h	48h	72h	3-10d	1wk	11-20d	2wk	3wk	21-45d	4wk	5wk	6wk	46-90d	8wk	12wk
<30	7.8	5.3	9.5		5.3		3.2			4.6				7.4	7.9
23-31	5.7		4.8		4.4			6.2		6.6	6.9	7.2			
30-31	11.5			7.7		7.5			7.8						
32-33	12.3			8.5		8.3			8.0						
33	9.3	7.8	7.3		7.1		7.8	8.8							
34				7.9											
34-35	12.4			10			10.5			9.3					
Term	12-72h 19			15.9		12.2			12.1				10.2		

Data from references 11, 15, 17, 27, 29.

concentrations declined to values significantly below those of cord blood at 1 and 2 weeks of age. In infants with birth weights less than 1,000 grams (Group 1) total T4 values were significantly lower than in the heavier and older preterm infants at 1 and 2 weeks of age. In all groups trends in free T4 values were qualitatively similar to the total T4 changes, although quantitively less. John and Bamford (15) reported reduced levels of both free T4 and free T3 in healthy preterm infants one to 10 days of postnatal age compared to full term babies. Whiteside et al (14) studied 22 premature infants less that 1,000 grams at birth and younger than 30 weeks gestational age (Figure 1). Total and free T4 levels fell to a nadir at two weeks. There was a surge of TSH secretion at two weeks of age followed by increase in total and free T4 values. Mercado et al (16) reported the incidence of hypothyroxinemia (defined as serum T4 value less than 5 ug/dL) in 108 infants delivered at 23-31 weeks gestation to be 84% at one week and 36% by six weeks of age. Thyroxine values were inversely proportional to gestational age during this postnatal study. Infants 23-28 weeks gestation had significantly lower T4, TBG, and T3 levels than did those of 29-31 weeks gestation (16).

In AGA premature well infants (>33 weeks gestation) serum T3 concentrations generally increase in the first 24-72 hours after birth, although the magnitude of the rise is dependent on gestational age and birth weight (7) (Table IV). Free T3 concentrations also rise during this period but remain less than in term infants (15). The free T3/T3 ratio does not differ between premature and term infants.

Between birth and 24-30 hours of age the correlation between T4 and T3 concentrations and gestational age increases while the correlation between T3 and T4 levels and birth weight does not change (17). In a cross-sectional study of a cohort of 280 infants with mean birth weight 1,330 gm and gestational age 30 weeks, Lucas et al (18) reported that plasma T3 concentrations were low during the first 3 days of life and doubled by 4 weeks of age. Median plasma T3 concentrations tended to fall during the first week in infants below 1,200 gms. During nursery hospitalization T3 levels rose 7 ng/dL/week in infants with birth weights <1,200 gms and 11 ng/dL/week in infants with birth weights >1,200 gms and achieved values >78 ng/dL after 8 weeks. In this study there

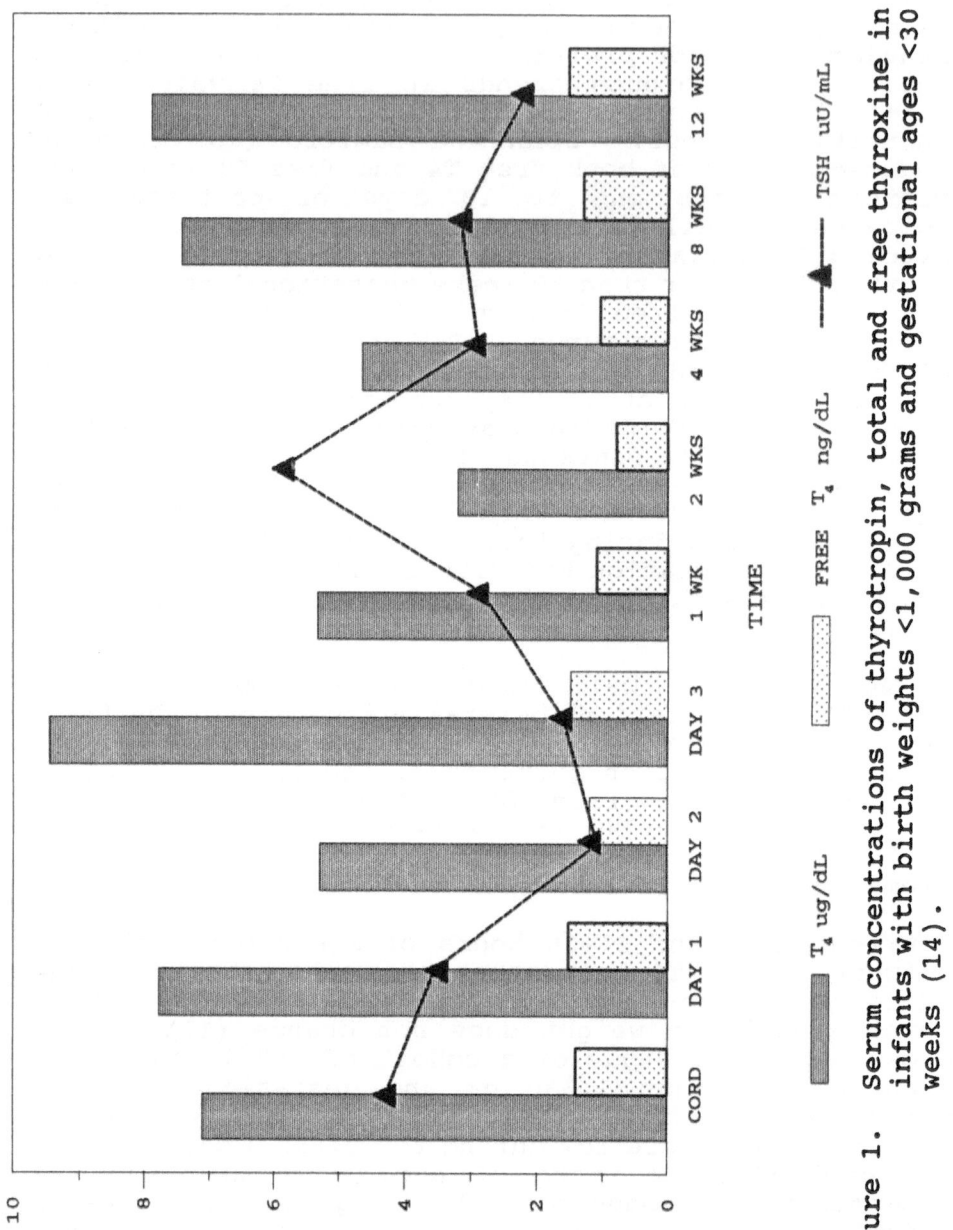

Figure 1.   Serum concentrations of thyrotropin, total and free thyroxine in infants with birth weights <1,000 grams and gestational ages <30 weeks (14).

TABLE IV. POSTNATAL CHANGES IN SERUM TRIIODOTHYRONINE CONCENTRATIONS (ng/dL) IN HEALTHY AGA PRETERM INFANTS

GESTATION	24 hours	48 hours	72 hours	1 week	2 weeks	3 weeks	4 weeks
27-36	65	99	57	87			
28.6	78		78	78		91	91
32.9	115						165
33.1	76	68	56	88	130	137	
33.2			5-8days 98				
34.2	12-72hours 103		3-20days 148				
Term	420	250	220				

Data from references 11, 13, 15, 20, 27, 31, 32.

was no association between plasma T3 concentrations and
necrotizing enterocolitis, somatic growth up to 18
months of age or later development of cerebral palsy.
However, babies whose nadir T3 concentration was below
20 ng/dL had, at 18 months corrected age, Bayley mental
and motor scores which were 8.3 and 7.4 points
respectively lower than those of infants with higher T3
concentrations, even after adjusting for a range of
antenatal and neonatal factors known to influence later
development. These data contrast with other studies
where   adverse   developmental   consequences   of
hypothyroxinemia   were   not   identified   (19,   20).
However, De Vries et al (21) reported that nerve
conduction velocity was delayed in preterm infants with
depressed T4 concentrations. The significance of this
observation to developmental outcome is uncertain.

Serum thyroglobulin levels rise on the first day of
life in preterm infants, then decrease significantly
after one week of age (12). The highest thyroglobulin
levels are found in infants of the youngest gestational
ages.   High thyroglobulin concentrations in preterm
infants may relate to a higher permeability of the
thyroid gland to the passage of this glycoprotein or to
delayed degradation of thyroglobulin by an immature
liver.

IV. THYROID FUNCTION IN THE ILL AND SGA PRETERM INFANT

Cord serum levels of T4, T3, rT3 and TSH are similar in
premature infants with and without respiratory distress
syndrome (RDS) (22, 23).   Prenatal administration of
glucocorticoids to mothers in preterm labor does not
affect   fetal   or   postnatal   thyroid   function   (23).
During the first two hours after birth the rise in TSH
values in infants with RDS is blunted compared to well
preterm infants (10, 21).   During the first week of
life preterm infants with RDS or other nonthyroidal
illnesses have lower T4, T3 and TSH concentrations than
do well preterm babies of similar gestational age;
values may remain below those of healthy term infants
for up to six weeks (7, 23, 24).   Kok et al (12)
reported that thyroglobulin levels were higher in the
first week of life in infants with RDS compared to well
AGA premature infants.   Serum TBG levels have been
reported to be low (25) or normal (10) in the sick
preterm infant.   Treatment with thyroid hormone does
not generally affect the clinical course of the well
hypothyroxinemic, preterm infant or of the sick neonate
or the long term outcome of the treated infant (19).

In 1981, Schonberger et al (26) reported that prophylactic administration of thyroxine and triiodothyronine to premature infants decreased their mortality rate, but this observation has not been replicated.

In well SGA, premature babies mean TSH, T4 and T3 values fluctuate in a manner parallel to those of AGA premature infants during the first few days of life. However, T3 values are signficantly lower at one week of age and both T3 and T4 are lower at two and three weeks of age in SGA than in AGA preterm infants (11).

IV. CONCLUSIONS

The fetal hypothalamic-pituitary-thyroid unit is active in utero and there is a defined sequence of function as the fetus matures. The major circulating fetal thyroid hormone is T4, while T3 values are low. Postnatally, in more mature preterm neonates (gestational age >30 weeks), the dynamic function of the hypothalmic-pituitary-thyroid unit is qualitatively similar to that of the term infant, but quantitatively less in that the TSH, T4 and T3 secretory surges are less dramatic. In gestationally immature (<30 weeks) preterm infants, thyroid hormone secretion is depressed postnatally increasing slowly after two weeks of age. In preterm infants with respiratory distress or other nonthyroidal illnesses, the secretion of TSH and thyroid hormone is also decreased transiently. The depressed function of the pituitary-thyroid axis in the very low birth weight infant or in the ill neonate is considered to be physiological and not an indication for treatment with thyroid hormone.

ACKNOWLEDGEMENT

The authors thank Ms. Sandy Jones for exceptional secretarial assistance.

REFERENCES

1. Fisher DA, Klein AH. Thyroid development and disorders of thyroid function in the newborn. N Engl J Med 304:702-712, 1981.

2. Fisher DA, Hobel CJ, Jarza R, Pierce CA. Thyroid function in the preterm fetus. Pediatrics 46:208-215, 1970.

3. Gorodzinsky P, Howard NJ, Ginsberg J, Walfish PG. Cord serum thyroxine and thyrotropin values between 20 and 30 weeks' gestation. <u>J Pediatr</u> 94:971-973, 1979.

4. Oddie TH, Fisher DA, Bernard B, Lamb KW. Thyroid function at birth in infants 30 to 45 weeks gestation. <u>J Pediatr</u> 90:803-806, 1977.

5. Greenberg AH, Czernichow P, Reba RC, Tyson J, Blizzard RM. Observations on the maturation of thyroid function in early fetal life. <u>J Clin Invest</u> 49:1790-1803, 1970.

6. Hirano T, Singh J, Srinivasan T, Pildes R. Postnatal thyroid function in low birth weight infants; a cross-sectional assessment of free thyroxine and thyroid hormone binding globulin. <u>Eur J Pediatr</u> 139:244-246, 1982.

7. Jacobsen BB. Thyroid function in infancy. <u>Danish Medical Bulletin</u>, 30:281-309, 1983.

8. Fisher DA, Dussault JH, Hobel CJ, Lam R. Serum and thyroid gland triiodothyronine in the human fetus. <u>J Clin Endocrinol Metabol</u> 36:397-400, 1973.

9. Isaac RM, Hayek A, Standefer JC, Eaton RP. Reverse triiodothyronine to triiodothyronine ratio and gestational age. <u>J Pediatr</u> 94:477-479, 1979.

10. Klein AH, Foley B, Kenny FM, Fisher DA. Thyroid hormone and thyrotropin responses to parturition in premature infants with and without the respiratory distress syndrome. <u>Pediatrics</u> 63:380-385, 1979.

11. Uhrmann S, Marks KH, Maisels MG, Kulin HE, Kaplan M, Utiger R. Frequency of transient hypothyroxinemia in low birth weight infants. <u>Arch Dis Child</u> 56:214-217, 1981.

12. Kok JH, Tegelaers WH, deVijlder JJ. Serum thyroglobulin levels in preterm infants with and without respiratory distress syndrome II. A longitudinal study during the first three weeks of life. <u>Pediatr Res</u> 20:1001-1003, 1986.

13. Jacobsen BB, Anderson HJ, Peitersen B, Dige-Peterson H and Hummer L. Serum levels of thyrotropin, thyroxine and triiodothyronine in full term, small for gestational age and preterm newborn babies. Acta Paediatr Scand 66:681-687, 1977.

14. Whiteside WH, Williams PR, Root AW, Strezlecki J, Curran JS, Kanarek K. Serum thyroxine (T4), free T4 (FT4) and TSH in premature infants less than 1000 grams during the first three months of life. Pediatr Res 18:179A, 1984.

15. John R, Bamford FJ. Serum free thyroxine and free triiodothyronine concentrations in healthy full term, preterm and sick preterm neonates. Ann Biochem 24:461-465, 1987.

16. Mercado N, Yu VY, Francis I, Szymonowicz W, Gold H. Thyroid function in very preterm infants. Early Hum Devel 16:131-141, 1988.

17. Bongers-Schokking JJ, Schopman WS. Thyroid function in healthy normal, low birth weight and preterm infants. Eur J Pediatr 143:117-122, 1984.

18. Lucas A, Rennie J, Baker BA, Morley R. Low plasma triiodothyronine concentrations and outcome in preterm infants. Arch Dis Child 63:1201-1206, 1988.

19. Hadeed AJ, Asay LD, Klein AH, Fisher DA. Significance of transient postnatal hypothyroxinemia in premature infants with and without respiratory distress syndrome. Pediatrics 68:494-498, 1981.

20. Chowdhry P, Scanlon JW, Auerbach R, Abassi V. Results of a controlled double blind study of thyroid replacement in very low birth weight premature infants with hypothyroxinemia. Pediatrics 73:301-305, 1984.

21. DeVries LS, Heckmatt JZ, Bunnin JM, Dubowitz LMS, Dubowitz V. Low serum thyroxine concentrations and neural maturation in preterm infants. Arch Dis Child 61:862-866, 1986.

22. Klein AH, Foley B, Foley TP, MacDonald HM, Fisher DA. Thyroid function studies in cord blood from

premature infants with and without RDS.   J Pediatr 98:818-820, 1981.

23. Franklin RC, Purdie GL, O'Grady CM.   Neonatal thyroid function: prematurity,   prenatal steroids, and respiratory distress syndrome. Arch Dis Child 61:589-562, 1986.

24. Abassi V, Merchant K, Abramson D.   Postnatal triiodothyronine   concentrations   in   healthy preterm infants and infants with respiratory distress syndrome.   Pediatric Res 11:800-804, 1977.

25. Jacobsen BB, Peitersen B, Hummer L.   Serum concentrations of thyrotropin, thyroid hormones and thyroid hormone binding proteins during acute and recovery stages of idiopathic respiratory distress syndrome.   Acta Paediatr Scand 68:257- 264, 1979.

26. Schonberger W, Grimm W, Emmrich P, Gempp W. Reduction of mortality rate in premature infants by substitution of thyroid hormones.   Eur J Pediatr 135:245-253, 1981.

27. Cuestas RA.   Thyroid function in healthy premature infants, J Pediatr 91:963-967, 1978.

28. Ginsberg J, Walfish PG, Chopra IJ.   Cord blood reverse T3 in normal, premature, euthyroid low T4 and hypothyroid newborns.   J Endocrinol Invest 1:73-77, 1978.

29   Diamond FB, Parks JS, Tenore A, Marino JM, Borgiovanni AM.   Hypothyroxinemia in sick and well preterm infants.   Clin Pediatr 18:555-561, 1971.

30. Hirano T, Singh J, Srinivasan G, Pildes R. Postnatal thyroid function in low birth weight infants; a longitudinal assessment of free thyroxine and thyroid hormone binding globulin. Acta Endocrinol 110:56-60, 1985.

31. Nagashima K, Onoda K, Suzuki S, Sakaguchi M, Kuroume T.   Reevaluation of thyroid function in low birth weight infants based on free triiodothyronine and triiodothyronine.   Biol Neonate 48:341-345, 1985.

32.   Erenberg A, Phelps DL, Lam R, Fisher DA. Total
       and free thyroid hormone concentrations in the
       neonatal period. Pediatrics 53:211-216, 1974.

USE OF TRH IN THE FETUS TO ADVANCE LUNG MATURITY

William F. O'Brien

University of South Florida
College of Medicine
Department of Obstetrics and Gynecology
Tampa, Florida

RESPIRATORY DISTRESS SYNDROME

Respiratory distress syndrome (RDS) secondary to prematurity is a leading cause of perinatal mortality. The incidence of RDS increases as the gestational age at birth declines and occurs in approximately two thirds of infants born at 29-30 weeks gestational age (1). The functional immaturity of the fetal lung is both anatomic and biochemical. The most obvious abnormality is insufficient production of surfactant by the Type II alveolar cells. Composed of proteins and phospholipids and packaged in granules secreted by the alveolar lining cells, surfactant lines the air-tissue interface of the alveoli and decreases surface tension. This decrease in surface tension allows the alveoli to remain expanded in the absence of continuous airway pressure.

In addition to surfactant deficiency the lungs of premature neonates demonstrate abnormalities in distensibility in a totally fluid environment. Thus therapies to ameliorate the pulmonary effects of preterm birth may affect surfactant production or the morphologic development of the fetal lung.

CORTICOSTEROIDS

The influence of corticosteroids on maturation of the fetal lung was discovered serendipitously during studies designed to investigate the initiation of labor in sheep. This knowledge was soon applied clinically. A multicentered trial conducted by the National

Institutes of Health comparing dexamethasone with placebo demonstrated a significant reduction in the incidence of RDS in the dexamethasone treated group (2). This effect was limited to non-white females with a gestational age at birth between 30 and 34 weeks. No effect was noted in infants delivered prior to 24 hours following the start of therapy. Controlled clinical trials of corticosteroids for the prevention of RDS have recently been reviewed by Crowley et al. (3). This overview demonstrates that the use of maternally administered corticosteroids reduces the occurrence of neonatal RDS with a typical odds ratio of approximately 0.5 and early neonatal death with a typical odds ratio of 0.6.

The mechanism of action of corticosteroids has involved both in vivo and tissue culture studies. Glucocorticoids bind to specific receptors in the cytoplasm and there is strong correlation between binding and production of phosphotidyl choline (PC), a major component of surfactant (4). This increase in production is due primarily to an increase in the activity of the enzyme which controls the rate limiting step in the production of PC, choline-phosphate cytidyltransferase (5).

The effect of corticosteroids on fetal lung maturation is not limited to surfactant production. Following exposure there is accelerated morphological maturation including increased air volume and thinning of the alveolar lining cells (6), increased collagen and elastin synthesis (7), an increase in beta-adrenergic receptors, and a decrease in protein leak into the alveoli (8).

THYROID HORMONES

Unlike corticosteroids, investigation of the role of thyroid hormones in the maturation of the fetal lung was stimulated by clinical as well as experimental observation. In a follow-up study of infants with RDS, Fisch et al noted a higher than expected incidence of hypothyroidism (9). Investigation of indices of thyroid function at birth in RDS and control neonates soon followed. Although the problem of conflicting variables led to some studies failing to demonstrate a statistically significant difference between the RDS and control groups (10), several studies demonstrated a significant decrease in umbilical cord levels of T4 and T3 along with an increase in TSH in infants with RDS (10).

In vitro and animal studies corroborated the possibility of an association. In the fetal lamb thyroidectomy at 95-99 days of gestation (term 145 days) results in a decrease in the lecithin to sphingomyelin ratio in tracheal fluid, less mature appearing Type II alveolocytes, and hypercellular lungs with thickened alveolar septae (11). Direct administration of T4 to the fetal rabbit results in stimulation of PC synthesis and accelerated morphological lung development (12). Importantly, stimulation of endogenous thyroid hormone synthesis by maternal administration of thyrotropin releasing hormone resulted in enhanced neonatal functional and morphological fetal lung maturation in the rabbit (13). Both T4 and T3 result in an increase in PC synthesis in organ culture (14), although the effect of T3 appears to be significantly greater, reflecting a greater affinity of nuclear binding sites (15).

In addition to the individual efficacy of either corticosteroids or thyroid hormones in organ culture, several lines of evidence point to an additive or synergistic effect. In organ culture the combination of dexamethasone and T3 results in greater stimulation of PC synthesis than that noted with either hormone alone (16). In the fetal lamb, direct infusion of T3 (17) or thyrotropin releasing hormone (18) along with cortisol resulted in a greater increase in PC synthesis and led to significant clinical response at a gestational age earlier than that at which cortisol alone has an effect. In the fetal rat maternal T3 administration augments the increase in fetal lung PC production following maternal betamethasone (19). Maternal TRH administration in rabbits augments the beneficial effects of prenatal corticosteroids and exogenous surfactant on neonatal ventilatory function (20).

INTRA-AMNIOTIC THYROXINE

Since T4, T3, and thyroid stimulating hormone cross the placenta poorly (21), the first studies involving humans utilized intra-amniotic installation of thyroxine via amniocentesis. In an early report installation of 250 ug of levothyroxine apparently resulted in an acceleration of fetal lung maturation as measured by comparison of amniotic fluid fluorescence polarization, a measure of surfactant activity, with historical controls (22). In another small study, however, instillation of T3 failed to produce an increase in umbilical cord blood concentration or improvement in outcome which the authors suggested may have been due to poor absorption (23).

The greatest experience with intra-amniotic thyroxine administration has been by Adamsons and his co-workers at the University of Puerto Rico (24). Between 1982 and 1988 they instilled 500 ug of T4 at weekly intervals between 26 and 31 weeks gestation in over 140 women at risk for preterm delivery. Results were evaluated by comparison of the LS ratio of the treated pregnancies to those of a non-treated group. In comparison to the values noted in untreated pregnancies, the LS ratio in the treated pregnancies demonstrated significant weekly increments. These increments, moreover, appeared to be greater in women who had received more than one injection of T4.

TRH

In contrast to T4 and T3, TRH easily crosses the placenta and maternal TRH administration results in an increase in fetal TSH, T4 and T3 (25,26). The potential for an easily administered regimen of maternal treatment utilizing a combination of corticosteroids and TRH has resulted in several trials comparing corticosteroid alone to corticosteroid and TRH. The first of these trials to be reported was conducted at the University of South Florida (27).

The study was conducted between November 1986 and November 1987 at the Tampa General Hospital. Women at high risk for delivering before 34 completed weeks gestational age were considered candidates. Pregnancies complicated by lethal congenital anomalies or demonstrating a LS ratio of 2.0 or greater were excluded. Volunteers were randomly assigned to receive either betamethasone (24 mg intramuscularly in two divided doses 24 hours apart) alone or betamethasone and TRH (400 ug intravenously every eight hours for six doses). The regimen was repeated weekly for two weeks if the preceding LS ratio remained below 2.0.

Staff blinded to the therapy recorded neonatal outcome with specific reference to oxygen and ventilator requirements which were used to categorize RDS severity into mild, moderate, and severe. An echoencephalogram was performed on the third day of life to determine intraventricular hemorrhage.

The effect of TRH therapy on maternal indices of thyroid function demonstrated significant stimulation. Following therapy there was a significant increase in total and free T4, and a reduction in TSH.   No difference in T3 was noted.

When cord blood levels at birth in the corticosteroid alone and corticosteroid and TRH infants born within one week of therapy were compared, there were no significant differences in T4 or T3 levels but a reduction in TSH in the TRH group was noted.   This probably reflects the transient nature of the response to TRH therapy.

Strong trends in favor of the combined treatment group were noted for several outcomes.   A LS ratio of greater than 2.0 was noted in 52% of the TRH group and 40% of the corticosteroid only group (p=0.06).   Although the difference was not statistically significant, RDS was less common in the TRH group (44% vs. 28%).   When absence of RDS in infants born within a weeks of therapy or a LS ratio greater than 2.0 were combined to evaluate success of therapy, a significant effect was noted with a greater percentage of the combined therapy group (61%) achieving success than the corticosteroid group (46%).

The potential for combined therapy to influence morphologic as well as biochemical maturation was demonstrated by comparison of the development of bronchopulmonary dysplasia. This long-term complication of pulmonary insufficiency at birth was significantly less common in the combined therapy group (8% vs. 24%).

When umbilical cord indices of thyroid function for infants in both group were examined, those with RDS demonstrated significantly lower levels of both total (6.32 ± 2.87 vs. 7.69 ± 2.19 ug/dL) and free T4 (0.70 ± 0.41 vs. 0.98 ± 0.49 ng/dL).

Although unpublished at this time, preliminary reports of a large, multicentered trial sponsored by the NICHD suggest similar findings with a decrease in the incidence of RDS and, as in our study, no noted adverse effects of TRH therapy.

SUMMARY

Maturation of the fetal lung is a complex process involving interactions between intrinsic biologic development of several cell types and hormonal triggering of biochemical and functional alterations. Catecholamines, corticosteroids, thyroid hormones and perhaps prolactin, all influence the transition from non-functional to functional pulmonary tissue, the most critical component necessary for extrauterine existence. The assimilation of clinical observation, in vitro experimentation, and animal studies has led to the potential of therapeutic intervention in a manner with minimal hazard to the fetus. The use of maternally administered TRH to reduce the incidence and severity of RDS, although still in the early stages of clinical application, appears to justify the long years of careful research in this area.

REFERENCES

1. Usher RH, Allen AC, McLean FH: Risk of respiratory distress syndrome related to gestational age, route of delivery, and maternal diabetes. Am J Obstet Gynecol 111:826-32 (1971).
2. Colaborative Group on antenatal Steroid Therapy: Effect of antenatal dexamethasone administration on the prevention of respiratory distress syndrome. Am J Obstet Gynecol 141: 276-86 (1981).
3. Crowley P, Chalmers I, Kerise MJ: The effects of corticosteroid administration before preterm delivery: an overview of the evidence from controlled trials. Br J Obstet Gynaecol 97:11-25 (1990).
4. Gross I, Ballard PL, Ballard RA, et al: Corticosteroid stimulation of phosphotidylcholine synthesis in cultured fetal rabbit lung. Evidence for de novo protein synthesis mediated by glucocorticoid receptors. Endocrinology 112: 829-37 (1983).
5. Post M: Maternal administration of dexamethasone stimulated choline-phosphate cytidyltransferase in fetal Type II cells. Biochem J 241:291-296 (1987).
6. Beck JC, Mitzner W, Johnson JWC, et al: Bethamethasone and the rhesus fetus: Effect on lung morphometry and connective tissue. Pediatr Res 15:235-40 (1981).
7. Schellenberg JC, Liggins GC: Growth, elastin concentration and collagen concentration of perinatal rat lung: Effects of dexamethasone. Pediatr Res 21:603-07 (1987).
8. Ikegamy M, Berry D, Elkay T, et al: Corticosteroids and surfactant change lung function and protein leaks in the lungs of

ventilated premature rabbits. J Clin Invest 79: 1371-8 (1987).

9. Klein AH, Folery B, Foley P, et al: Thyroid function studies in cord blood from premature infants with and without RDS. J Pediatr 98:818-20 (1981).

10. Cuestas RA, Lindall A, Engel RR: Low thyroid hormones and respiratory-distress syndrome of the newborn: Studies on cord blood. N Engl J Med 295:297-301 (1976).

11. Erenberg A, Rhodes ML, Weinstein MM, Kennedy RL: The Effect of fetal thyroidectomy on ovine fetal lung maturation. Pediat Res 13:230-5 (1979).

12. Wu B, Kikkawa M, Orzalesi E, et al. The effect of thyroxine on the maturation of fetal rabbit lungs. Biol Neonate 22: 161-8 (1973).

13. Devaskar U, Nitta K, Szewczyk K, et al: Transplacental stimulation of functional and morphologic fetal rabbit lung maturation: Effect of thyrotropin-releasing hormone. Am J Obstet Gynecol 157:460-4 (1987).

14. Gross I, Wilson CM, Ingleson LD, et al: Fetal lung in organ culture. III. Comparison of dexamethasone, thyroxine, and methylxanthines. J Appl Physiol 48:872-7 (1980).

15. Ballard PL, Hovey ML, Gonzales LK: Thyroid hormone stimulation of phosphotidylcholine synthesis in cultured fetal rabbit lung tissue. J Clin Invest 74:898-905 (1984).

16. Gross I, Wilson CM: Fetal lung in organ culture. IV. Supra-additive hormone interaction. J Appl Physiol 52: 1420-5 (1982).

17. Schellenberg JC, Liggins GC, Manzai M, et al: Synergistic hormonal effect on lung maturation in fetal sheep. J Appl Physiol 65:84-100 (1988).

18. Liggins GC, Schellenberg JC, Manzai M, et al: Synergism of cortisol and thyrotropin-releasing hormone in lung maturation in fetal sheep. J Appl Physiol 65:1880-4 (1988).

19. Gross I, Dynia DW, Wilson CM, et al: Glucocorticoid-thyroid hormone interactions in fetal rat lung. Pediatr Res 18: 191-6 (1984).

20. Ikegami M Jobe AH, Pettenazzo A, et al: Effects of maternal treatment with corticosteroids, T3, TRH, and their combinations on lung function of ventilated preterm rabbits with and without surfactant treatments. Am Rev Respir Dis 136:892-8 (1987).

21. Fisher DA, Lehman H, Lackey C: Placental transport of thyroxine. J Clin Endocrinol 24:393-400 (1964).

22. Mashiach S, Barkai G, Sack J, et al: The effect of intra- amniotic thyroxine on fetal lung maturity in man. J Perinat Med 7:161-70 (1979).

23. Schreyer P, Caspi E, Letko Y, et al: Intra-amniotic triiodothyronine instillation for prevention of respiratory distress syndrome in pregnancies complicated by hypertension. J Perinat Med 10:27-33 (1982).

24. Romaguera J, Zorrilla C, de la Vega A, et al: Responsiveness of L-S ratio of the amniotic fluid to the intra-amniotic administration of thyroxine. Acta Obstet Gynecol Scand 69: 119-122 (1990).

25. Roti E, Gnudi A, Braverman L, et al: Human cord blood concentrations of thyrotropin, thyroglobulin, and idothyronines after maternal administration of thyrotropin-releasing hormone. J Clin Endocrinol Metab 53: 813-7 (1981).

26. Moya F, Mena P, Heusser F, et al: Response of the maternal, fetal, and neonatal pituitary-thyroid axis to thyrotropin-releasing hormone. Pediatr Res 20:982-6 (1986).

27. Morales WJ, O'Brien WF, Angel JL, et al: Fetal lung maturation: The combined use of corticosteroids and thyrotropin-releasing hormone. Obstet Gynecol 73:111-6 (1989).

DIVERSE ABNORMALITIES OF THE C-ERBAβ THYROID HORMONE RECEPTOR GENE IN GENERALIZED THYROID HORMONE RESISTANCE

Stephen J. Usala, M.D., Ph.D.

Department of Medicine
East Carolina University School of Medicine
Greenville, North Carolina 27858-4354

Barry B. Bercu, M.D.

Department of Pediatrics
University of South Florida College of Medicine
Tampa, Florida 33612

Samuel Refetoff, M.D.

Thyroid Study Unit
Departments of Medicine and Pediatrics
University of Chicago
Chicago, Illinois 60637

CLINICAL DIVERSITY OF GENERALIZED THYROID HORMONE RESISTANCE

The syndrome of generalized thyroid hormone resistance (GTHR) was first described in 1967 in a kindred, G, with elevated free thyroid hormones and absence of the typical clinical features of hyperthyroidism (1). The proband, a 6-year-old girl, demonstrated stippled epiphyses, dysmorphic features (bird-like facies, pigeon

Supported in part by USPMS Grants DK 15070 and RR 00055 (S.R.) and DK42807-01 (S.J.U.)

breast, and winged scapulae) and deaf-mutism. The syndrome was transmitted as a recessive trait and affected members were the product of a consanguineous union. As children, affected members had intelligence quotients within the ranges normally seen in hearing-impaired individuals (1,2). Although minimal delay of bone age was observed in affected members, final adult height was above the parental mean. Interestingly, affected members showed a paradoxical increase of serum TSH in response to the administration of suppressive doses of T3 (3) and no significant effect of antithyroid drugs on the level of TSH (2). This constellation of clinical findings in the original kindred--stippled epiphyses, somatic abnormalities, and deaf-mutism--has never been reported in other kindreds with GTHR. However, less severe hearing defects, learning disabilities and growth retardation are not uncommon.

The syndrome in the majority of other kindreds with GTHR segregates as an autosomal dominant disorder and the patients generally have no dysmorphic features (4,5). Inheritance appears to be clearly recessive in 10% of cases and somatic abnormalities have been described in 5%. Parameters of thyroid hormone action in the tissues such as bone, brain, liver, heart, and pituitary, as well as the level of basal metabolism, have been assessed in many kindreds (1-8). There is a spectrum of refractoriness to thyroid hormone action among different tissues within a given patient with GTHR, and different kindreds have variable patterns of tissue resistance. The differences in phenotype among the various kindreds suggested a heterogenous genetic defect.

GTHR Is Linked to c-erbAβ

It was originally hypothesized that patients with familial GTHR harbored abnormalities of the level of a putative thyroid hormone receptor (2). Attempts to test this hypothesis with T3-binding studies utilizing patient fibroblasts and lymphocytes were sometimes suggestive of affinity or number defects but were not definitive (9,10).

Progress in understanding thyroid hormone action occurred with the identification of two different c-erbA proto-oncogenes that were putative thyroid hormone receptor genes (11,12). The c-erbAα gene on chromosome 17 (α-receptor) and the c-erbAβ gene on chromosome 3 (β-receptor) coded for proteins that *in vitro* had the affinity and specificity for thyroid hormones predicted from *in vivo* studies of the thyroid hormone receptor. The relationship between the c-erbAβ receptor gene and thyroid hormone resistance was first demonstrated in the laboratory of B. D. Weintraub in a large kindred, A, where the syndrome segregated as a dominant disorder (13). Highly informative restriction fragment length polymorphisms were found at the c-erbAβ locus and were shown to be tightly linked to GTHR in Kindred A. Tight linkage between GTHR and the c-erbAβ gene has now been demonstrated in multiple kindreds with GTHR of variable phenotypes (7,8,14). These linkage studies suggested that different kindreds were very likely to have different genetic abnormalities of c-erbAβ. Four kindreds have been found where c-erbAβ is not linked to GTHR (S. Refetoff, unpublished data). In these kindreds the c-erbAα receptor would be the best candidate for the genetic abnormalities causing GTHR.

Dominant Negative Forms of c-erbAβ

Different point mutations have recently been identified in the T_3-binding domain of c-erbAβ in kindreds with dominantly-inherited GTHR. These mutant c-erbAβ receptors belong to a functional class of proteins called dominant negative inhibitors, where a mutant form is co-expressed with the wild-type and disrupts normal activity. In the case of GTHR with a dominant negative mutation, a single mutant allele results in the inhibition of the function of normal thyroid hormone receptors from the wild-type β-allele and α-alleles. In a kindred, Mf, a single guanine to cytosine replacement in codon 345 resulted in a glycine to arginine change (15). This mutation was found in only one of two alleles of two affected patients belonging to the same family (Table 1).

The Mf receptor was synthesized *in vitro* and did not bind T_3. Furthermore, transfection experiments have shown that this mutant receptor not only fails to activate thyroid hormone responsive genes but also interferes with the transactivating activity of the normal c-erbAβ and c-erbAα receptors (16). Two different point mutations were found in Kindred D and Kindred A: a G→C at nucleotide 1305 resulting in the replacement of the normal glutamine 340 with histidine and a C→A at nucleotide position 1643 resulting in a proline to histidine replacement at codon 453. In contrast to the Mf receptor, the kindred A receptor synthesized *in vitro* did bind T_3 but with a fivefold reduced affinity (17). In a kindred, S, the nucleotides (CAC) at positions 1295-1297 were deleted (18). This deletion resulted in the deduced loss of threonine 337. This unusual mutation lies in proximity of those found in kindreds Mf and D. This clustering may indicate a domain of frequent mutations for the syndrome of GTHR. This region of the receptor appears to be crucial for T_3-binding since the *in vitro*

Table 1. Dominant Negative Mutations in c-erbAβ

Kindred	Mutation		T$_3$-Binding[a]
	Nucleotide	Codon[b]	
Mf	1318 G→C	345 GLY→ARG	None
A	1643 C→A	453 PRO→HIS	Reduced
D	1305 G→C	340 GLN→HIS	?
S	1295-1297 Deletion CAC	337 Deletion THR	None

a   T$_3$-binding measured with *in vitro* synthesized mutant receptors. Kindred Mf (ref. 15); Kindred A (ref. 7,17); Kindred D (ref. 8); · Kindred S (ref. 18).

b   Codon position defined by the co-ordinates in reference 21. Codon positions in references (7,8,18) for mutations A,D, and S defined by co-ordinates in Weinberger *et al*. (ref. 12).

synthesized mutant receptor of kindred S did not bind detectable quantities of ^{125}I-labeled T$_3$ (18).

The diverse structure and function of these mutant receptors are almost certainly responsible for the observed differences in phenotype. More than ten other point mutations in the T$_3$-binding domain of c-erbAβ have been isolated in kindreds with GTHR. (S. Usala, S. Refetoff, unpublished data; B. Weintraub, personal communication).

Homozygous Defects of c-erbAβ in Man

The molecular defect in the original Kindred G has now been identified (ref. 19, Table 2). A major deletion in the c-erbAβ gene of affected members of Kindred G was first suspected by the inability to amplify the exons encompassing the T$_3$-binding domains. Deletion of all coding sequences of the c-erbAβ gene in both alleles of affected members of this family was confirmed by Southern blotting and *in situ* hybridization. The fact that members with only one functional β-receptor allele are clinically and biochemically normal and have normal TSH and thyroid hormone levels supports the notion that mutant c-erbAβ receptor genes act in a dominant negative fashion and not solely through lack of activity. Furthermore, any thyroid hormone action in the homozygous patients of Kindred G is presumably mediated by the c-erbAα1 receptor suggesting that it must play a crucial role in brain development.

A single patient with GTHR who had free thyroid hormones and TSH levels, much higher than other affected members of his kindred (Kindred S) or other kindreds has been described (ref. 20, Table 2). This patient was born at 35 weeks gestation with a birth weight of 1480 grams, compatible with intrauterine growth retardation. During the neonatal period the patient had signs of hyperthyroidism, including exophthalmos, thyromegaly, and tachycardia. Now at 3 11/12 years of age the patient manifests profound growth retardation, marked delay in bone age, profound psychomotor retardation, but remains tachycardic and underweight. This patient was the product of a consanguineous union of two affected members and is homozygous for the deletion of threonine-337. Unlike the situation in Kindred G, the mutant c-erbAβ receptor of Kindred S acts in a dominant negative manner and heterozygous members are also affected, albeit less severely.

The TSH and thyroid hormone levels in heterozygous and homozygous affected members of kindreds G and S are compared in Table 2. Of particular interest is the fact that obligate heterozygotes from family G showed no abnormalities while those from Kindred S had clearly elevated serum thyroid hormone levels which failed to suppress their TSH. More importantly, the homozygous subject of Kindred S expressing the mutant c-erbAβ was more severely resistant to thyroid hormone than the corresponding subjects with complete c-erbAβ deficiency. Indeed, his serum TSH level was 20-fold higher despite 2 to 5-fold elevation of thyroid hormones above the mean level of homozygous subjects from Kindred G. These results clearly support the antagonistic function of c-erbAβ when not activated by the hormone ligand. Clinical trials are in progress to measure TSH secretion as a function of thyroid hormone levels in the homozygote of Kindred S.

The advances in molecular genetics have enabled the indepth study of the mechanisms of thyroid hormone action in man. Further investigation of patients with receptor-dependent GTHR will provide information on the relative role of the β- and α-receptors.

Table 2. Comparison of Thyroid Function Tests in Heterozygous and Homozygous Individuals from Kindred G and Kindred S

Subject	Molecular Defect	T$_4$ (nmol/l)	T$_3$ (nmol/l)	FTI (%)[a]	TSH (mU/L)
Kindred G - mother father	Heterozygous β-receptor deletion	116 128	1.81 --	-21 -18	3.1 1.8
Kindred S - mother grandfather	Heterozygous deletion-mutation; Threonine 337 in β-receptor	244 196	4.58 3.52	+96 +53	1.0 2.2
Kindred G - proband brother brother	Homozygous β-receptor deletion	307 257 344	12.75 4.53 4.75	+130 +88 +125	5.8 5.5 1.6
Kindred S[b]- proband	Homozygous deletion mutation; Threonine 337 in β-receptor	653	20.07	+1017	389
Normal Range		64-154	1.38-2.84[c]	-56-0	0.5-4.0

References for Kindred G (1,2) and Kindred S (20)
a Free thyroxine index above (+) or below (-) the upper limit of normal.
b Age 3 weeks
c Upper limit of normal in prepubertal children is 3.22 nmol/L

REFERENCES

1. S. Refetoff, L.T. DeWind, L.J. DeGroot. Familial syndrome combining deaf mutism, stippled epiphyses, goiter, and abnormally high PBI: possible target organ refractoriness to thyroid hormone. *J. Clin. Endocrinol. Metab* 27:279-294 (1967).
2. S. Refetoff, L.J. DeGroot, B. Bernard, L.T. DeWind. Studies of a sibship with apparent hereditary resistance to the intracellular action of thyroid hormone. *Metabolism* 21:723-756 (1972).
3. S. Refetoff, L.J. DeGroot, C.P. Barsano. Defective thyroid feedback regulation in the syndrome of peripheral resistance to thyroid hormone. *J. Clin. Endocrinol. Metab.* 51:41-45, (1980).
4. S. Refetoff. Syndromes of thyroid hormone resistance. *Am. J. Physiol.* 243:E88-E98 (1982).
5. J.A. Magner, P. Petrick, M. Menezes-Ferreira, B.D. Weintraub. Familial generalized resistance to thyroid hormones: report of three kindreds and correlation of patterns of affected tissues with the binding of [125I] triiodothyronine to fibroblast nuclei. *J. Endocrinol Invest* 9:459-69 (1986).

6. R.C. Smallridge, R.A. Parker, E.A. Wiggs, K.R. Rajagopal, H.G. Fein. Thyroid hormone resistance in a large kindred: physiologic, biochemical, pharmacologic, and neuropsychologic studies: *Am. J. Med.* **86**:289-96 (1989).

7. S.J. Usala, G.E.Tennyson, A.E. Bale, R.W. Lash, N. Gesundheit, F.E. Wondisford, D. Accili, P. Hauser, B.D. Weintraub. A base mutation of the c-erbAβ thyroid hormone receptor in a kindred with generalized thyroid hormone resistance. *J. Clin. Invest.* **85**:93-100 (1990).

8. S.J. Usala, J.B. Menke, T.L. Watson, J. Berard, W.E.L. Bradley, A.E. Bale, R.W. Lash, B.D. Weintraub. A new point mutation in the T3-binding domain of the c-erbAβ thyroid hormone resistance is tightly linked to generalized thyroid hormone resistance. *J. Clin. Endocrinol. Metab.*, (1991) (in press).

9. M.M. Menezes-Ferreira, C. Eil, J. Wortsman, B.D. Weintraub. Decreased nuclear uptake of [125I]-Triiodo-L-thyronine in fibroblasts from patients with peripheral thyroid hormone resistance. *J. Clin. Endocrinol. Metab.* **59**:1081-1087 (1984).

10. K.I. Ichikawa, I.A. Hughes, A.L. Horwitz, L.J. DeGroot. Characterization of nuclear thyroid hormone receptors of cultured skin fibroblasts from patients with resistance to thyroid hormone. *Metab. Clin. Exp.* **36**:392-399 (1987).

11. J. Sap, A. Munoz, K. Damm, Y. Goldberg, J. Ghysdael, A. Leutz, J. Beng, B. Vennstrom. The c-erbAβ protein is a high-affinity receptor for thyroid hormone. *Nature* (Lond.) **324**:635-40 (1986).

12. C. Weinberger, C.C. Thompson, E.S. Ong, R. Lebo, D.J. Gruol, R.M. Evans. The c-erbA gene encodes a thyroid hormone receptor. *Nature* (Lond.) **324**:641-46 (1986).

13. S.J. Usala, A.E. Bale, N. Gesundheit, C. Weinberger, R.W. Lash, F.E. Wondisford, O.W. McBride, B.D. Weintraub. Tight linkage between the syndrome of generalized thyroid hormone resistance and the human c-erbAβ gene. *Mol. Endocrinol.* **2**:1217-20 (1988).

14. H.G. Fein, K.D. Burman, Y.Y. Djuh, S.J. Usala, R.C. Smallridge. Linkage between the syndrome of generalized thyroid hormone resistance (GTHR) and the human c-erbAβ gene is present in multiple kindreds. *Endocrinology Suppl.* **122**:T-1 (abstract) (1989).

15. A. Sakurai, K. Takeda, K. Ain, P. Ceccarelli, A. Nakai, S. Seino, G.I. Bell, S. Refetoff, L.J. DeGroot. Generalized resistance to thyroid hormone associated with a mutation in the ligand-binding domain of the human thyroid hormone receptor β. *Proc. Natl. Acad. Sci. USA.* **86**:8977-8981 (1989).

16. A Sakurai, T. Miyamoto, S. Refetoff, L.J. DeGroot. Dominant negative transcriptional regulation by a mutant thyroid hormone receptor β in a family with generalized resistance to thyroid hormone. *Mol. Endocrinol.* 4:1989-1994 (1990).

17. S.J. Usala, F.E. Wondisford, T.L. Watson, J.B. Menke, B.D. Weintraub. Thyroid hormone and DNA binding properties of a mutant c-erbAβ receptor associated with generalized thyroid hormone resistance. *Biochem. Biophy. Res. Commun.* 171:575-580 (1990).

18. S.J. Usala, J.B. Menke, T.L. Watson, F.E. Wondisford, B.D. Weintraub, J. Berard, W.E.C. Bradley, S. Ono, O.T. Mueller, B.B. Bercu. A homozygous deletion in the c-erbAβ thyroid hormone receptor gene in a patient with generalized thyroid hormone resistance: isolation and characterization of the mutant receptor. *Mol. Endocrinol.* (1991) (in press).

19. K. Takeda, S. Balzano, A. Sakurai, L.J. DeGroot, S. Refetoff. Screening of 19 unrelated families with generalized resistance to thyroid hormone for known point mutations in the thyroid hormone receptor β gene and the detection of a new mutation. *J. Clin. Invest.* (1991) (in press).

20. S. Ono, I.D. Schwartz, A.W. Root, B. Bercu. Evolution of hypothalamic-pituitary-thyroid function in a suspected homozygotic child from a kindred with generalized resistance to thyroid hormone. *Clin. Res.* (Abstract) 38:48A (1990).

21. A. Sakurai, A. Nakai, L.J. DeGroot. Structural analysis of human thyroid hormone receptor β gene. *Mol. Cell. Endocrinol.* 71:83-91 (1990).

INDEX